I0500356

Mining Surface Arrangements
Ore Dressing and Milling, Sampling Ores, Roasting Ores, The Cyanide Process

by Colliery Engineer Company

with an introduction by Kerby Jackson

This work contains material that was originally published in 1899.

This publication was created and published for the public benefit, utilizing public funding and is within the Public Domain.

This edition is reprinted for educational purposes and in accordance with all applicable Federal Laws.

Introduction Copyright 2017 by Kerby Jackson

Introduction

It has been years since the Colliery Engineer Company released his important publication "Surface Arrangements". First released in 1899, this important volume has now been out of print since those days and contains important information about ore extraction that has been unavailable to the mining community since those days, with the exception of expensive original collector's copies and poorly produced digital editions.

It has often been said that "*gold is where you find it*", but even beginning prospectors understand that their chances for finding something of value in the earth or in the streams of the Golden West are dramatically increased by going back to those places where gold and other minerals were once mined by our forerunners. Despite this, much of the contemporary information on local mining history that is currently available is mostly a result of mere local folklore and persistent rumors of major strikes, the details and facts of which, have long been distorted. Long gone are the old timers and with them, the days of first hand knowledge of the mines of the area and how they operated. Also long gone are most of their notes, their assay reports, their mine maps and personal scrapbooks, along with most of the surveys and reports that were performed for them by private and government geologists. Even published books such as this one are often retired to the local landfill or backyard burn pile by the descendents of those old timers and disappear at an alarming rate. Despite the fact that we live in the so-called "Information Age" where information is supposedly only the push of a button on a keyboard away, true insight into mining properties remains illusive and hard to come by, even to those of us who seek out this sort of information as if our lives depend upon it. Without this type of information readily available to the average independent miner, there is little hope that our metal mining industry will ever recover.

This important volume and others like it, are being presented in their entirety again, in the hope that the average prospector will no longer stumble through the overgrown hills and the tailing strewn creeks without being well informed enough to have a chance to succeed at his ventures.

Kerby Jackson
Josephine County, Oregon
March 2017

PREFACE

The International Library of Technology is the outgrowth of a large and increasing demand that has arisen for the Reference Libraries of the International Correspondence Schools on the part of those who are not students of the Schools. As the volumes composing this Library are all printed from the same plates used in printing the Reference Libraries above mentioned, a few words are necessary regarding the scope and purpose of the instruction imparted to the students of—and the class of students taught by—these Schools, in order to afford a clear understanding of their salient and unique features.

The only requirement for admission to any of the courses offered by the International Correspondence Schools is that the applicant shall be able to read the English language and to write it sufficiently well to make his written answers to the questions asked him intelligible. Each course is complete in itself, and no textbooks are required other than those prepared by the Schools for the particular course selected. The students themselves are from every class, trade, and profession and from every country; they are, almost without exception, busily engaged in some vocation, and can spare but little time for study, and that usually outside of their regular working hours. The information desired is such as can be immediately applied in practice, so that the student may be enabled to exchange his present vocation for a more congenial one or to rise to a higher level in the one he now pursues. Furthermore, he wishes to obtain a good working knowledge of the subjects treated in the shortest time and in the most direct manner possible.

In meeting these requirements we have produced a set of books that in many respects, and particularly in the general plan followed, are absolutely unique. In the majority of subjects treated the knowledge of mathematics required is limited to the simplest principles of arithmetic and mensuration, and in no case is any greater knowledge of mathematics needed than the simplest elementary principles of algebra, geometry, and trigonometry, with a thorough, practical acquaintance with the use of the logarithmic table. To effect this result, derivations of rules and formulas are omitted, but thorough and complete instructions are given regarding how, when, and under what circumstances any particular rule, formula, or process should be applied; and whenever possible one or more examples, such as would be likely to arise in actual practice —together with their solutions—are given to illustrate and explain its application.

In preparing these textbooks, it has been our constant endeavor to view the matter from the student's standpoint, and to try and anticipate everything that would cause him trouble. The utmost pains have been taken to avoid and correct any and all ambiguous expressions—both those due to faulty rhetoric and those due to insufficiency of statement or explanation. As the best way to make a statement, explanation, or description clear is to give a picture or a diagram in connection with it, illustrations have been used almost without limit. The illustrations have in all cases been adapted to the requirements of the text, and projections and sections or outline, partially shaded, or full-shaded perspectives have been used, according to which will best produce the desired results. Half-tones have been used rather sparingly, except in those cases where the general effect is desired rather than the actual details.

PREFACE

It is obvious that books prepared along the lines mentioned must not only be clear and concise beyond anything heretofore attempted, but they must also possess unequaled value for reference purposes. They not only give the maximum of information in a minimum space, but this information is so ingeniously arranged and correlated, and the indexes are so full and complete, that it can at once be made available to the reader. The numerous examples and explanatory remarks, together with the absence of long demonstrations and abstruse mathematical calculations, are of great assistance in helping one to select the proper formula, method, or process and in teaching him how and when it should be used.

The numerous questions and examples, with their answers and solutions, which have been placed at the end of each volume, will prove of great assistance to all who consult the Library.

Two of the volumes of this library, of which this is the first, deal with the metallurgy of gold, silver, copper, lead, and zinc. In the present volume the following subjects are treated: Surface arrangements at reduction works, ore dressing and milling, sampling ores, roasting and calcining ores, and the cyanide process. The subject of ore sampling treats both of mechanical and hand sampling. In Ore Dressing and Milling the reduction and concentration of ores and also the amalgamation of gold and silver ores are treated. The papers on the Cyanide Process are very thorough and complete, and every precaution was taken to insure that the information given was accurate and practical. This volume, together with the other volume treating on the same subject, forms the most thoroughly up-to-date and practical work that has yet appeared on the metallurgy of the metals specified.

The method of numbering the pages, cuts, articles, etc. is such that each subject or part, when the subject is divided into two or more parts, is complete in itself; hence, in order to make the index intelligible, it was necessary to give each subject or part a number. This number is placed at the top

PREFACE

of each page, on the headline, opposite the page number; and to distinguish it from the page number it is preceded by the printer's section mark (§). Consequently, a reference such as § 16, page 26, will be readily found by looking along the inside edges of the headlines until § 16 is found, and then through § 16 until page 26 is found.

INTERNATIONAL TEXTBOOK COMPANY.

CONTENTS

CONTENTS

CONTENTS

CONTENTS

SURFACE ARRANGEMENTS AT REDUCTION WORKS

INTRODUCTION

1. Definitions.—Reduction plants are usually distinguished as *wet* and *dry*. In the first instance the plants are supposed to use water and chemicals for reduction purposes, while in the second they use fire to obtain metals from the ore. The **dry process** would include magnetic concentration and the concentration of minerals by air, were there at this time any other than experimental plants existing for the latter. The **wet process** is made to include all those plants that reduce the bulk of ores by means of water, whether chemicals enter into the reduction or not. **Concentration**, from a metallurgical standpoint, is defined as a separation of ore or metal from its containing rock; whether water or heat is the more convenient and suitable agent for concentration will depend on the location of the mill and the character of the ores, for in some cases one method would answer while the other would be entirely unsatisfactory and too costly.

2. Hydrometallurgy.—The process of reducing metallic ores by means of liquids is termed **hydrometallurgy.** Hydrometallurgy includes both concentration by means of water and reduction by means of chemical solutions. It also includes amalgamation and such methods as are termed combination processes. Hydrometallurgy, as it

§ 24

For notice of copyright, see page immediately following the title page.

is understood today, embraces both chemical and mechanical concentration wherein water enters as a factor.

3. Hydraulicking is a term given to mining, transporting, and concentrating metal mineral by means of water. The term to within a short period referred entirely to the operation of concentrating gold by means of a stream of water, which mined the ground and gravel containing the gold, transported the same, and allowed the gold to become massed at one point and thus made recoverable. This principle has been extended to mining other minerals, such as iron, phosphate, rock, zinc ore, and platinum, and the term should no longer be limited to gold mining, but it should be extended to such operations as use the same stream of water for mining, transporting, and concentrating minerals.

4. Dredging for minerals approaches hydraulicking as far as concentration by water enters into the operation, but differs from it materially in that the excavating is accomplished by other means than water alone, such as centrifugal pumps or dredging buckets. While water in some measure assists the dredging apparatus in excavating, nevertheless it would not transport and concentrate the mineral, as is done in hydraulicking, without the use of a water elevator. Hydraulic elevators, such as are used in the West for placer mining, belong to dredging rather than to hydraulicking.

5. Hydrometallurgical Apparatus.—Various mechanical devices are used to concentrate minerals by washing them free from dirt or other impurities. Such machines are not classified with hydraulicking, as they are not mining machines; but although they wash and concentrate minerals, they come under the head of hydrometallurgical apparatus. Mechanical devices, such as jigs, trommels, log washers, hydraulic classifiers, and other similar machines, are not to be placed in a separate class, as they may be made adjuncts to several operations that differ completely in their ultimate method of obtaining the desired results. This character of hydrometallurgical apparatus is described in *Ore Dressing and Milling* and will not be specifically explained here; but it is well to understand that such apparatus may be used for concentrating secondary products of milling, as well as the primary products of mining.

6. A second class of hydrometallurgical apparatus includes wet crushing as a preliminary operation, followed by sizing and concentration. Water forms the medium through which the decrease in barren mineral becomes possible. This operation is treated in *Ore Dressing and Milling* and includes such machines as jigs, trommels, spitzkasten, cone classifiers, bumping tables, concentrators, buddles, and vanners.

Concentration by the use of such machines is for the purpose of obtaining mineral matter in small bulk and then subjecting it to some other treatment. The milling ores thus treated are usually sulphuret and lean ores, which must be concentrated to a small bulk and treated at some distant point. Concentrates containing sulphur are more easily and economically reduced when freed from barren vein material, and in some cases if the ore were not concentrated no values could be obtained.

This system is used to separate lead from zinc in the Missouri-Kansas zinc districts of the United States, where zinc and lead are found in the same ore, and as they do not alloy they cannot be concentrated by fire so as to obtain each separately without making the cost of the concentration exceed their value. The only other means left, then, is a system of hydrometallurgy that will separate the two as far as possible, owing to their difference in specific gravity. This process is not practical when pyrite is associated with blende and galena. The same system is also practiced with the native copper ores in Michigan, where, on account of the difference in specific gravity between the rock and the native copper, they are successfully separated. The process was attempted at the Huston Mines, near Nace, Virginia, in order to separate manganese from limonite, but on account of the nearness of the minerals in their specific gravities, it was a complete failure.

7. A third hydrometallurgical process includes wet crushing with amalgamation, such as is usually practiced in stamp milling.   This process may be extended to include pan and barrel amalgamation.   In case there is any considerable loss of gold from incomplete amalgamation, the process may be followed by that previously described or the fourth process.

The operation is hydrochemical in some instances, but not in the sense of lixiviation, for chemicals are not added in the pan process to extract the precious metals, but for the purpose of keeping the almalgam quick and preventing its sickening, flouring, or becoming incorporated with copper.

When the mercury becomes sickened it refuses to act upon the gold, and hence the gold is not recovered by this process.   If, therefore, it is possible to keep the mercury bright and active, as may be done by adding chemicals, a larger percentage of precious metals can be recovered.

8. It often happens in stamp milling that only about 40 per cent. of the precious metals in the ore is recovered by amalgamation, and it is seldom that the recovery exceeds 75 per cent.   In the case, then, of a $12 ore, the value remaining after amalgamation would be $7.20, if the recovery was but 40 per cent.   If by grinding the tailings in pans containing mercury a further percentage of the values may be recovered at a cost that will prove profitable, it is generally done, provided this system of recovery is cheaper and more profitable than some other.   Again, if the ore will leave a profit after concentration and shipping expenses to the smelters are deducted, that method will prove advisable provided pan amalgamation does not yield as large returns. In some cases lixiviation may prove more remunerative for a secondary operation than the methods set forth; in any case, the method adopted should be tested by practical experiments, which will also in a great measure be dependent on local conditions.

It is not good metallurgical engineering to adopt an expensive process that in a short time may become worthless on account of such a change in the character of the ore as to require its abandonment for some other; yet the West is dotted with just such near-sighted experiments, which would not have occurred had some metallurgical engineer been called in to treat the ore experimentally before the mill was erected.

9. A **fourth hydrometallurgical process** consists in the reduction of minerals by chemical solutions. The factors entering into such processes, if not well known and recognized, will cause failure. It is seldom that an ore can be obtained which does not require some preliminary treatment before leaching, and in some cases the preliminary treatment becomes a part of the chlorination and Russell processes. In nearly every case, except that mentioned, where tailings are treated, crushing precedes leaching. In case the ore is refractory, roasting must precede cyaniding; and it is an absolute necessity where chlorination is practiced on sulphurets. There are other chemical processes that have been conditioned to the metals they are to extract, such as the leaching of lead, zinc, and copper ores. So far, they have not come into general use because the expense connected with the recovery of the last-named metals by leaching their ores, with other necessary treatment, amounts to more than the value of the metals after recovery.

Lixiviation is defined as a process by which a soluble alkali or saline compound is extracted from an earthy mixture by washing out. As this does not cover the case of such metals as native gold, silver, or copper, the definition must be adjusted to include the term **leaching,** by which is meant the separation of soluble matter by percolation or drainage. The distinction between lixiviation and leaching would then be that the former was a process and the latter a part of that process. A further distinction can be drawn from the fact that lixiviation includes dissolving with special solutions, while leaching is a matter of draining off the matter dissolved.

10. Ore Dressing and Milling.—The definitions given show that the four different hydrometallurgical processes mentioned require different milling methods and that their subdivisions also require different apparatus to carry out the particular processes involved. As an illustration, the cyanide mill requires very different arrangements and machinery from a chlorination mill, but since nearly every kind of apparatus that is used in hydrometallurgy has been discussed in *Ore Dressing and Milling*, it is unnecessary to go into the details of their construction and workings. In the case of that metallurgical apparatus which is to be described under particular headings, such as tanks, barrels, pumps, filter presses, etc., it would simply be an unnecessary repetition to describe them at this time. Their description is given under the processes in which they are employed and which are included in this Course.

11. Roasting furnaces are described in *Roasting and Calcining Ores*. These form a very important adjunct to milling and smelting arrangements, few metallurgical plants being able to do without them. They are, therefore, mentioned, but not described in detail.

12. The question of **water supply** enters into the subject of mill location, and if not always for power, at least for milling and smelting purposes. In some cases it may be the only means available for power, and in any case no mill or smelter can be run successfully without a supply of water. The subject of water supply for power has been fully discussed in *Hydraulics and Hydraulic Machinery*.

13. Economic Arrangements.—In case there is not sufficient water to furnish power, but fuel for steam boilers is available, the position in which the steam boilers are to be placed at the mill is one worth consideration. The situation of the engine room will be a matter of some moment for the reason that line shafting and belting must, as a usual thing, extend to various remote parts of the mill. The more distant it is from the power, the more difficult it becomes to maintain and keep it in repair. If it is possible to arrange a mill so that fuel may be delivered direct to the boiler house, considerable labor will be saved, and if by the expenditure of $1,000 the labor of one man can be saved, it will prove economical in the end. At small works, where it costs 20 cents per ton to load, cart, and unload coal, and this is in most instances a low figure unless done by contract, a saving of $146 per year can be accomplished when 2 tons are used daily, provided the delivery track can be laid directly to a trestle adjacent to the boiler room.

14. In the arrangement of a mill, ore delivery is probably one of the most important items. If it is possible to place the mill near the mine whose ore is to be treated, an ideal location would be a side hill, provided the ore could be delivered at the top of the mill. In a situation of this kind gravitation can be made to assist and thus avoid the expense of raising the material. Advantages of this kind would not always apply to those mills that use driers before crushing and screening.

Calcining furnaces are sometimes built high up in a mill, but there is less danger from fire and they are generally better located with regard to fuel and general handling of hot material if they are on the lowest floor of the mill. Even in case it is advisable to drop the ore to the drying furnace, there still remains certain advantages if a side hill is used for a mill site, as it usually furnishes a tailing dump, besides affording easy methods for dealing with slimes, tailings, and exhausted liquors.

15. Should the ore be delivered to the mill in lumps, so that it requires crushing, it is considered by some to be a disadvantage to have the mill on a side hill. This objection arises from the fact that some one once placed a crusher at the top of the mill, without first making allowance for vibrations and thrust from loaded ore bins, besides not properly anchoring the crushers. Some will smile that such claims should be advanced as a drawback to placing the crusher at the top of the mill; nevertheless, the position of the crusher is one that requires serious thought and much careful expert work, in order to obtain a firm foundation and prevent excessive vibrations. If the crusher or the stamps cause a water tank to vibrate in unison with them, the building will be seriously affected and weakened to its foundations.

16. In case the mill is to be built on comparatively level land, the location should be chosen first with preference to the ore supply and then with reference to transportation of products and the delivery of fuel. In case it is to be a custom mill, similar preferences should be shown. The disposition of tailings is very important, but, like the water supply, it is considerably more flexible than the delivery tracks at a mill. The ideal location for a mill that must receive ore from different mines would be the brow of a hill, where the water supply might reach it by gravity, ore be delivered into bins from railroad cars, and tailings leave the mill in such manner that the expense of handling them would be practically nothing. There are instances where mills have been very favorably located, but it is simply impossible to find every advantage in one mill.

17. In case the mill can be run by water-power furnished by some stream in the vicinity, it may be advisable to transport ore to a mill so located. On the other hand, it may be advisable to locate the mill near the mine and transmit the power generated by water to the mill. This second arrangement would cause a great loss of power, but it may so happen that the location of the water-power was practically inaccessible to transportation facilities and mill construction. There are several instances on record where the power has been carried to the mill in the form of electrical energy.

LAWS RELATING TO WATER RIGHTS

DITCHES AND WATER

18. The laws passed by the United States Congress to encourage mining and assure to çitizens agrarian rights are given in the following sections as far as they have bearing upon water rights, mill sites, etc.

"Whenever, by priority of possession, rights to the use of water for mining, agricultural, manufacturing, or other purposes, have vested and accrued, and the same are recognized and acknowledged by the local customs, laws, and the decisions of the courts, the possessors and owners of such vested rights shall be maintained and protected in the same; and the right of way for the construction of ditches and canals for the purposes herein specified is acknowledged and confirmed; but whenever any person, in the construction of any ditch or canal, injures or damages the possession of any settler on the public domain, the party committing such injury or damage shall be liable to the party injured for such injury or damage."—*Sec. 9, Acts of Congress (A. C.), July 26, 1866.*

19. Excepted in Patent.—"All patents granted or preemption or homesteads allowed shall be subject to any vested and accrued water rights, or rights to ditches and reservoirs used in connection with such water rights as may have been acquired under or recognized by the preceding section."—*Sec. 17, A. C., July 9, 1870.*

20. Claims Subject to Ditches and Flumes.—"All mining claims now located, or which may be hereafter located, shall be subject to the right of way of any ditch or flume for mining purposes, etc. Provided always, that such right of way shall not be exercised against any location duly made and recorded, and not abandoned, etc. without the consent of the owner, except by condemnation, as in the case of land taken for public highways. And provided further, that such ditch or flume shall be so constructed that.

the water from such ditch or flume shall not injure vested rights by flooding or otherwise."

21. Ditch Rights.—"Ditch rights are located and a notice posted, after which a certificate of ditch and water rights is made out by the locator and sworn to before a notary public. The ditch should be staked and work commenced and prosecuted with reasonable diligence, otherwise the record and notice amount to nothing. Different States have different forms upon which the locator of a ditch is to record his statements. As these can be obtained from the Secretary of State, they are not given here."

DUMPS

22. "The **right to dump** is but little, if at all, affected by statutory regulation, and the right to dump, as of necessity or by custom, across lower claims, has been looked upon as custom or subject to only nominal damages. The exception in such cases would be where damage was done to mining operations on the claims below or to improved lands. The dump, when placed upon another's claim, is considered real estate by law; hence, tailings or piles of lean ore dumped upon another claim belong to that claim and cannot be removed by the miner from whose mine it came."—*Morrison.*

MILL SITES

23. A **mill site** for purposes incidental to mining may be located on government lands, provided the United States statutes are complied with. There are, under United States laws, two classes of mill sites, which comprehend two classes of mills—private and public. The first class, that is, a mill site with lode, being of a private nature, is of greater interest to the miner than the second class, or the mill site that is intended to serve as a place for erecting a custom mill. Such sites can be patented in a manner similar to lode claims.

24. How to Patent Mill Sites.—"Where non-mineral land, not contiguous to the vein or lode, is used or occupied by the proprietary of such vein or lode for mining or milling purposes, such non-adjacent surface ground may be embraced and included in an application for patent for such vein or lode and the same may be patented therewith, subject to the same preliminary requirements as to survey and notice as are applicable to veins and lodes; but no location hereafter made on such non-adjacent land shall exceed 5 acres, and payment for same must be made at the same rate as fixed for the superficies of the lode.

"For filing claim, a charge of $10 is made; for mill-site survey and platting, $30; for mill site, including United States survey with a lode, $15. The owner of a quartz mill or reduction works not owning a mine in connection therewith may also receive a mill site, as provided in Sec. 15, A. C., May 10, 1872."

LOCATION AND RECORD

25. American Practice.—Mill sites are located by posting a notice in some conspicuous place on the claim, after which the locator records his notice with the proper county or State authority in which his mill site is located.

LOCATION NOTICE

"I claim the Juanita mill site (600 feet northeast by 200 feet southwest) as staked on this ground. Date of location, September 4, 1901.

"G. C. MUNSON."

RECORD LOCATION CERTIFICATE

"*To All Whom These Presents May Concern:* Know ye that I, G. C. Munson, County of Arapahoe, Commonwealth of Colorado, do hereby declare, and publish as a legal notice to all the world that I have a valid right to the occupation, possession, and enjoyment of all and singular that tract or parcel of land not exceeding 5 acres, situate, lying, and being in the Empire Mining District, County of Clear Creek, State of Colorado, bounded and described as follows, to wit: The Juanita mill site, beginning at corner No. 1, from which a line north 30°, east 100 feet, etc., to place of beginning. Together with all and singular the hereditaments and appurtenances thereunto belonging or in anywise appertaining.

"G. C. MUNSON.   [SEAL.]

"Witness my hand and seal, this fourth day of September, A. D. 1901."

26. **Non-Mineral Affidavit.**—It is sometimes customary to make a non-mineral affidavit that no portions of the mill site contain minerals. In such cases, on two days' notice parties who own the land make affidavit to that effect; the claimant is not required to file his own affidavit to the same effect. The claimant must, however, make affidavit in regard to his being a citizen of the United States and finally publish his notice. After the application for survey has been made and the land surveyed, the patent will be given for the mill site, provided everything has been regular and the law complied with. Where a mill site is applied for in connection with a lode, a second affidavit (according to some authorities) of stated form, and which can be obtained by applying to the proper official, is required. This latter affidavit is a brief of mill site used for mining or milling purposes. In case a patent has been granted for a mill site and mineral should afterwards be discovered, the probabilities are that the patent would be good to cover both mill site and mineral, it having been received in fee simple.

27. **Staking.**—The locator of a mill site, before he makes claim to any portion of the land, will proceed to mark its boundaries by means of suitable stakes, or, if stakes are not possible, by means of monuments of rock or stone. A stone or boulder properly marked fully answers the purpose of a stake, or even a pile of stones in such places where stakes cannot be driven. A stake is set at each angle of the claim, marked with the name of the mill site and the number of the corner. Whenever it is possible, one corner should be tied to some natural landmark, government survey, or if that is not possible, to a natural permanent monument. If on account of the precipitous ground it is impossible to set the stake where the claim corners, a witness stake suitably marked to designate the position of the corner should be set at the nearest available point along the lines of the survey.

BRITISH COLUMBIAN MILL SITE

28. British Columbian laws require that the land for a mill site be unoccupied public land, and as far as known, not to contain mineral. British Columbian statutes entitle the owner of a mill site to surface rights only, reserving all minerals that may be subsequently discovered on the land, together with the right to enter the property and mine such minerals for the government and its licensees. They also require that the mill site shall be as nearly square as possible.

Aside from the difference already mentioned, the British Columbia statutes are very similar in intent to those of the United States, as in the United States the area of a mill site is limited to 5 acres. The corners are marked by legal posts, with the notice on each post, stating, first, the name of the locator; second, the number of his Free Miner's Certificate; third, the intention, within 60 days from date of notice, to apply for the land as a mill site; and fourth, the date of notice.

29. Application for Lease.—Having properly located and staked his mill site, the locator, within the 60 days specified, applies to the Provincial Land Surveyor for a lease of the property, and on depositing duplicate plans and making affidavit as to the location of the claim, is granted a lease for 1 year. If in that time the lessee has placed or constructed machinery or done other work on the property for mining or milling purposes to the value of $500, he can obtain a Crown grant (equivalent to the United States patent) to the mill site at the expense of $5 per acre. The interpretation of the term "mining purposes" is not so broad as in the American practice. It includes only the erection of machinery and buildings for transporting, reducing, crushing, and sampling ores, or for the transmission of power for working mines.

PLANT FOR TREATMENT OF ORE

30. General Considerations.—In a matter of this kind, the first thought should be the supply of ore. Mills have been erected almost simultaneously with the operation of breaking ground at a mine, irrespective of the quantity and quality of ore that the mine would produce. In some cases, the mines proved to be merely pockets, so that all moneys expended for mills have been entirely lost. In other cases, there was not sufficient ore for one-third or one-fourth the capacity of the mill erected, and yet again there have been mills built at great expense to treat ore by a certain process, which proved complete failures, since the process was not applicable to the ore. It sometimes happens that the character of the ore completely changes with the depth, in which case the method of treatment must change in order to work economically and satisfactorily.

31. The erection of a plant for the reduction of ore is a matter of considerable importance to mine owners and is not one to be guessed at or to be the subject of experimentation by ordinary mine managers. The West is full of good little mines that have failed to pay the owners interest on the capital that they have invested, although interest and capital have been made from the mine. The natural inference when one regards such results coolly is that the returns have been misapplied, and upon an expert examination it has been found that usually they were squandered in some character of unsuitable mill or process, and in many instances the profits were lost in the tailings.

32. The first general thought relative to the erection of a mill is the question of a suitable ore supply. The second thought is the kind of power most readily available. The third, the location of the mill. The last, the character of the mill suitable for treating the ore mined.

Assuming that we have a free-milling ore, the question of the mill to be used is narrowed down to the stamp mill, as that, although primitive, has been ascertained by costly experiments to be the one best suited for that ore. This being the case, there are two classes of mills to choose from: The California mill with heavy stamp (weighing with stem, head, and shoe 950 pounds) and short, quick drop and the Gilpin County, or Colorado, stamp mill, with light stamp (600 pounds) and long, slow drop. These two mills, from their different modes of working, require different mortars —one a low discharge, the other a high discharge. The arguments for and against these mills have been long, sometimes bitter, without either side gaining an advantage; but with the disappearance of the old guard, there seems to have been a compromise, in which the tendency is towards heavy stamps and quick drop—that is, from 60 to 95 drops per minute.

33. Arrangement of Buildings.—Mill buildings should be arranged with reference to the ore coming from the mine, the handling of material in the mill, the fuel supply, railroad connections, and tailings dump. The plant, if simply intended for coarse-crushing ore, which is subsequently to be picked for shipment, should be at the mine, since in that locality dumping ground for lean ore and rock is usually available and transportation of barren stuff is avoided. If such a plant is located so as to have railroad connections, the work may be conducted in the cheapest manner possible, advantage being taken of gravity to load the cobbed ore directly into cars from storage bins. In instances where the mine is almost inaccessible for machinery or erection of proper milling machinery, the ore must be trammed to the mill.

The mill may be a simple crushing plant, but if of large capacity, it should have a steady delivery of ore, possibly

FIG. 1

obtained through the use of a wire tramway, such as that shown in Fig. 1. .

The illustration is of a wire tramway whose length is 9,000 feet, with a span of 1,173 feet across the town of Wardner, Idaho. Wire-rope tramways are more expensive to operate than railways; however, in some situations they are imperative, railway construction being out of the question. They may be made to work up and down hill, across ravines or rivers, and are very useful adjuncts to mining.

34. Picking Belts.—Simple crushing mills, where ore is assorted for shipment, should be provided with a picking belt. The Robbins picking belt, shown in Fig. 2, has proved successful both as a conveyer and as means for assorting

and cobbing the ore.   The ore is cobbed with a hammer on
the belt as it travels along.   One of these belts is said to
have conveyed 350,000 tons of heavy crystalline ore, in

FIG. 2

pieces about 3 inches in diameter, at the Franklin, New
Jersey, Zinc Mines.

By careful hand picking, the value of rich mineral in ore
has been raised 26 per cent., a matter of considerable importance, since if ore transportation is $2 per ton, $2 will be
saved on every 4 tons shipped.   One advantage of a movable picking belt lies in its serving the purpose of a conveyer.   Another case is that of the Ferreira Company,
in South Africa.   The value of the ore as it comes from the
mine is $17.17.   After assortment the ore is worth $25.78,
the increased value, due to sorting, being $8.61.   The increase
in value leaves but $.84 per ton in the waste rock, although
the latter is 36 per cent. of the ore mined.

35. It is advisable that mill buildings should be erected with a view to economy; nevertheless, they should not be so arranged as to increase fire risks. At some plants the storage bins are connected with the head-gear above the shaft. This is a faulty arrangement, provided the shaft is perpendicular, for then the head-house must be above the shaft, and in case of fire the miners are cut off from escape. Many lives have been lost by this arrangement of buildings.

Fig. 3 shows the plant at the Mother Lode Mine, Anaconda, British Columbia. The head-frame and gear *a* is

FIG. 3

seen in the background. The ore when dumped at the head-house passes over the grizzly bars, the finer dropping on to a 12-inch belt conveyer, which carries it to the ore bins *b*, shown in the foreground. The coarse pieces are delivered to the crusher, located in the building *c* in the back of the illustration and to the right of the head-frame. A 36-inch sorting conveyer, 111 feet between centers and located in building *d*, receives the ore as it falls from the crusher and carries it to the bins. The waste and lean ore

to be discarded is assorted by hand labor and dropped into
chutes, which deliver it to a 16-inch conveyer *e* having 540 feet
between centers, which carries it to the rock dump. This
conveyer may be seen to the left, connecting the ore bins
with the dump.

CONSTRUCTION OF MILLS

36. Masonry.—The supervision of mill construction is
generally one of the many duties intrusted to the metallurgical engineer. The design, as well as the kind of mill
to be constructed, is a matter usually left for the metallurgist's decision; therefore, he should be able to draw up
plans and specifications in order to estimate the cost of construction. For this purpose he must know the cost of labor
and materials delivered at the mill site, and which involve
an estimate of the cost of transportation, a matter varying
widely in the mountainous districts. The cost of masonry is
particularly difficult to estimate in a new country, for at
times, with good stone quarries near by, it may be economy to
import stone or bricks from a distance. Masonry should not
exceed $10 per cubic yard, and from that price it can be made
to taper down to $1.50. In estimating the cost of buildings
and masonry, labor should not exceed 60 per cent. of the
cost of the materials used, otherwise something is wrong.
This rule one may say is flexible enough to fit into any district where labor is dear or cheap, as in such situations materials are correspondingly high or low in price. Masonry
depends on the price of stone and labor. In some instances
stone can be quarried and delivered to the masons for 75 cents
per cubic yard. Under such favorable circumstances, the
wall in place should not cost over $1.50 per cubic yard.
The above figure for masonry is exceptional; a usual figure
for a good cement mortar wall, pointed and well bound, is
about $2.50 per cubic yard. The stones used for mill foun-
dations are not dressed, that is, cut to size, but are usually
faced and split by the mason as he lays them, in order to
keep a line and bind the wall.

37. Masonry at metallurgical works is usually confined to building foundations and engine beds, although at times it extends to vats, chimneys, and the construction of the entire plant. When constructing simple foundation walls for heavy buildings and loads, a good rule to follow, either for temporary or permanent masonry, is to give the stones plenty of binding material, such as a mixture of two-thirds cement and one-third sand, and also to have all spaces well grouted and spalled. Headers and corner stones should be properly laid to strengthen the bond and course joints should be properly broken. The batter given to foundation walls is about 1½ inches per foot, from the surface up, and while such walls should go to bed rock, it is not necessary to batter them below the surface, unless it is desired to give them an extra wide base. A firm foundation wall whose top is less than 20 inches wide cannot be readily built. Walls 18 inches wide and possibly less can be constructed for dwellings, but this will not answer for heavy mill construction. Sometimes it becomes necessary when erecting mills on side hills to build retaining walls which answer as building foundations and at the same time keep the earth from sliding or moving the building.

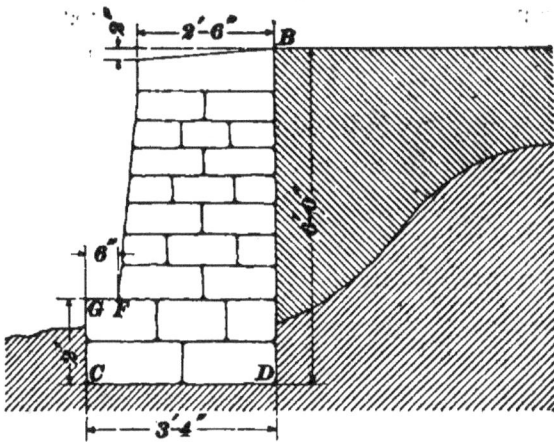

Fig. 4

38. Retaining Walls.—The usual form of a retaining wall is shown in Fig. 4. There is no fixed rule for determining the dimensions of retaining walls, but the one given below will in all probability meet every requirement.

Rule.—*When the backing is loose, a wall of first-class large stones laid in mortar should have a base C D equal to one-third its vertical height; a wall of bricks laid in mortar should have a base of two-fifths its vertical height.*

Retaining walls must have a firm, wide foundation and must be constructed in the manner explained for dams in *Hydraulics and Hydraulic Machinery.* In case they are built up against shaly or other rock, they should be well backed by loose stones, and in instances where the rocks carry water, all spaces should be filled with cement to hold back the water or a drain made to carry the water away from the masonry.

Probably the latter plan will be better in case the wall is placed at the foot of a high hill or where the water is likely to come from a high elevation and create a great pressure against the wall. If, however, there are other avenues of escape for the water, which is easily determined by examination of the strata, it will not matter if the backing is made water-tight.

39. Guarding Against Frost.—Where freezing occurs, the back of the wall should be sloped, as shown at *a b*, Fig. 5, and smoothly finished to lessen the hold of the frost, which might otherwise displace the masonry. The foot of the slope *b* should be at the frost line, usually about 2 feet below the surface in moderate climates and from 4 to 6 feet in climates above the forty-fifth parallel of latitude. High altitudes will also affect the frost line.

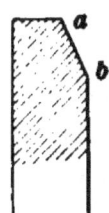

FIG. 5

40. Having proportioned a retaining wall by the rule given, its stability may be increased by stepping it in back, as shown in Fig. 6, without adding to the volume of masonry.

The offsets are determined as follows: Through *c*, the middle point of the back, draw any line *f g*. From *f* erect the perpendicular *f h*. Divide *g h* into any even number of parts, in. this instance four, and draw

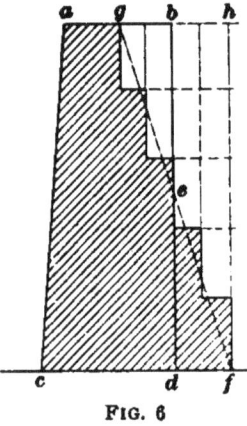

FIG. 6

through these points of division lines parallel to *f h*. Then divide *f h* into one greater number of equal parts than *g h*, and through these points of division draw lines at right angles to *f h*, forming the offsets shown in the figure. By increasing the thickness of the wall at the base, the center of gravity is lowered and the stability consequently increased. The backing included between the lines *g h* and *f h* exerts only vertical pressure against the offsets, which tends greatly to prevent the overturning of the wall. The theory of retaining walls differs little from that of dams and consequently is not repeated here, the student being referred to *Hydraulics and Hydraulic Machinery*.

- - -

FRAMING OF TIMBER STRUCTURES

41. Trestles.—As most reduction works require railroad tracks, and as it is frequently necessary to construct trestles, either leading to bins or dumps or during the regular construction of any mine railroad, illustrations are given for framing some simple forms of trestles. Fig. 7 illustrates the various parts of a trestle; Fig. 8 two forms of bents and a side elevation of a portion of a pile bent trestle. The various parts are numbered and their names given in the following list:

1, a bent framed; *2*, a pile bent; *3*, cap piece; *4*, cross-tie; *5*, dapping or gaining; *6*, guard rail; *7*, jack stringer; *8*, longitudinal brace; *9*, mortise; *10*, mudsill; *11*, packing block; *12*, packing bolts; *13*, a pile given a batter, an inclined brace; *14*, a vertical, plumb, or upright pile; *15*, a vertical, plumb, or upright post; *16*, a post with a batter, or inclined; *17*, sill; *18*, stringer; *19*, sway-brace; *20*, tenon.

The portion of the illustration at (*a*), Fig. 7, shows the arrangement of a pile bent trestle, while that at (*b*) shows the arrangement of a framing bent trestle. In case the trestle

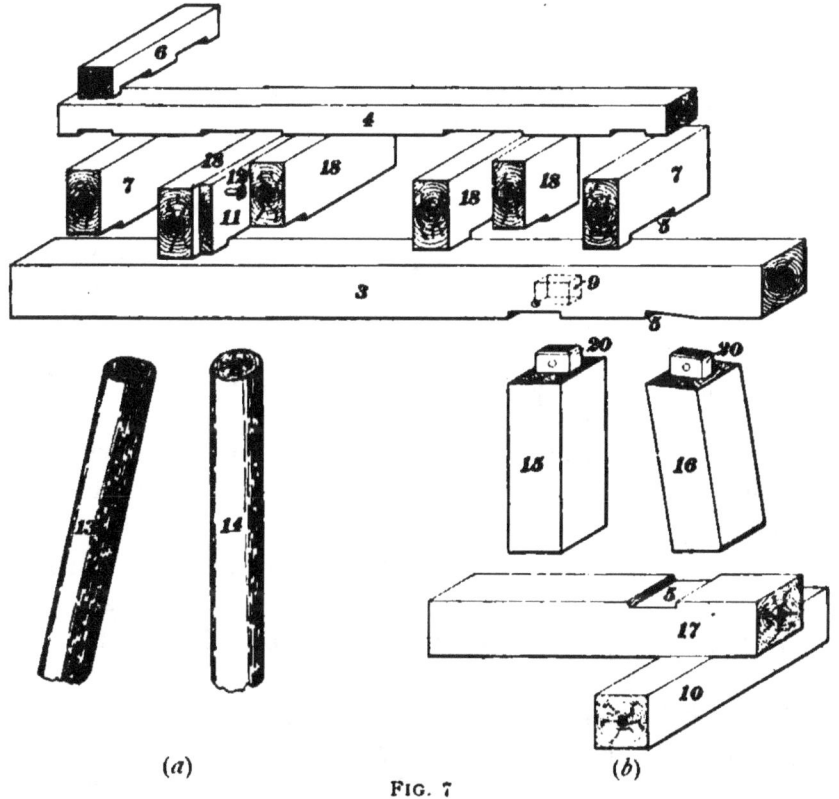

(*a*) (*b*)

FIG. 7

is not very high and piles are used, they may be driven vertically, but it is always best to have the outer piles driven at an angle so as to form batter braces. Fig. 9 illustrates a

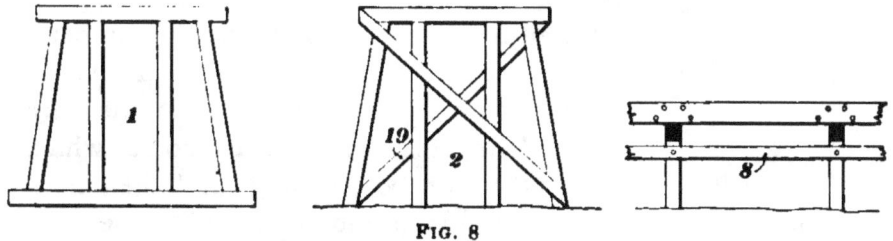

FIG. 8

pile having a tenon formed on the upper end to receive the cap. When this method of securing caps is used, a hole is drilled through the cheeks of the mortise in the cap and through the tenon.   It is well to have the hole in the cheeks of the mortise so placed that when the pin is driven through two holes it will tend to draw the cap down on to the top of the pile.

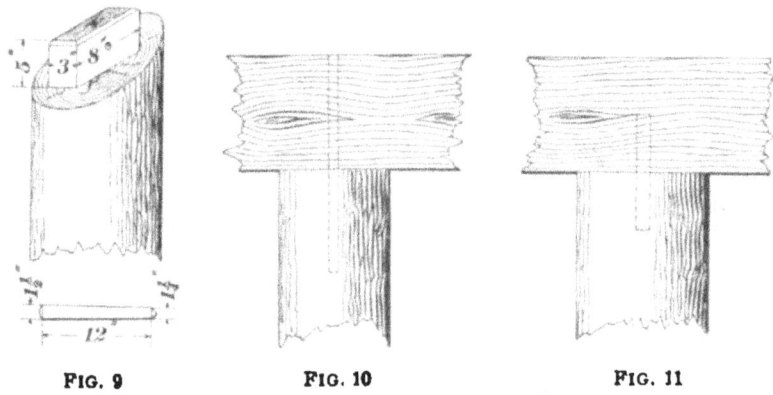

FIG. 9   FIG. 10   FIG. 11

The pin used for this purpose is commonly called a **treenall,** and should be made of hardwood, locust wood if possible, and slightly tapered, as shown in the lower part of the illustration. Sometimes the caps are not mortised and tenoned on to the piles, but may be secured by means of drift bolts, as shown in Fig. 10, or by means of dowels, as shown in Fig. 11.

42. **Split Caps.**—Another arrangement is shown in Fig. 12.   This is called the "split" cap; in place of using one 10″ × 10″ timber, two 5″ × 10″ timbers are employed and the top of the pile is cut as shown in the illustration.   The timbers can be seen at *a* and *b*, while *c* is a tenon, the full width of the pile, that is allowed to project up between the timbers. No notches are cut in the timbers where they rest on top of the piles, but they are secured in place by means of a bolt *d*, which passes through both timbers and the tenon.

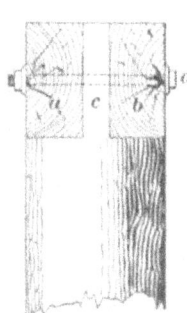

FIG. 12

Some of the advantages of this method of framing the caps are as follows:

1.   On account of the smaller size of the cap pieces, it is possible to obtain better timber.

2. Repairs can be made with greater ease than where caps are mortised and tenoned or fastened with drift bolts to the top of the piles, for either of the caps can be removed and replaced without interfering with traffic or cutting any portion of the timber work.

43. Framed Bents.— Where it is not possible to drive piles and form pile bents, framed bents are used. Fig. 13 illustrates a framed bent in which all the timbers are simply secured by means of drift bolts.

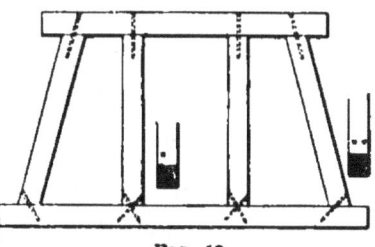

FIG. 13

44. Foundations for Sills.—The sill of the bent should always be placed upon some form of foundation. This may be composed of timber mudsills, as shown at *10*, Fig. 7, but it is a better practice to construct stone or masonry walls under the sills and to see that the latter are well bedded. When masonry is used as a foundation for sills, care must be taken to see that the stones are well laid; it is never good practice to construct these foundations of round stones laid up like rubblework, for the constant passage of trains over the trestle is liable to break up such a foundation.

45. Placing Timbers. — The batter braces should have a uniform angle of 3 inches per foot. Fig. 14 illustrates the method of framing on the ends of batter braces

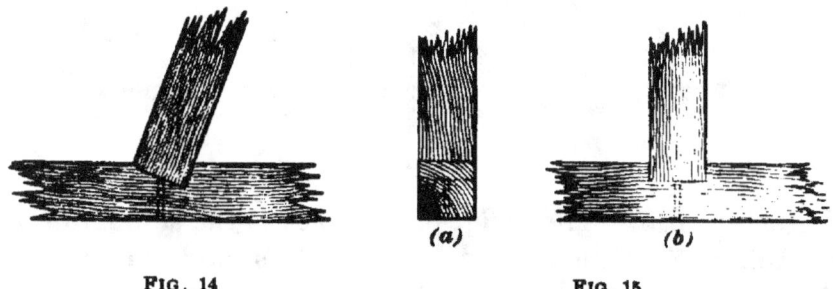

(a)   (b)

FIG. 14   FIG. 15

and posts in framed bents and also shows a drainage hole bored in such a manner that any water collecting under the jointing will immediately flow out through the drain and thus reduce the tendency that timbers have to rot. It may be well to mention here that green oak timbers or wet oak timbers spiked or bolted with iron soon decay in the vicinity of the iron. Fig. 15 (*a*) and (*b*) show the method of mortising and tenoning the legs to bents.

46. Illustration of a Framed Bent.—Fig. 16 is a dimensioned drawing showing a timber bent as used on one

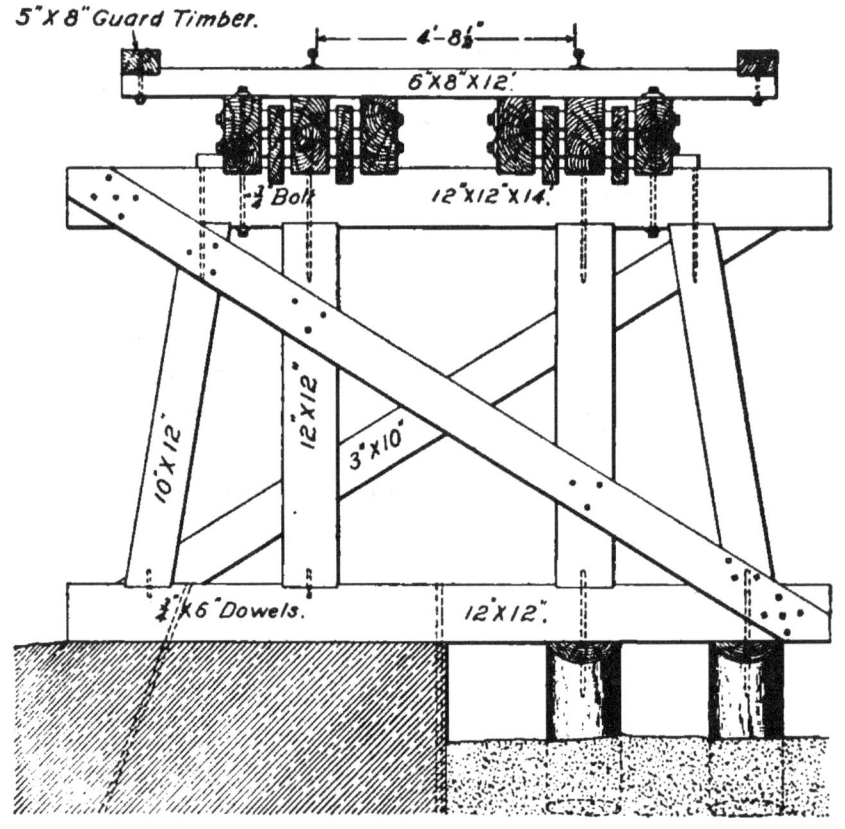

FIG. 16

line of railroad. The gauge of the track is the standard for the United States, that is, 4 feet 8½ inches, and the dimensions on the drawing fully explain the various parts.

47. Elevation of Outer Rail.—Where the trestle comes in a curve on the railroad track, if it is intended that the cars should move at any considerable speed, it is necessary that the outer rail be elevated. This may be accomplished by wedge blocks placed on the top of the cap and under the stringers; usually, however, the cross-ties are cut wedge-shaped to give the desired elevation.

48. Framing Buildings.—Small buildings about mills can be framed of small stuff, without any special framing, the pieces simply being spiked together to form a balloon frame, which is covered with either siding or corrugated iron.

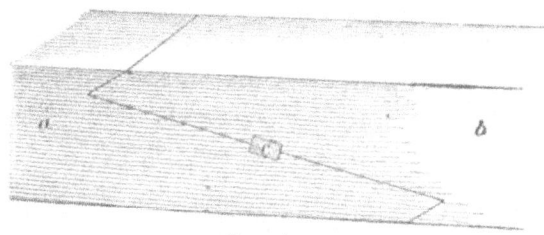

FIG. 17

When it becomes necessary to build somewhat heavier structures, some form of framing may be used. The different joints used in framing are all similar to those illustrated in connection with trestlework and consist mainly in the use

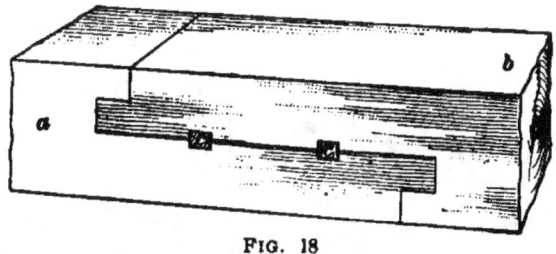

FIG. 18

of tenons or notching the timbers together. When timbers must be joined in the direction of their length, special joints or a different method may be necessary. When two timbers are joined without an increase of size, it is called a "scarfed" joint.

Fig. 17 illustrates one form of scarfed joint in which the timbers *a* and *b* are joined as illustrated, and are held in place by means of the key *c*. Fig. 18 illustrates another form of scarfed joint, which is better adapted for resisting end

thrusts and in which the timbers *a* and *b* are held together
by two keys *c* and *c*. Usually scarfed joints are strengthened

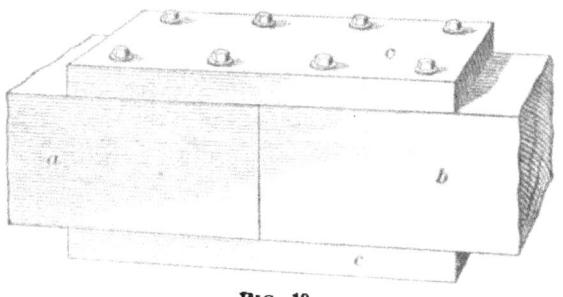

FIG. 19

by means of iron or wooden plates called *fish-plates*, and
when these are set into the wood in such a manner as not

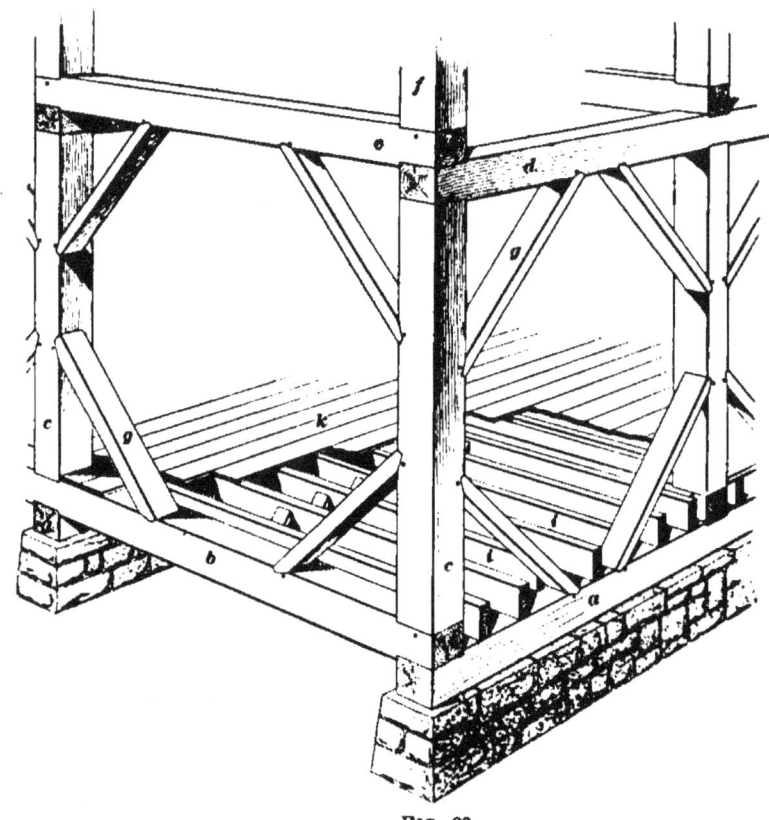

FIG. 20

to increase the size of the timber, it is still a scarfed joint;
but when the pieces used for joining the timbers are simply

bolted on to the outside, as illustrated in Fig. 19, the joint becomes a pure fished joint, and the plates c and c are called fish-plates.

49. Heavy Framing by Cutting Joints.—One form of heavy framing is illustrated in Fig. 20. The heavy sill timbers a are secured to the timbers b by notching into each timber, as shown in the illustration, and also by the use of drift bolts. The posts c are fastened to the timbers b by means of tenons and treenails, or pins. The timbers for the second floor e and d are united by notching and by drift bolts, while the posts f for the next upper story are secured by tenons and treenails. The braces g are notched into the posts and sills and secured by heel tenons and pins.

Fig. 21 illustrates a tenon on the end of one of the braces. The floors may be formed by using a joist i the same depth as the timbers b and notching them on to the sills a to the same depth the timbers b were notched down. After this the floor k can be laid on top of the joist and the timbers b. Where

FIG. 21

the braces do not extend the full width of the post, they can be placed either flush with the outside of the building, as illustrated, or centrally on the timbers. The manner of placing the braces flush with the outside as illustrated has the advantage that whatever form of siding is used, it will be secured to both the posts and the braces, and will thus aid in stiffening the building.

50. Heavy Framing Without Cutting Joints.—Fig. 22 illustrates another system of framing which is to a large extent on the balloon-frame order, for it has no mortise-and-tenon joints and very little framing of any kind. The sill timbers a and b may be notched together and secured by means of drift bolts ; the posts c may be slightly notched into the timbers b and secured by drift bolts. The braces g are simply pieces of plank cut and spiked as shown in the illustration, the pieces *h* being spiked against the posts or sills in such a manner as to fill the space between the ends of the braces. Where the floor joists *m* rest upon the sill *a*, it may be necessary to use a narrower piece between the braces, as at *i*, or to saw notches into the piece *i*, into which the joist can be placed. The timbers *d* and *e* for the upper

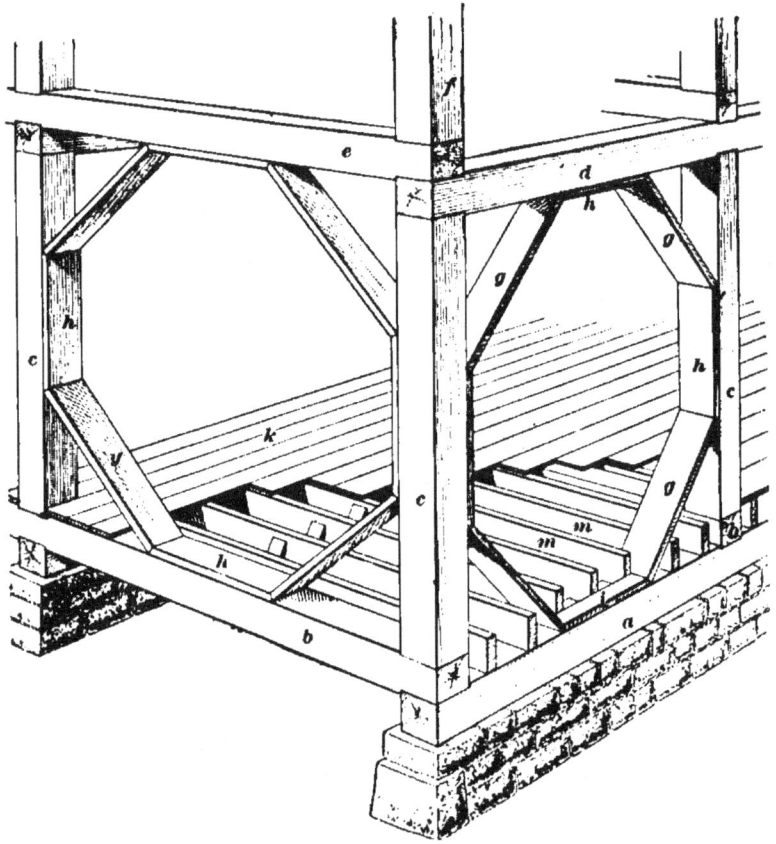

FIG. 22

floor of the mill are fastened together and to the posts *f* and *c* by slight notching and drift bolting. In some cases no notching is done, the posts simply being sawed off square and secured by large spikes or drift bolts. The floor in this style of construction is laid exactly as in the previous case and is shown at *k*.

51. Corbels.—Where it is necessary to join sills or large horizontal timbers on top of posts, corbels may be used, as shown in Fig. 23. The corbel a is bolted to the two timbers b and c, and the post d is usually tenoned into the corbel, while the post a may have a wide tenon and be secured to the timbers b and c by means of two treenails or pins. By making the corbels fairly long, they will help support the timbers b and c, thus relieving them of a portion of

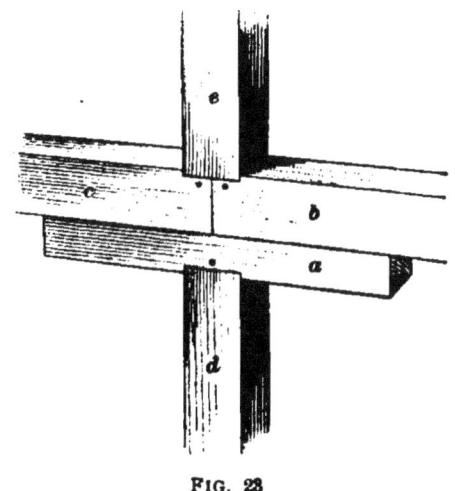

FIG. 23

the weight they would otherwise have to carry. On this account, corbels are sometimes used whether joints occur above the corbels or not.

THE STAMP MILL

52. The gold stamp mill has been described in *Ore Dressing and Milling.* Fig. 24 shows one, in cross-section, situated on a side hill, for the convenience of receiving its ore by tram cars that dump over a grizzly. The ore that passes over the grizzly passes to a crushing floor and thence through a crusher to the ore bin, where it again mingles with the ore that passed through the grizzly. From the ore bin, the ore passes through a gate into an automatic ore feeder, which supplies a stamp battery composed of five stamps. The crushed ore passes from the stamp-battery mortar, as pulp, over amalgamating plates; then from the launder at the foot of the plate it passes to concentrating tables, an arrangement not used with free-milling ores, but quite natural with a rather refractory ore.

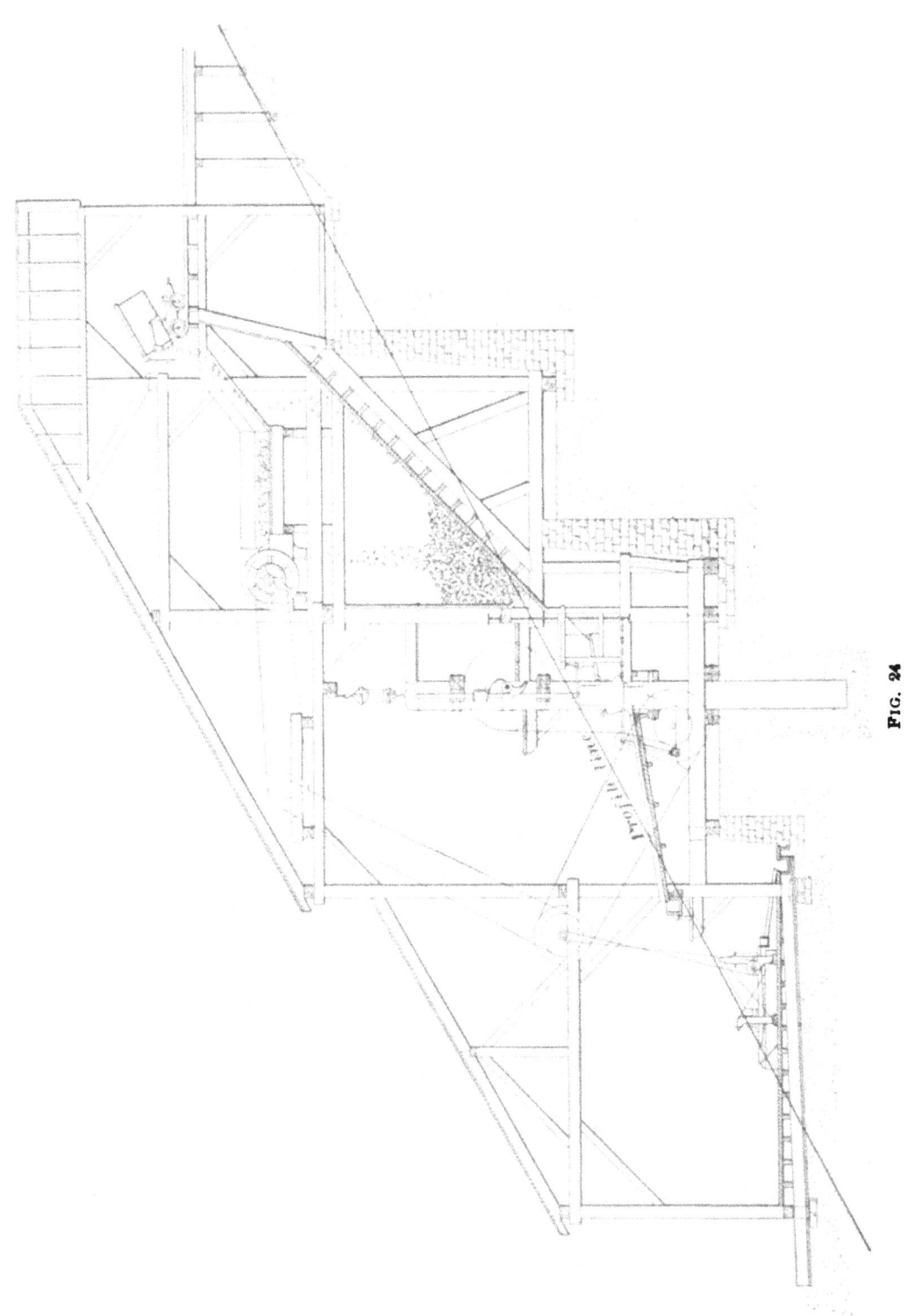

Fig. 24

Fig. 25 shows an elevation of a double gold stamp mill, by which it is meant that the stamps are placed

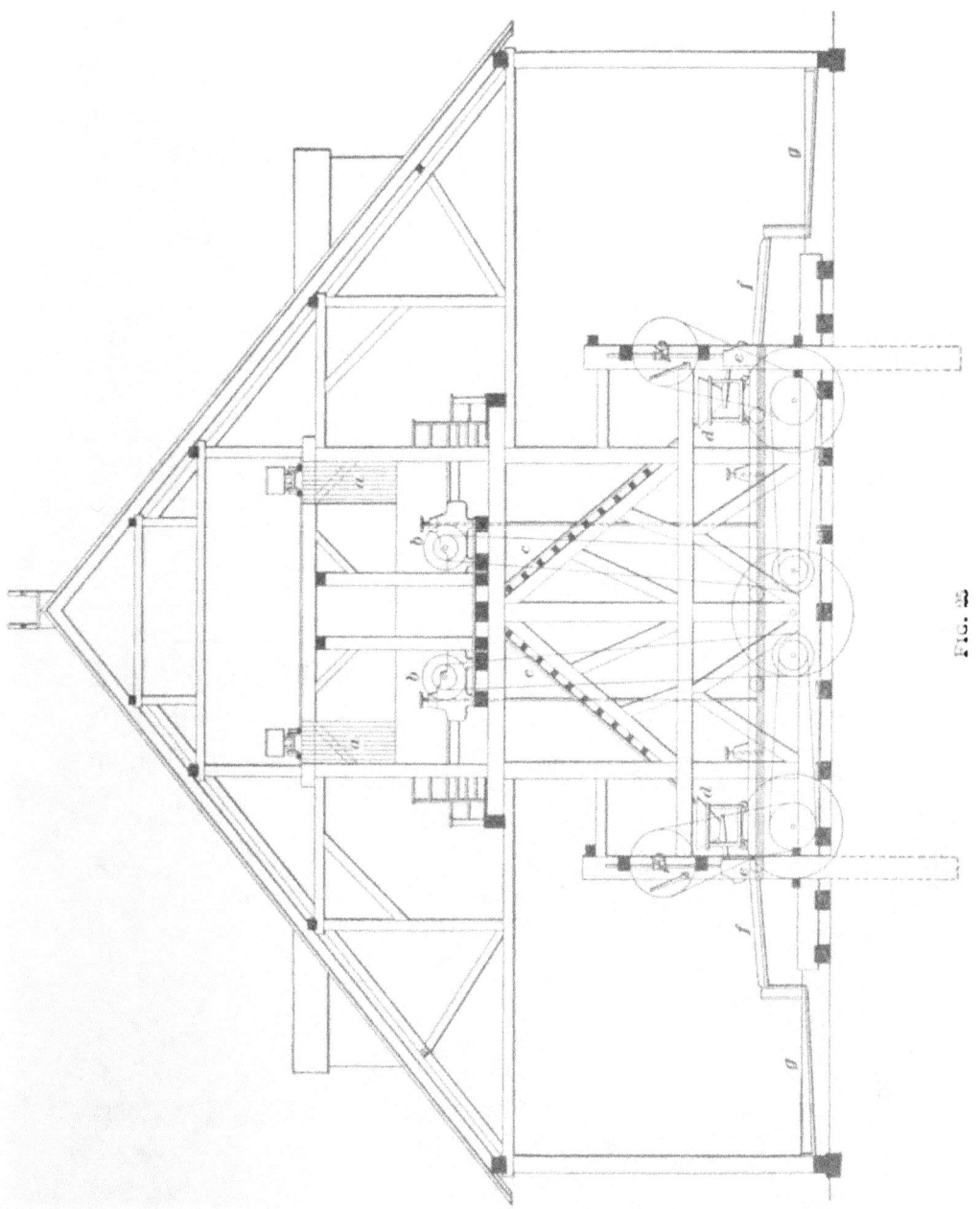

in double rows, in order to increase the number of stamps and lessen the lengths of single-line shafting. *a* are the grizzlies, *b* the rock crushers, *c* the ore bins, *d* the automatic ore feeders, *e* the stamps, *f* the plates, *g* the launder

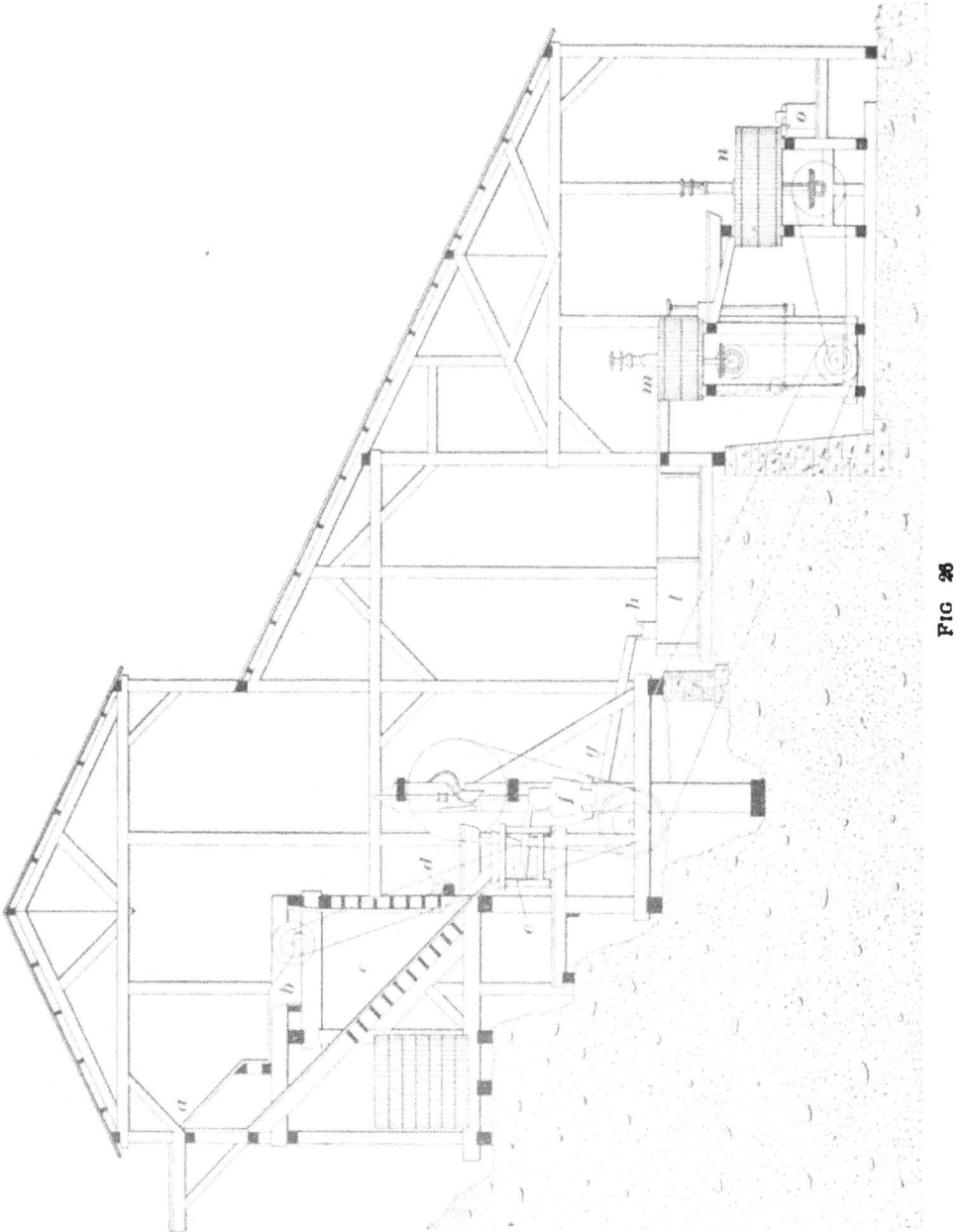

Fig. 26

leading to the settling tank or pond. Where a large number of stamps are to be used and the ground is comparatively level, double stamp mills are preferable to single stamp mills.

53. Silver stamp mills do not differ from gold stamp mills until after the pulp reaches the settling tanks. At this point the excess of water is drained and the thick pulp shoveled in regular charges into amalgamating pans, in which it is worked several hours in order to obtain as much of the precious ·metals in the form of amalgam as possible. The pan digestion being completed, the contents are run into large settlers, where the quicksilver, in the form of amalgam, and the free mercury settle to the bottom. The quicksilver and amalgam are separated from the rest of the pulp by a trap and the amalgam is finally separated from impurities by treatment in the clean-up pan.

A wet-crushing silver mill, such as has just been described, is shown in Fig. 26. In the figure, *a* is the grizzly, *b* the rock crusher, *c* the ore bin, *d* the ore gate, *e* the automatic ore feeder, *f* the stamp battery, *g* the plate, *h* the launder leading from the amalgamating plate to the settling tank *l*, *m* the amalgamating pan, *n* the settling pan, and *o* the mercury and amalgam trap.

54. Dry Crushing Silver Mill.—This character of mill is intended for refractory ores that must be roasted prior to amalgamation. It differs from the mill described by having a drier placed between the ore crusher and the stamps, which in this instance crush dry. The ore may be run in chutes, lined with sheet iron, from the driers to the automatic feeder. The pulverized product from the stamps is carried by conveyers to bucket elevators, which discharge into the hopper of the roasting furnace. The ore in the furnace is desulphurized and also chloridized by the addition of salt, thus preparing the pulp for what is known as barrel amalgamation. In Fig. 27 is shown a dry-crushing silver mill, in which *a* are the grizzlies, *b* the crusher, *c* the ore bin, *d* the drier, *f* the stamp battery of 20 stamps, *g* the screw conveyers, one on each side of the battery, the mortar

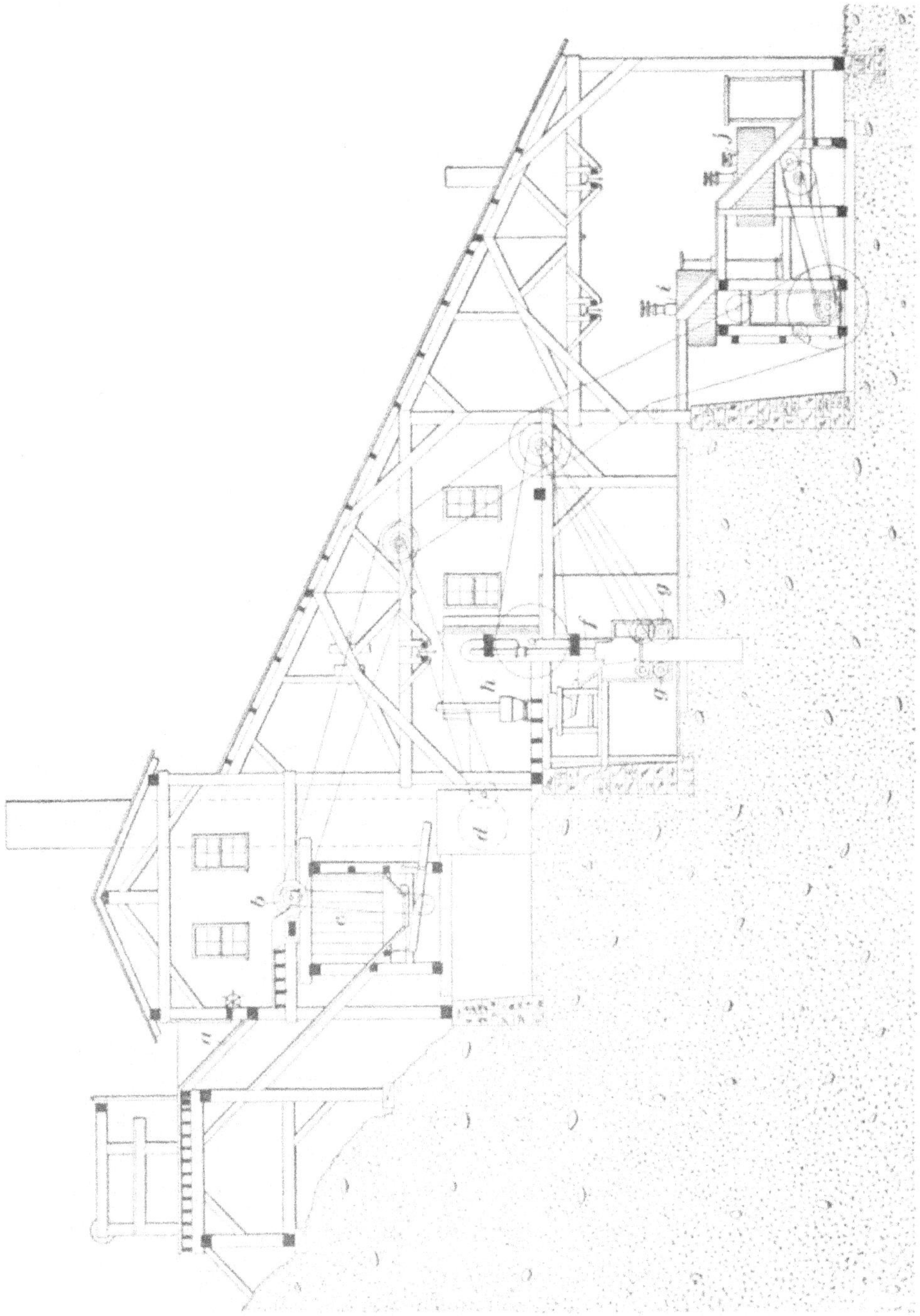

FIG. 27

FIG. 28

in this case being of the double discharge pattern; *h* the furnace, *i* the amalgamating pans, *j* the settlers.

At some dry-crushing mills the ore is crushed by rolls and properly sized for the furnace by means of rotary screens. One great objection to this method is the dust created by the several crushings and resizings, but it produces an increased amount of fine ore in a given time, though at the expense of power.

55. Concentrating mills are for preparing ores for subsequent metallurgical treatment. They may be divided into two classes: One reduces the bulk by the elimination of worthless gangue; the other not only does this, but it also separates those minerals that have different specific gravities, for instance, zinc and galena. Usually water is the separating medium, but lately a form of air jig has been quite successful as a

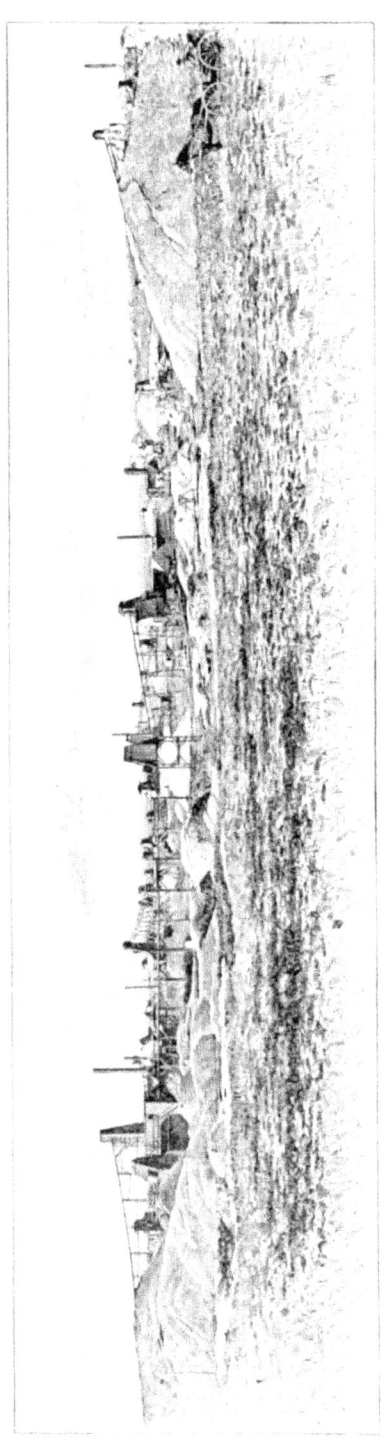

FIG. 28

concentrator. The concentrating mill has been so thorougly discussed in *Ore Dressing and Milling* that it will not be taken up here in detail.

In Fig. 28 is shown a zinc plant in the Joplin, Missouri, district. In a few instances in this ore field it is possible to find a hillside, so as to take advantage of gravity, but in most instances the ore is hoisted from the shaft high enough in the head-house to permit its being trammed directly to the top of the mill, which is frequently 1 story high. From the crusher room in the top of the mill the ore passes downwards during the various operations of concentration until it reaches the ground floor. In case concentration is not complete, it is again elevated to the top of the building and again descends through the various processes by gravity. In some mills, portions of the ore is elevated three or four times, but as a rule two or three times is sufficient after it once descends from the crusher room. When concentration has been completed, the tailings go to the waste pile. Water is furnished from the creeks, when they are accessible, and from the mines when no other source is available; in the latter case the water is impounded and used over and over again, if necessary. Centrifugal pumps raise the water from ponds to tanks situated in the buildings or elevated on towers. Creek or well water when not contaminated with sulphate of lime is used for steam production. In many instances, however, spring water and city water must be used and in extreme cases water is hauled in barrels several miles to the mill.

56. In Fig. 28 the head-houses are seen at *a*, the extreme right and left, the mill *b* being located between them and, as shown, connected by trestles *c* over which tram cars move. In the foreground is seen a waste rock pile *d*; to the extreme left, part of an ore pile *e*. The tower *f*, back of the mill, contains an elevator for raising the tailings, which are run out automatically from it on special cars, whose tracks are supported by the skeleton towers *g*. The tailings are automatically dumped, as shown by the white pile *i* back of the head-house on the left. Between the rock pile in the foreground and the mill is a settling pond, which may be distinguished only by the bank *m*, which serves to impound the water and slimes. The slimes that accumulate are sometimes quite rich in mineral and are treated in separate mills, called **sludge mills.** In Fig. 29 is shown a group of mills and head-frames at the Mastin diggings, Galena-Joplin district.

CONVEYERS

57. Elevators and Conveyers.—Since the introduction of elevating and conveying machinery, the handling of material at reduction works has been revolutionized. The cost of handling material has been lessened, thereby increasing profits and making it possible to treat and handle larger quantities of material.

The term conveying and hoisting machinery includes bucket conveyers, wire-rope tramways, elevated railways, cableways, and cantilever cranes. These various devices have numerous modifications to condition them to the work to be performed. The introduction of this class of machinery has made it possible to place reduction works on ground that would at one time have been considered unfit for the purpose. Occasionally mills are seen almost entirely hemmed in by waste rock and tailing piles, so that it looks as if they must be covered over if work is to be continued. This state of affairs is now past, owing to the improved machinery for handling the waste tailings and rock. Conveying and hoisting machinery is used inside the mills as well as outside; moreover, its flexibility permits it to be applied to new and various conditions that are continually arising.

58. Chain-belt conveyers were probably the first in the field and have held their own in most instances against new devices intended in a measure to supersede them.

FIG. 30

In Fig. 30 is shown a Jeffrey ore conveyer as used at a cyanide-process mill for filling the vats. The carriers are so arranged that they may dump their buckets into any vat or any particular place in the vat. Bucket elevators for unloading vats and delivering tailings to conveyers that run to the dump are also made. The same elevators can be made to deliver into tram cars. Where the roasting furnace is separated from the mill, conveyers may be used to move the ore to the place it is to be treated, thus avoiding the

FIG. 31

otherwise necessary haulage, loading, and unloading. Fig. 31 shows a refuse conveyer much used in the anthracite fields of Pennsylvania, where it handles veritable mountains of refuse termed **culm.**

59. Hoisting and conveying machinery is sometimes used at smelters for the purposes of conveying ore, fuel, and flux to the furnaces. Fig. 32 shows an elevator building containing a bucket elevator, which is 60 feet between

centers and intended to lift granulated slag and water to a
height that will permit them to flow to the dump or deliver
them to the dump pile through troughs carrying scraper

FIG. 32

conveyers.   Sufficient water, with a slight inclination, will
move the slag along and a considerable saving may be
effected by getting rid of furnace slag in this way.   Besides,

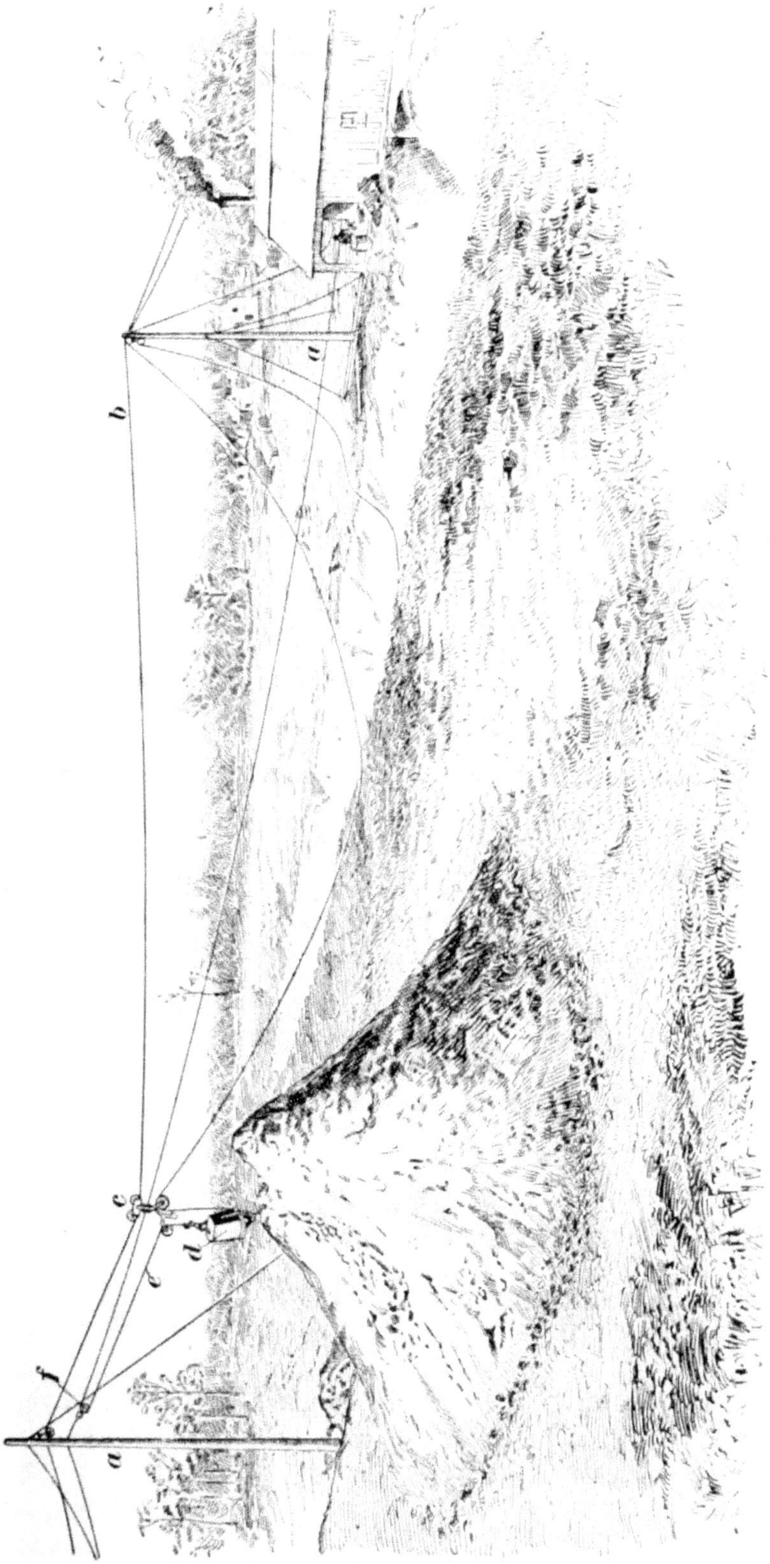

FIG. 88

the necessity of a slag dumping ground is obviated if there is a good-sized stream or river to dump into.

In the figure, *a* is a hole leading to a sump which answers as an elevator boot and is kept flushed with water. The slag is dumped into this well from slag carts *b* and, becoming granulated, is carried by the elevator buckets *c* to the bucket discharge in the house *d*. The water and slag that the buckets discharge flow together down the trough *e* to the slag pile or river, as the case may be.

60. The Ligerwood cableways have two towers, between which is stretched a wire rope. In Fig. 33, the poles *a* will answer for towers. Upon the wire rope *b* a carriage *c* runs back and forth, carrying with it a bucket *d*. The traveling

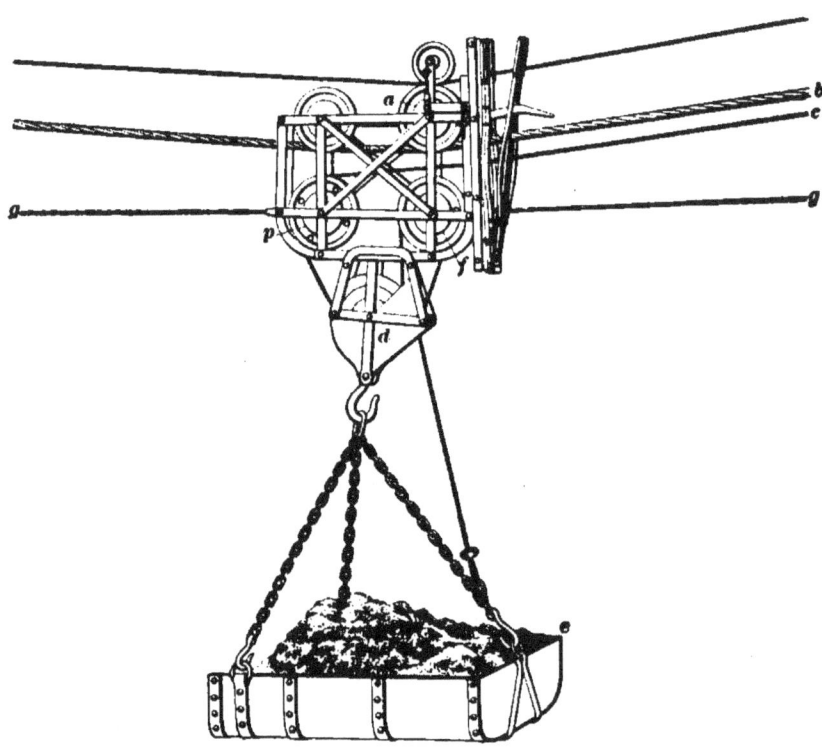

FIG. 34

rope *e* is attached to one end of a bucket and passes over a pulley *f* and then between the span and over another pulley leading to the engine room. It is given two or three turns around the drum, then passes up to another pulley, and so on back to the bucket, making virtually an endless rope. This rope moves the carriage back and forth on the cable b.

The carriage a, shown in Fig 34, is another arrangement for dumping refuse. The carriage runs on rope b; there is a fall rope c, which is connected with the engine at one end and with the bucket e at the other. The fall rope passes half around pulley p, then half around the block d, up to pulley f, and then, as shown, down to the bucket at e. The traveling rope g moves the carriage back and forth to the dump, and when it is desired to dump holds it there. The rope c then comes into play and tilts the bucket, allowing the contents to fall on the dump pile. The bucket is next pulled into its normal position and moved back for more refuse by the traveling rope g moving the carriage a.

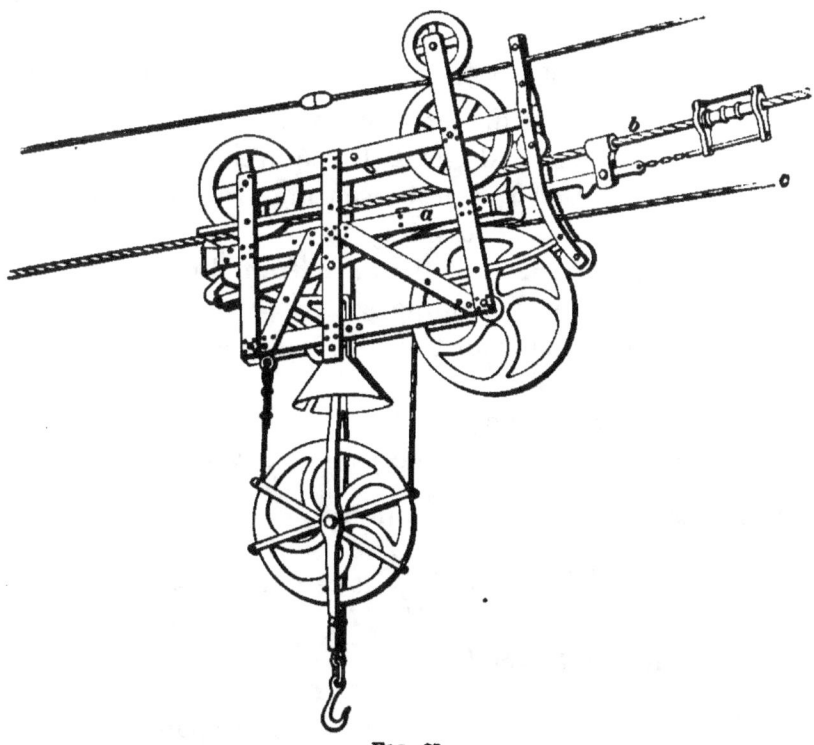

FIG. 35

In Fig. 35 is shown another carriage a, which travels down an inclined rope b. The loads are raised by means of the fall rope c, which also acts as a haul rope and brings them from a lower level to a higher.

61. Rope Sag Calculations.—While a cable may be made sufficiently taut to answer as a runway for the carriage and car, nevertheless there will be a certain amount of deflection that must not be neglected, otherwise it will

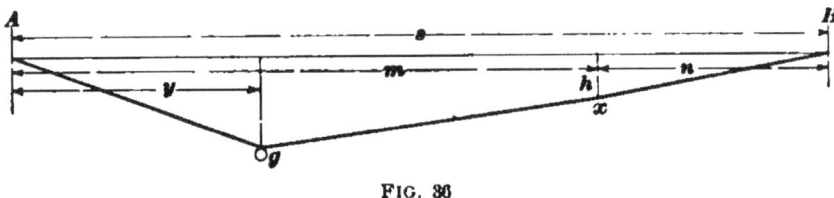

FIG. 36

necessitate the shortening of the span or require the tower away from the mill to be the higher. This deflection can be estimated if the length of the span, the weight of the rope, and the weight of the load are known. In Fig. 36

let s = span or the distance $A B$ between supports;

  m and n = arms in feet into which span is divided by a vertical through required point of deflection x, m representing arm corresponding to loaded side;

      y = horizontal distance from load to support corresponding with m;

      w = weight of the rope per foot in pounds;

      g = load;

      t = tension;

      h = required deflection at any point x.

Deflection due to rope alone:

$$h = \frac{m\,n\,w}{2\,t} \text{ at } x, \text{ or } \frac{w\,s^2}{8\,t} \text{ at center of span.}$$

Deflection due to load alone:

$$h = \frac{g\,n\,y}{t\,s} \text{ at } x, \text{ or } \frac{g\,y}{2\,t} \text{ at center of span.}$$

If $y = \frac{s}{2}$, $h = \frac{g\,n}{2\,t}$ at x, or $\frac{g\,s}{4\,t}$ at center of span.

If $y = m$, $h = \dfrac{g\,m\,n}{t\,s}$ at x, or $\dfrac{g\,m}{2\,t}$ at center of span.

Total deflection:

$$h = \frac{w\,m\,n\,s + 2gny}{2\,t\,s} \text{ at } x, \text{ or } \frac{w\,s^2 + 4\,g\,y}{8\,t} \text{ at center of span.}$$

If $y = \dfrac{s}{2}$, $h = \dfrac{w\,m\,n + g\,n}{2\,t}$ at x, or $\dfrac{w\,s^2 + 2\,g\,s}{8\,t}$ at center of span.

If $y = m$, $h = \dfrac{w\,m\,n\,s + 2\,g\,m\,n}{2\,t\,s}$ at x, or $\dfrac{w\,s^2 + 2\,g\,m}{8\,t}$.

If tension is required for a given deflection, transpose t and h in the above formulas.   (*Trenton Iron Company.*)

WEIGHT OF WIRE ROPE

62. To obtain the weight per foot of wire rope of any diameter,

let w = weight per foot;
             D = diameter of the rope in inches.

Then

For ordinary wire rope with hemp core, $w = 1.57\,D^2$.
For ordinary wire rope with hemp core, $w = 1.7\,D^2$.
For patent locked-iron rope, $w = 2.5\,D^2$.
For solid round bar, $w = 2.62\,D^2$.

THE HUNT ELEVATOR AND AUTOMATIC RAILWAY

63. The **Hunt elevator and automatic railway** is operated by gravity, needing neither steam, horse, nor manual power. It requires power, however, to raise material to the car. This car, Fig. 37, is arranged to run on a railway track, as shown, and is discharged by means of the tripping block a, placed on the track where the load is to be dumped. The sides are not fastened to the car, but to each other, so that if one is unfastened both are. The load is thus evenly discharged and without danger of overturning the car, although the gauge is but 22 inches.   The bottom of the car has a center ridge so arranged that the material runs entirely out

FIG. 37

when the sides are unfastened.   The bearings on some of these cars are so made that the car runs around a 30-foot radius with comparative ease.

FIG 38

64.   In Fig. 38 is shown the car at work removing coal from a boat to the dump pile.   *a* is the trolley and automatic

FIG. 39

bucket that raises the tailings to the dump car *b*; above the car is a pocket *c* into which the bucket dumps automatically. From the pocket the ore slides directly into the car, which runs down a narrow-gauge track *d*, laid upon the trestle, to the dumping ground *f*, after it has been loaded and started by the man at the loading chute.

The chief peculiarity consists in storing the energy that the loaded car creates in descending the inclined track, so as to return the car to the loading chute after it has emptied itself. This is accomplished by the car picking up a cable which is attached to the weight *g* in its journey down the incline. The car raises the weight *g* only a limited distance, and that by a gradual movement, so as to prevent strains on the various parts as far as possible. When the car has dumped, the weight falling gives the car sufficient momentum to run up the plane to the chute, after which it is ready to receive another load. This arrangement is probably best suited to unloading vessels or cars, but may, by having movable towers, be applied to almost any character of loading and unloading.

65. Another method of disposing of tailings is shown in Fig. 39. This arrangement is shown in the **cantilever crane,** that when used on the Chicago drainage canal had a length of 353 feet over all and a maximum height of 80 feet for a dump. It was arranged to travel upon a truck whose wheel base was 37 feet. This truck ran upon a portable track, so that the entire structure could be moved with comparative ease. The cars or buckets could be loaded, then raised by a fall rope, and hoisted up the incline to be dumped automatically. With such an arrangement, a dump pile 80 feet high could be made and continued indefinitely along the canal.

66. **Centrifugal pumps** are sometimes used to raise the exhausted tailings from leaching vats. When used for this purpose, care must be taken to give the bends of the tail-pipes and the delivery pipes as large a radius as possible to prevent friction. When the pump shown in Fig. 40 is stationary, so as to require long suction pipes, the latter must be air-tight and given a gentle rise towards the pump. They will elevate 50 feet to a launder large quantities of sludge, but are probably most efficient when elevating about 20 feet. As the construction of centrifugal pumps was fully described in *Hydraulics and Hydraulic Machinery*, Part 4, only the work that can be done by them will be discussed here. Ordinarily a pulp consisting of 2 pounds of water to 1 pound of solid matter can be handled by a centrifugal pump when there is but a slight head to be

FIG. 40

overcome. This quantity of water must be augmented, as the elevation is increased above 10 feet, to from 3 to 10 pounds of water to 1 of mineral matter. It is to be understood that mineral matter of a coarse nature is more difficult to raise than fine, but that there is a limit to the density of the sludge for good work, no matter how fine the sands may be, and this limit is its mobility. The necessity for having the foot-valve of a centrifugal pump well submerged in the sludge arises from the fact that the tailings sand will not otherwise enter the tail-pipe and the pump will draw only water.

The objections to the use of centrifugal pumps for this purpose is the large quantity of material that must be handled to attain the desired end, the high power required to drive them, and their low efficiency. Their advantage, however, is that they are capable of handling very large quantities of material in a comparatively short time. Their use effects a saving in time, but a loss of power.

67. In South Africa the ore is stamped and amalgamated previous to cyaniding. The tailings from the stamp mills are conducted in launders to large **tailing wheels,** sometimes 60 feet in diameter, which lift the tailings in buckets on their inside peripheries and dump them into a launder leading to the leaching vats.

Fig. 41

An illustration of this wheel and the method of filling cyanide vats is shown in Fig. 41; *a* is the flume carrying tailings to the pit, which are lifted therefrom by buckets *b*, in the wheel, and raised by the wheel so as to dump into the launder *c*. The boxes *d* are **V**-shaped spitzkasten to catch any sulphides in the tailings. Large wheels for handling tailings were probably first used at the Lake Superior Copper Mines, although the idea no doubt originated in Cornwall, England.

68. The **Chinese pump** shown in Fig. 42 may be used for handling tailings and water with probably more economy than any of the wheels, pumps, or other devices so far described, provided it is intended to charge the lixiviation vats by launders and with tailings. These pumps require comparatively little power to run them.

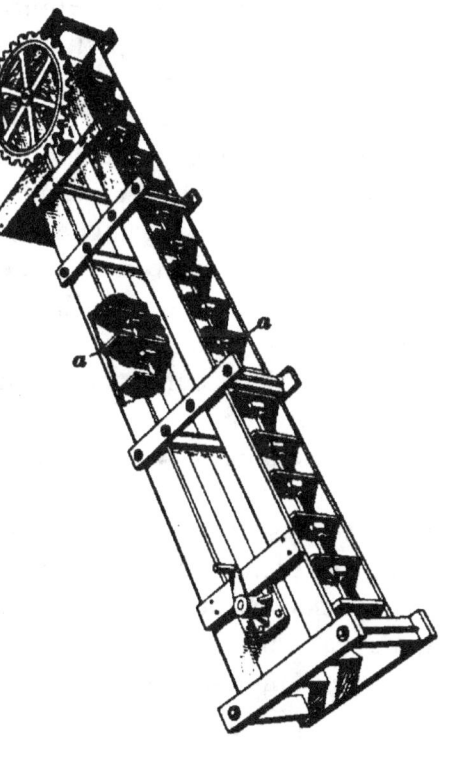

FIG. 42

SMELTING PLANTS

69. General Considerations.—Smelting plants must be located with reference, first, to water supply; next, fuel and flux; and lastly, ore. Furnaces cannot be worked without water; it is needed for boilers and the modern economical furnaces, which are constructed with water-jackets. The supply of water required for furnaces will be found in Art. **88.** Fuel and flux are both necessary items and furnaces should be located so as to have, if possible, railroad transportation for these materials; or if that be not possible, to have good wagon roads at least. Custom smelters are usually situated where cheap fuel can be obtained, for, as the ore must come to them, they care naught for the location of mines. This is due to the fact that the miners must pay the freight on ore and the smelters the freight on the coal and the flux. Private smelting plants can in some instances afford to pay more freight per ton for fuel than for ore, from the fact that much more ore is handled at the furnace than fuel.

70. The **ideal furnace location** is one where water is abundant and where fuel, flux, and ore can be delivered with a short haul into the stock piles without extra handling. This, of course, requires a system of railway tracks and trestles connecting with the main haulage railways. It was customary in the past and is so yet in some instances to regard 36- and 42-inch gauge tracks as the proper width for furnace delivery tracks. This idea is erroneous, for it is possible to put in a standard railroad track which has a gauge of 4 feet 8½ inches where it is possible to put in the above. If narrower than standard-gauge tracks are used, it causes the rehandling of the fuel and sometimes of the flux and ore. The idea prevailed in the past pretty generally that curves on narrow-gauge roads could be given more curvature, so as to take up less room. The wheel base on two-wheel mine cars of any size is nearly the same as for four-wheel trucks on narrow-gauge cars and the latter is about the same as on broad-gauge cars. The small difference in curvature required for standard-gauge car trucks will not take up much more room than ordinary mine cars, such as would be used to transport ore to the furnace a mile or more.

71. Ideal furnace locations are not always obtainable; but whenever it is possible, locations should be picked where nearby ground will furnish a dumping place for slag. Granulating slag and having it floated away by some stream has been spoken of as an economical method of getting rid of a troublesome furnace product. When such locations are not convenient, but water is abundant, granulating slag and then floating it, as described in Art. **59,** to some depression in the ground near by is to be recommended. Again, the molten slag may be granulated and elevated to dump piles by some of the methods described for the disposal of tailings. Lastly, the slag may be removed to dumping grounds in a molten condition and used for filling up depressions in the surface, if convenient, and if not, for creating slag heaps.

72. Arrangement of Furnace Plants.—The location of the plant having been determined, the next step is the arrangement of buildings, stock piles, and furnace flues with reference to the furnaces and roasters.

The furnaces and roasters must be located with regard to the quantity of ore that may be delivered and also to any possible increase that may occur to double the output in the future. The actual number of furnaces erected should not exceed the output of the mines, but space should be left for additional furnaces and stock piles, should the ore come in greater abundance in the future than the original furnaces can manage.

Flues, water, and steam supply should be considered with an eye to an increase in the number of furnaces, and should in each case, whether an increase occurs or not, be in excess of the actual requirements of the plant. The reasons for this are obvious: For instance, if one boiler should give out, there would be too little steam; or if an áuxiliary engine of some kind were needed, there would be too little steam power; or, again, in case a furnace were to work badly, it might require an additional blast pressure to cure the trouble; finally, it is economy to have an overabundance of steam supply, rather than a shortage or just enough. The same method of reasoning will apply to keeping an additional furnace in the plant, even though it is idle three-quarters of the time, also reserve engine power, blowers, and water supply. The first cost in such cases will be greater, but it is, nevertheless, cheaper in the long run, because fixed charges about furnaces do not decrease when something goes temporarily wrong, and in furnace work it is a steady output that keeps down expenses.

73. While the arrangement of a furnace plant will depend somewhat on the location, it will also depend in detail on the metallurgist in charge. Details of requirements will necessarily vary, but general details will not.

74. If the plant is to be permanent, the buildings should be of stone, brick, and iron; but if it is a temporary affair, the buildings may be constructed of light framing and rough boards. Sometimes corrugated iron is used for sidings, but it is not as durable as wood; neither will an iron or steel roof last as long as a first-class shingle roof. But a smaller fire risk, which even then is large, ofttimes causes the management to favor the metal construction.

In case the capital necessary for a large furnace plant is not available and must be derived from profits, any temporary structure will suffice; but while the building may be ramshackle, the furnace and appliances should be first class in every respect. The general plan in this latter case should be thought out as if the structures and entire plant were to be erected at once.

75. The Ore Beds.—The space for ore beds must be large enough to permit one to be building while the other is being drawn upon to supply the furnaces. Two beds, each containing 3,000 tons of ore, are sometimes planned. In such cases all the furnaces in blast are fed from one pile while the other is being made up. In case furnaces are running on one character of ore, ore beds are not necessary, ordinary stock piles being sufficient; but where furnaces are receiving ore from many mines, bedding piles are a necessity. Sulphides and carbonate ores are kept in separate piles, the former usually being dumped in close proximity to the roasting furnaces, from which they are transported to the bedding floor and spread. The carbonate and oxidized ores are dumped nearer the furnace and, consequently, the bedding floor. The beds should be protected from snow and heavy rains.

The manner in which the ores arrive is of much importance; that is, whether they come in large or small consignments, at regular or irregular intervals. In case all ores arrive during the summer months, large stock piles may be necessary; this calls for a series of railroad trestles and tracks, if the furnaces are to run throughout the year. In case the ore comes in small quantities at regular intervals, the various ores can be carried, unless they need roasting, direct to the bedding pile.

The question of room for the bedding floor requires particular attention only when various ores are to be mixed and smelted.

76. Location of Smelting Plant.—Whenever a hillside is decided upon for a furnace site, the questions to be considered are, can one terrace be obtained for the ore supply, another for the products of the furnace, and yet another for the slag dump? With insufficient fall, the main stress will be laid on having at least two terraces—one for the ore floor and one for the furnace floor. In case but one terrace is possible, preference should be given to the slag dump; the trestles should then be raised to such a height that the stock piles or at least the fuel piles will be above the top of the furnace. In some instances the furnaces must be situated on perfectly level ground, which requires elevators for nearly every product going into and coming out of the furnace.

77. Handling Materials.—The principal aim at smelting plants should be to simplify the handling of materials, both loading and unloading, as well as charging. This is done by handling the materials to be moved from one place to another in such a manner that they will fall from and into trucks by gravity. The runways between loading and unloading points should not be long, but still they should not be so short as to crowd and prevent the free movement of apparatus. Tracks, elevators, and scrapper lines should all be arranged with this object in view.

WATER SUPPLY FOR STAMP MILLS

78. Water for stamp mills should be free from grease and should contain as little mine water as possible, if that is acid, as is usually the case. The average quantity of water used per stamp per hour in California is about 190 gallons and about 2,600 gallons per ton of ore crushed. The average fineness to which this ore is crushed is 40 mesh. The average quantity of water used per stamp per hour in Colorado is given as 125 gallons, and for each ton of ore stamped about 2,400 gallons. Clayey ore requires more water than harder ores, while the degree of fineness to which the product is crushed also decides the quantity of water that must be used.

79. Loss of Water.—The water mentioned in the above cases is that necessary for stamping where the water runs to waste. For instance, in Colorado a 20-stamp mill would, if each stamp pulverized 1 ton of ore, consume 48,000 gallons per day. It is possible by means of settling ponds to economize in the use of water and use the water over and over again in the battery, less the loss from absorption by the ore and evaporation. The absorption by ore is considered to be 66 gallons per ton for hard quartz and 96 gallons per ton for soft ores. The total loss of water will in no case be less than 25 per cent.

80. Water for the Combination Process.—In this instance, the stamps will require as much water as formerly, say, from 125 to 190 gallons per stamp per hour. Each amalgamating pan will require at least 125 gallons per hour and each settler 65 gallons per hour. These quantities are subject to variations and may be more or less, according to the character of the ore being milled.

81. Leaching processes require water in considerable quantities. Even though the weak and exhausted solutions are used over and over again by being standardized, the fact remains that a 25-per-cent. loss from absorption and evaporation will occur under the most favorable circumstances. On the Rand, South Africa, it is estimated that the tailings after cyaniding retain 30 gallons of water per ton, while the "slimes" contain 240 gallons per ton. The Russell process of lixiviation requires from 60 to 75 gallons of water per ton of ore.

82. All leaching processes require an easy-flowing pulp and then water for wash purposes. In case the ore being treated is silicious and comparatively coarse, not more than 80 gallons per ton will be required, while if heavy concentrates are being treated, probably as high as 120 gallons per ton may be required. The quantity of water required for the treatment of any ore should be determined by experimental tests, and the quantity used will not vary much from those tests.

83. In the **chlorination process**, assuming the quantity of water found to be 80 gallons per ton for lixiviating and to make an easy-flowing pulp, the wash water required will be 140 gallons per ton of ore. This quantity of water is for barrel chlorination and is double that required for tank chlorination, where only enough water is required at first to make the ore wet and porous. The wash water in both cases will probably be the same, from 80 to 140 gallons per ton of ore.

84. Water for the cyanide process varies according to the character of ore being treated. The quantity of water for weak and strong solutions, also for wash water, will probably approach 240 gallons per ton of ore treated, assuming 80 gallons sufficient to form an easy-flowing pulp. For each additional wash 80 gallons per ton of ore in the vat will be required, all tons being those of 2,000 pounds.

85. Water for Concentrators.—The quantity of water that will be necessary for concentrating machines, such as Frue vanners, will be 1.5 gallons per minute of clear water and from 1.5 to 3 gallons per minute with the pulp from the stamps. The quantity of water required for jigs is variously given and the same for classifiers,

The only true way to determine the quantity for any ore is to experiment with the ore and machines. Where there is an unlimited water supply, not much attention need be given to the water that concentrators of this class require, but where water is to be bought, it is good policy to ascertain the quantity required at some mill in the vicinity, or if none, to experiment before jumping at conclusions in regard to the machines.

WATER FOR SMELTING PLANTS

86. This subject is of considerable importance for modern smelting practice, as the quantity of water required by a water-jacketed smelting plant depends on local conditions. The direct object in supplying water to a water-jacket furnace is to prevent the jacket from burning, if of wrought iron or steel, or from cracking if of cast iron. The water should be discharged from the jackets below the boiling point, in order to prevent steam generating in the jackets. No more coke should be supplied to the furnace than is sufficient to smelt the flux and charge, otherwise more water will be required per minute to reduce the surplus heat to normal. The water supply will also depend on the blast pressure, for if that is too strong the heat will be moved upwards towards the *tunnel head*, and less water may be needed. Again, if the ore smelted is of such a nature that it clings to the sides and forms a coating, unless this coating is a good conductor of heat less water will be required than if the jacket walls were perfectly clean.

87. Hoffman states that a furnace 36 in. × 92 in. at the tuyeres, making a silicious calcareous slag, requires under normal conditions 11 gallons of water per minute. This figures out 15,840 gallons per day, which is considerably less than furnace builders recommend. It must be understood that the water supply here spoken of is for silver-lead furnaces.

In blowing-in and blowing-out furnaces the quantity of water must be increased. Mr. Hoffman suggests doubling the regular quantity, or 22 gallons per minute. It must be explained in this connection that the jackets during the warming-up period or "blowing-in" are subjected to a very strong, hot coke fire with little ore to reduce the heat, and that just before the furnace is stopped or "blown out" ore is not charged into it, but coke, until all the material in the furnace has been drawn off, when the furnace is allowed to gradually cool off.

88. Peters, in his book on "Modern Copper Smelting," gives the following quantities of water for furnaces properly managed :

Hearth Area. Square Feet	Water per Hour Normal Running. Gallons
3.0	460
5.0	600
7.0	950
9.5	1,100
12.5	1,300
18.0	1,500
24.0	1,800
30.0	2,000
36.0	2,200

The above table is, of course, assuming that the water is delivered to the furnace at 60° F. Should the temperature of the water delivered to the furnace be 120° instead of 60°, the quantity of water must be increased, for a cool jacket depends on the temperature or heat units per pound of water, and hence the quantity of water necessary for the furnace. Assuming the water left the furnace at 200° F., the heat units per pound of such water will be 168.7. If the water is delivered at 60° F., the number of heat units it will contain will amount to 28.12; the furnace will then have been cooled 140.58 heat units by each pound of water. Supposing the water to have entered at 120° F., it will contain 88.06 heat units per pound, and the water leaving the furnace with 168.7 heat units will have cooled but 80.64 heat units. Referring now to the Peters table, it will be found that if 460 gallons of water were required for the furnace when it was delivered at 60° F., 800 gallons will be required when the water delivered is 120° F. to accomplish the same cooling effect.

ORE DRESSING AND MILLING

(PART 1)

ORE DRESSING

ORE DRESSING MACHINERY

ROCK BREAKERS

1. Classification.—Under the heading of ore dressing are included those processes by which the miner prepares his ore for milling, smelting, or sale. The old method of cobbing ore is practiced in the mine and on the dump, consequently does not enter into this discussion, since it is presumed that ore has been assorted before it reaches the mill. There are two classes of machines used in ore dressing: one class break coarse material and are called rock crushers; the other class crush fine material, that is, pieces smaller than ½ inch in diameter, and are termed fine crushers.

It is true fine crushers sometimes approach pulverizers in their product, but it is not usual or desirable to pulverize ore to impalpable powder for milling purposes, consequently pulverizers are seldom needed.

2. Rock Breakers.—Ore as it reaches the mill is usually in lumps varying in weight from a few ounces to many pounds. To reduce the size so that fine crushers can work it finer, it is usual to employ a class of machines known as rock breakers. These machines are limited to two classes, termed jaw and gyratory crushers, on account of the movement of their parts·in performing work.

§ 25

For notice of copyright, see page immediately following the title page.

3. Blake Crusher.—The Blake crusher shown in Fig. 1 has the entire frame cast in one piece. The swinging jaw *b* is pivoted at *h*, so that its greatest movement is

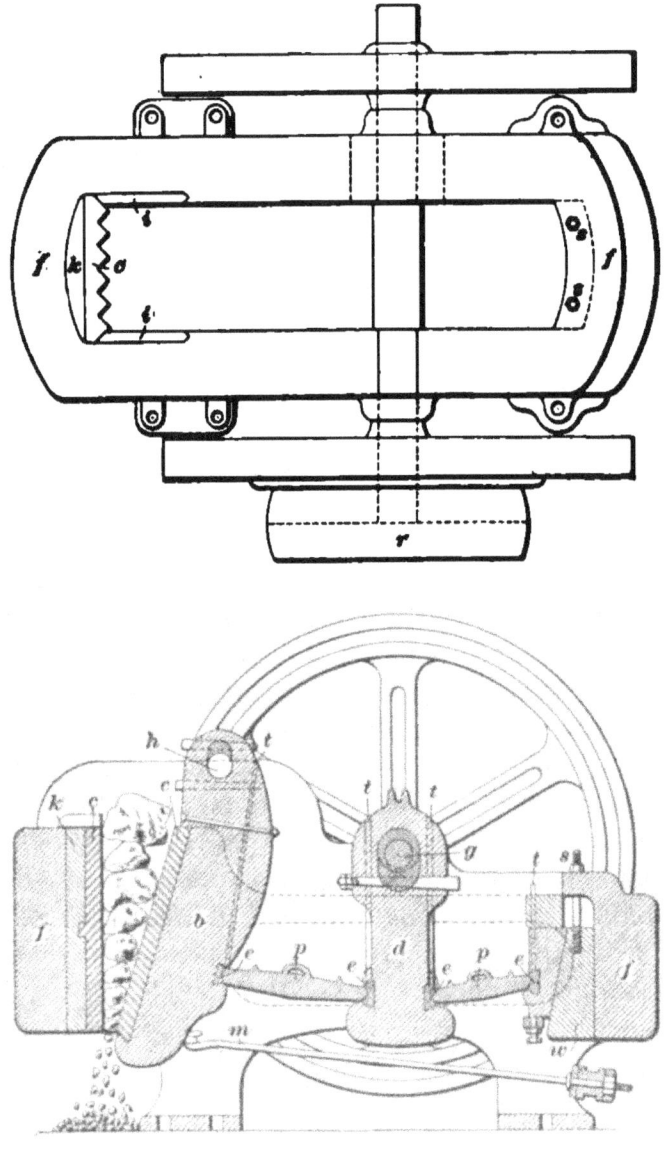

FIG. 1

at the discharge opening. At each full throw of the jaw the discharge opening varies, thereby causing a corresponding variation in the size of the product, and the only control the operator can have over the latter is to set the jaw by means of the wedge *w* and setscrew *s*, thus fixing the maximum size of rock that shall pass through the opening. The pulley *r*, which travels at the rate of from 225 to 250 revolutions per minute, operates an eccentric on the main shaft *g*. As the pitman *d* is raised and lowered with each revolution of the eccentric, the toggle plates *p* impart a reciprocating motion to the jaw *b*. The steel bearings *e* are lubricated through the tubes *t*. A tension rod *m* connects the jaw *b* with a coiled steel or rubber spring, insuring a rapid return of the jaw after each stroke. The jaw plates shown at *c* are usually cast with vertical corrugations, in order to give them a larger crushing area and better bite. The stationary jaw *k*, against which the plate *c* is held by the check plates *i*, is bedded in zinc ¼ inch thick, directly against the frame.

TABLE I

THE BLAKE CRUSHERS

Size of Receiving Capacity	Trade Number of Crusher	Weight of Heaviest Piece	Total Weight	Extreme Dimensions			Proper Speed	Horsepower Required
				Length	Breadth	Height		
Inches		Lb.	Lb.	Ft. In.	Ft. In.	Ft. In.		
3 × 1½	Laboratory	40	100	1 1	0 6	0 10	250	¼
6 × 2	One	560	1,200	2 10	2 1	2 3	250	4
10 × 4	Three	1,800	4,900	4 0	3 3	3 9	250	6
10 × 7	Five	3,800	8,000	5 1	3 9	4 5	250	8
15 × 9	Eight	7,400	15,500	6 6	5 0	5 11	250	15
15 × 10	Nine	7,800	16,000	6 6	5 5	5 11	250	15
20 × 6	Ten	5,300	11,200	5 3	2 11	4 6	250	15
20 × 10	Ten	8,100	18,300	6 10	5 9	5 11	250	20
12 × 30	Sixteen	14,200	33,000	7 10	8 4	6 4	250	30
15 × 30	Twenty	14,200	35,000	7 10	8 4	6 4	250	30

4. Dodge Crusher.—The Dodge crusher shown in Fig. 2 has an eccentric e on the shaft g which gives a rocking motion to the lever l. The lever is cast in one piece with the

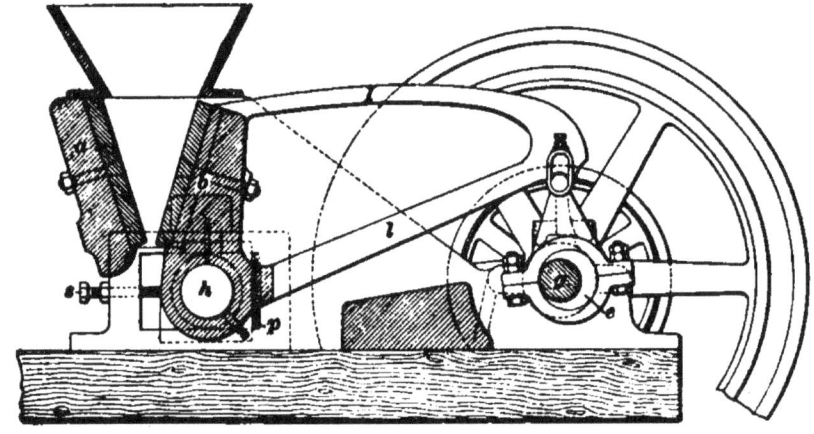

FIG. 2

jaw b and pivoted on the shaft h. The stationary jaw a, as in the Blake crusher, is cast as part of the frame. The jaw shaft rests in sliding boxes, provided with setscrews s to regulate

TABLE II

THE DODGE CRUSHER

No.	Size of Jaw Opening. Inches	Diameter of Pulleys. Inches	Width of Belt Used. Inches	Horsepower Required	No. Tons per Hour. Nut Size	Revolutions per Minute	Weight Complete
1	4 × 6	20	4	2 to 4	½ to 1	275	1,200
2	7 × 9	24	5	4 to 8	1 to 3	235	4,300
3	8 × 12	30	6	8 to 12	2 to 5	220	5,600
4	10 × 16	36	8	12 to 18	5 to 8	200	12,000

the size of the product. The setscrews should always be set tight, with locknuts set, while crushing. Packing strips p are used between the frame and the shaft boxes to take up wear. The shaft makes from 200 to 350 revolutions per minute. This crusher has the discharge opening near the center of motion of the movable jaw, hence the variation in the discharge opening is so slight that a practically uniform product is obtained. This feature of the Dodge crusher has led to its adoption as an intermediate crusher between the coarse and the fine crushers in large mills. The vibration caused by jaw crushers is sometimes destructive to the building. They should be therefore set on timbers or foundations independent of the rest of the framework, especially where there are leaching tanks; otherwise they may throb in unison with the crushers.

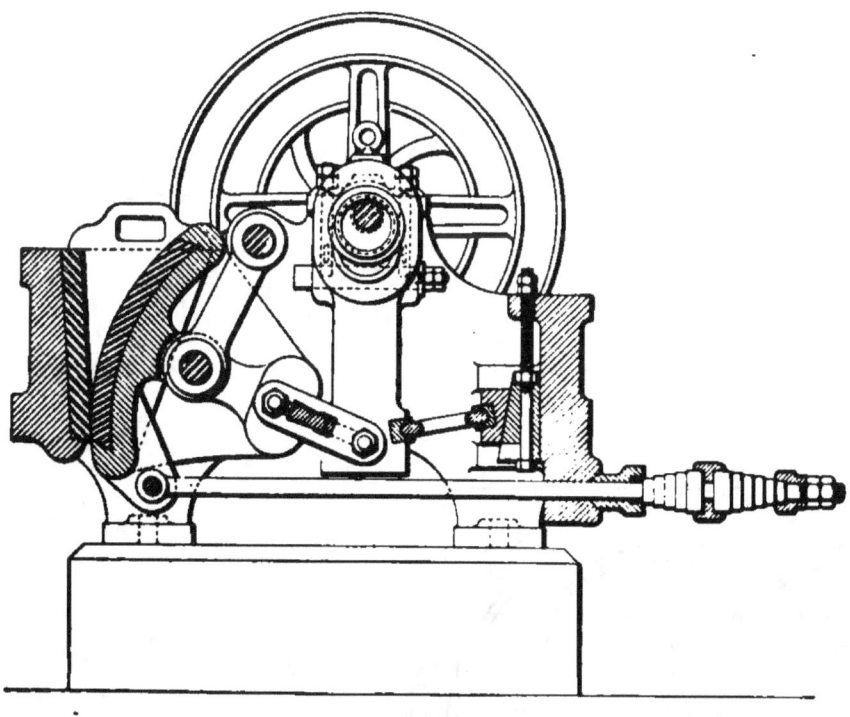

FIG. 3

5. The capacity of the Blake crusher is much larger, on account of the variable discharge opening, which reduces the tendency of the ore to clog; but at the same time this renders crushing to any degree of fineness and uniformity impossible. The Blake crusher is therefore used where the

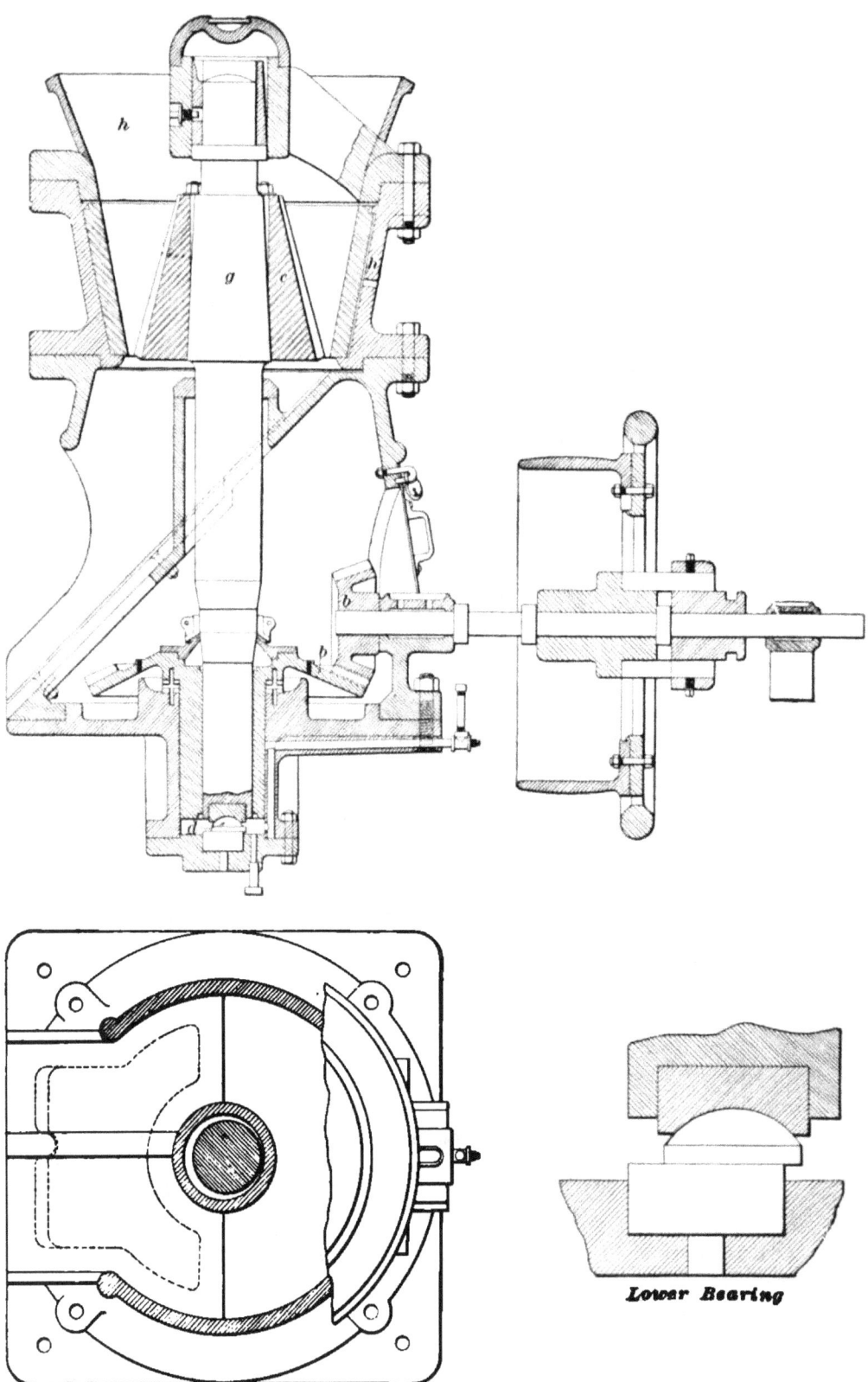

FIG. 4

Lower Bearing

amount of material handled is large, while the Dodge type is limited to mills of moderate capacity, or to secondary breaking in large mills, as previously mentioned.

6. Roll-Jaw Rock Breaker. — In the Schranz rock breaker, shown in section in Fig. 3, the motion, instead of being reciprocating, as in the types previously described, is rocking. The same principle is employed in the **Sturtevant roll-jaw crusher.** The advantages are that owing to the peculiar motion of the movable jaw the material is crushed with comparative ease and the product is approximately sized. The disadvantages are that as the discharge is small there is danger of blocking the machine and their capacity is limited.

7. Multiple-Jaw Crushers. — For special work, multiple-jaw crushers are used. The movable jaw in such machines consists of a number of segments, each of which is worked independently. They are but little used in gold or silver milling, but give a uniform product which makes them desirable crushers for lean magnetic iron and zinc-blende ores, especially where magnetic concentration is practiced and it is not desirable to break the mineral crystals.

8. Gyratory crushers have a large capacity and continuous action. They consist of a heavy cast-iron frame, the upper portion of which is a conical hopper h, shown in Fig. 4. The hopper is lined with a ring of steel, against which the material is crushed by a conical steel head c which fits on a shaft g, the bottom of which is placed in an eccentric bearing so that the amount of space between the hopper and head varies as the head rotates. The material is dumped into the receiving hopper and when crushed passes downwards through the machine and out the spout shown in the cut.

The advantages of this style are that the large pieces of material are received at the top of the jaws, where the motion is the least and the leverage or purchase greatest, thus reducing the work necessary in this heavy preliminary crushing. The relative movement between the crushing members is maximum at the discharge opening, but the amount of this movement is so small that the product is approximately sized. The fact that the maximum movement is at the point of discharge assures a free discharge. There is practically no jerking imparted to the building by the gyratory crushers. Their capacity is very great and besides a large-sized material may be dumped into the hopper *h* directly from the cars. For small capacity, a gyratory crusher is more expensive than a large crusher. Sometimes where there are large amounts of material to be crushed, a large gyratory crusher is used as a secondary crusher after jaw crushers of the Blake pattern, the product from the jaw crushers ranging from 3- to 6-inch cubes and that from the gyratory crushers from $1\frac{1}{2}$- to $2\frac{1}{2}$-inch cubes.

These crushers are made of a capacity varying from 1 ton to 200 tons per hour. They have not come into general use in the West, owing to prejudice against any new and untried machinery, but wherever they have been introduced they have been found to answer as well as any.

FINE CRUSHING MACHINERY

9. Crushing Rolls.—The degree of fineness to which crushing is carried depends upon the screen the ore particles are to pass through. Ore is said to be fine crushed when the particles will pass through a one-half-inch-diameter screen and from this point to an impalpable powder.

The principal representative of this type of machine is the ordinary Cornish roll, having a fairly wide face and rather small diameter. The diameter of these rolls was kept down for a number of years on account of the fact that cast-iron shelves properly chilled could not be obtained in large sizes and were expensive and hard to handle. With the advent of the rolled-steel shells it became possible to employ larger diameters and higher speeds. Rolls of the Cornish type vary from 4-inch face and 9-inch diameter to 16-inch face and 42-inch diameter. The distinctive feature of the Cornish roll is a comparatively wide face, compared with the diameter, and a rather slow peripheral speed. Many of the modern Cornish rolls are provided with rolled-steel shells, especially when employed for very fine crushing, owing to the fact that these shells are of a more uniform texture, work more evenly, and can be worn much finer before being discarded and can be trued up with less difficulty than is the case when chilled iron is employed. The large-diameter rolls with narrow faces are usually employed as finishing rolls, while the rough rolls have wider faces and are not intended for fine crushing, but to crush the product coming from the rock breakers.

10. Theoretical Capacity of Rolls.—The amount of material that can pass between any pair of rolls is proportional to the number of square feet of working surface passing per minute. Hence, the capacity of wide rolls may be increased by increasing the speed, or the

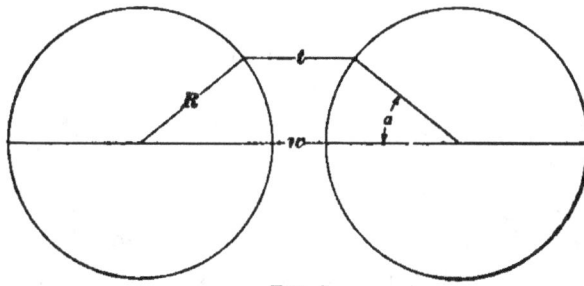

Fig. 5

same capacity may be obtained by reducing the face and increasing the speed. If the distance between the contact points of the material with the rolls be t, Fig. 5, the distance between the crushing face of the rolls be w, the angle be as shown in the figure, and R be the radius of the roll, then $R = \dfrac{t - w}{2(1 - \cos a)}$. The amount of material Q that may be crushed by a pair of rolls in a given time is equal to one-fourth of a layer whose length is the circumference of a roll multiplied by the number of revolutions, whose width is the length of the rolls, and whose thickness is equal to the space or distance between the rolls. Or, $Q = \dfrac{d \pi n l w}{4}$, where d = diameter of rolls; π = 3.14;

$n =$ number of revolutions in a given time; $l =$ length of rolls; $w =$ space between rolls; and one-fourth $=$ coefficient to allow for the irregular feeding of the material and the space between the pieces. The Denver Engineering Works Company gives the following formulas for the capacity of crushing rolls:

$T =$ tons per hour;
$R =$ revolutions per minute;
$S =$ mesh of screen in inches.

For $14'' \times 27''$ rolls, $T = 7.725 \, R \, S$; for $16'' \times 36''$ rolls, $T = 11.775 \, R \, S$; for $12'' \times 20''$ rolls, $T = .327 \, R \, S$. The pressure on the bearings necessary to crush ore depends directly on the face width, and hence if the capacity can be kept the same and the face width decreased, it is evident that there will be less pressure on the bearings and less loss in friction.

11. High-Speed Rolls.—The difficulty experienced with the old Cornish rolls in keeping the bearings cool when crushing hard rock led to the introduction of high-speed, narrow-faced rolls for certain classes of work. One objection to running small diameter rolls fast is that the larger pieces of ore have a tendency to dance off the face of the rolls rather than be crushed, while the bite is better when the speed is slower. The advantages of high-speed, narrow-faced rolls are greater capacity for a given bearing pressure, less loss of power from friction, and less dancing of the ore on the roll face, owing to the fact that the angle of approach between the surface of large rolls is more acute than with rolls of a smaller diameter.

High-speed, large-diameter rolls will also handle coarser material and hence make a greater range of reduction than small-diameter rolls.

One disadvantage of high-speed rolls is the tendency to hammer and pulverize the ore, and this feature, with brittle minerals, may be a detriment. In general, it may be stated that for crushing to any definite size with the least possible production of very fine material, rolls are the best form of machinery on the market. For crushing brittle material fine, quite slow speeds may give the best results. The accompanying table gives some facts in regard to the crushing-roll practice of several manufacturers, data having been taken from information furnished by them.

TABLE III

CRUSHING ROLLS

Name	Size. Inches	Peripheral Speed in Ft. per Min.	Spring Pressure in Lb. per In. of Face Width	Character of Rolls
Frazer & Chalmers...	24 × 8 36 × 16	600–1,500	4,000 for hard quartz	Cornish
Frazer & Chalmers...	44 × 5 56 × 8	2,200–2,300		Narrow face, high speed
Earle C. Bacon......		1,000		Cornish
Sturtevant Mill Co..	16 × 3 27 × 5	3,000		Special Centrifugal
E. P. Allis Co.......	20 × 12 26 × 14 30 × 14 36 × 14	800		Cornish
E. P. Allis Co.......		1,885		Narrow face, high speed
Colorado Iron Works	20 × 12 27 × 14 36 × 16 40 × 16	600	4,000 for hard rock 4,800 for very hard rock	Cornish
Denver Engineering Works Co.........	16 × 10 to 42 × 16	350–1,000	3,500–4,500	Cornish
Gates Iron Works...	9 × 4 26 × 15 36 × 15	470–850	2,266–3,333	Cornish

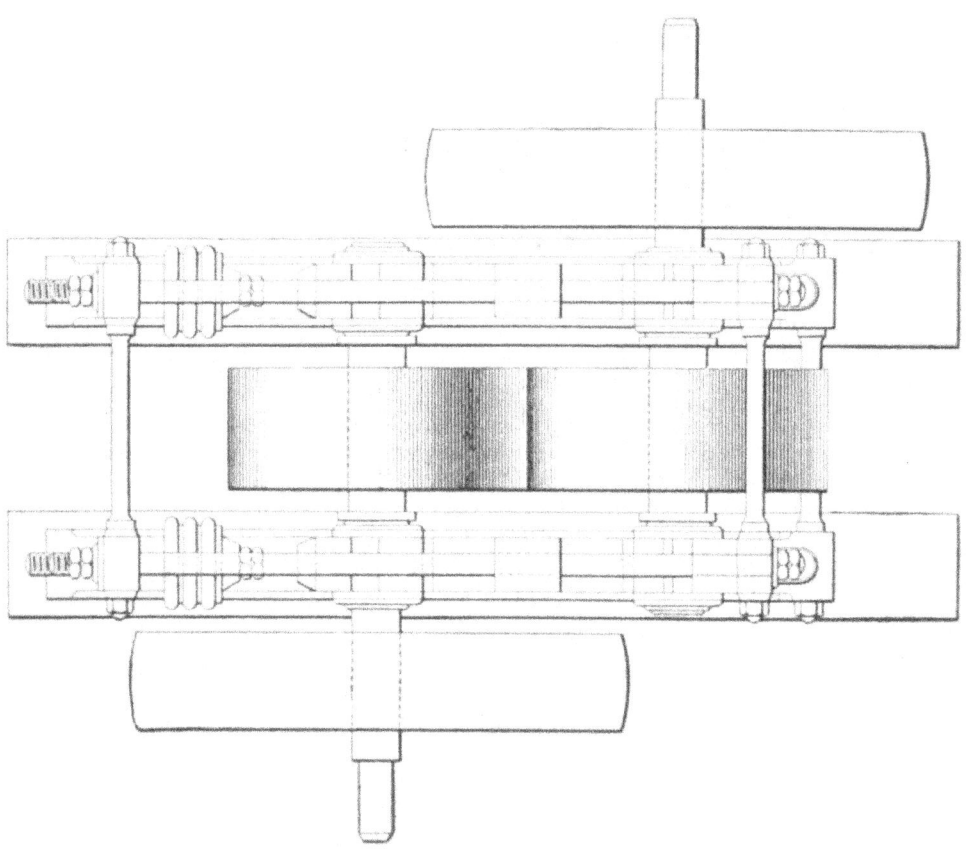

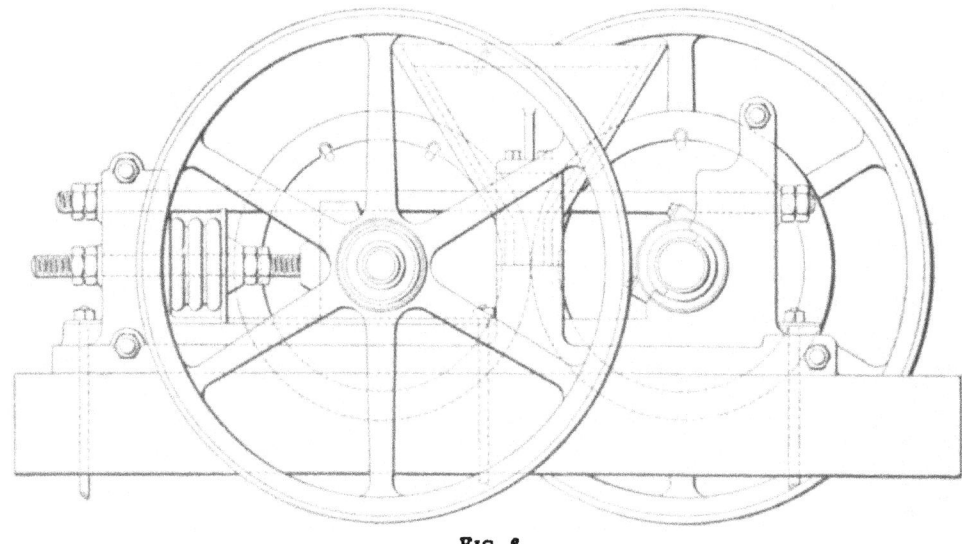

FIG. 6

12. Construction of Rolls.—The ordinary form of crushing rolls, shown in Fig. 6, consists of two cylinders set in a strong cast-iron frame, with their axes parallel and in the same horizontal plane. One roll has fixed journal bearings, cast as a part of the frame. The other roll has adjustable bearings arranged to slide horizontally and take up wear and also prevent rupture of the machine in case some very hard substance, such as a hammer head, should accidentally get into the ore and pass into the rolls. The adjustable roll is held in position by rods which terminate in some safety device, so that in case of such an accident as mentioned, the pressure will force the rolls apart and allow the obstruction to pass through. This was formerly done by means of cast-iron " breaking cups," used as washers for the adjusting rods. When the pressure became excessive, the cups would break and allow the roll to slide back.

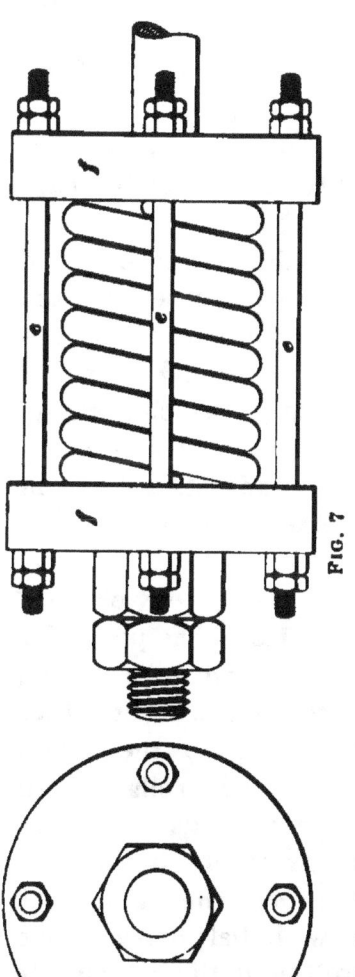

Fig. 7

The cups were inexpensive, but whenever one broke the rolls had to be stopped, and frequently the whole mill had to wait until it could be replaced.

Springs placed in frames *f*, shown in Fig. 7, compressed, say, to 25,000 pounds and then fastened in place by means of nuts on the rods *c*, so that they cannot relax, are now generally used for crushing rolls by placing them as washers on the rods that draw the adjustable roll into position. The roll is thus held in place by a force of 50,000 pounds (25,000 pounds on each side).

In the **Rogers rolls** the axes of the rolls are not in the same horizontal plane, but are in a plane inclined at an angle of about 45°, so that the upper roll practically rests upon the lower while the machine is in operation. The upper roll is an idler, being driven by friction from the lower roll. It is set in a sliding bearing and is held in position by its weight alone, no rods or springs being used except in the small sizes, in which the weight of the roll is not sufficient to perform the crushing satisfactorily. Rolls of this type are only adapted to crush fine ore particles, as they would have to be built inconveniently heavy in order to get weight enough in the idler roll to satisfactorily crush coarse ore.

One of the latest improvements in the construction of rolls dispenses with the sliding journal, its place being taken by the swinging pillow-blocks b, shown in Fig. 8, which are pivoted at p and carry the adjustable roll in fixed bearings g.

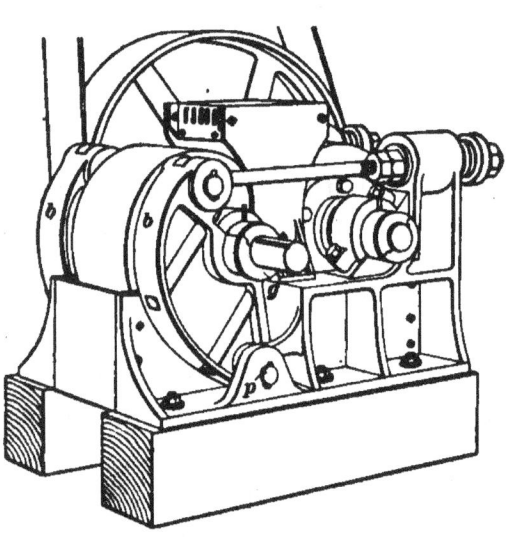

FIG. 8

13. Power of Rolls.—The power for driving may be applied to the shafts of either roll or to the shafts of both rolls. With geared rolls it is almost invariably divided between the two, but with belt-driven rolls it is frequently applied to the fixed roll, and the adjustable roll is driven by friction. It is preferable, however, to have both rolls driven directly, particularly when crushing comparatively coarse material, as this reduces the slip and also avoids the heavy blow resulting from the sudden starting of the idler roll while the driving roll is at full speed—something that is very apt to happen when idler rolls become stopped by clogging. When both rolls are driven directly by belts, the pulleys are placed on opposite sides of the machine, one being driven by a crossed belt and the other by an open belt from the same countershaft. This practice secures the reversed motions.

14. The wear on roll faces is necessarily very great, so that for economy they are made as interchangeable chilled-iron or wrought-steel shells which fit on over a turned core, usually tapered. When the roll shells become so worn that they can no longer be used for fine crushing, they may be used on the coarse crushing rolls until worn out, or the faces may be turned down and the rolls readjusted. With chilled-iron shells the amount which can·be thus removed is limited to the depth of the chilling, but steel shells may be turned down until they are ¾ of an inch thick; below this they will not be safe. When a shell is worn out, it should be removed and replaced by a new one.

15. Application of Rolls.—While rolls may be used for any purpose for which approximate sizing is necessary, by far their largest application is in preparing ore for the jigs, lixiviation, and roasting. As before mentioned, they are frequently used, instead of the jaw crusher, for secondary crushing, even when uniformity is not a desideratum, or for roasting. Dry-crushing rolls must be completely covered, or "housed," with wood or sheet iron, to confine the dust. It is usual in the case of dry crushing for lixiviation to calcine the ore, as that makes it more friable and prevents ore sticking to the face of the rolls. The product is better for crushing and also leaching, but the dust is correspondingly increased.

16. Dimensions of High-Speed Rolls.—In Fig. 9 is shown an elevation of a high-speed roll, with a section of the housing *a* thrown back to show the roll *b*. This, it will be seen, is a narrow roll. The rigid roll, being the main driver, is supplied with two heavy pulleys *c* of large diameter, while the adjustable roll receives its motion from a smaller pulley *d*. A test of 56″ × 8″ rolls showed that they could easily crush *300 tons* of hard granite in 24 hours, when material was fed at 1½-inch ring and rolls set ¼ inch apart;

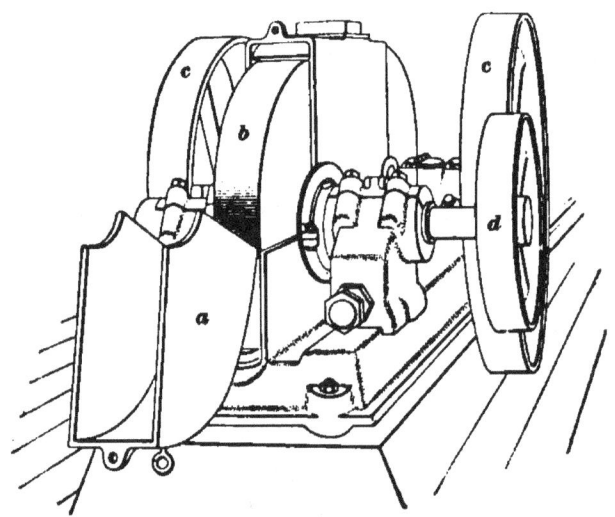

FIG. 9

of the product in this case, 10 per cent. ran over ¼ inch, 90 per cent. passed ½-inch screen, 60 per cent. through ¼-inch, 37½ per cent. through ⅛-inch, 27 per cent. through ₁⁄₁₆-inch, and 12 per cent. through a 30-mesh screen.

WEIGHTS, DIMENSIONS, ETC., FOR HIGH-SPEED ROLLS

Size of Rolls in Inches	Size of Large Pulleys in Inches	Size of Small Pulleys in Inches	Revolutions per Minute	Total Weight Roll Complete Without Housing	Additional Weight for Housing
44 × 5	50 × 7	30 × 6	200	15,500	1,550
56 × 8	72 × 9	36 × 7	150	28,000	2,400

17. Speed of Crushing Rolls.—The most advisable speeds for different sizes of crushing rolls, to give the most satisfactory product and best economy in operation, are based on data furnished by extensive investigation assuming that every size of ore particle fed to the roll should have a certain speed.

18. Theoretical Capacity of Crushing Rolls.—Fig. 10 is a diagram which shows the theoretical capacity in cubic feet and tons of ore crushed per hour for 14″ × 27″ crushing rolls, through screens from 4 to 40 mesh. Referring to the diagram, suppose it be desired to find the cubic feet of ore which will be crushed to 10 mesh when the rolls are running 90 revolutions per minute. Find the figure *90* in the left-hand column of figures marked "Revolutions of roll shells per minute;" follow the horizontal line, passing

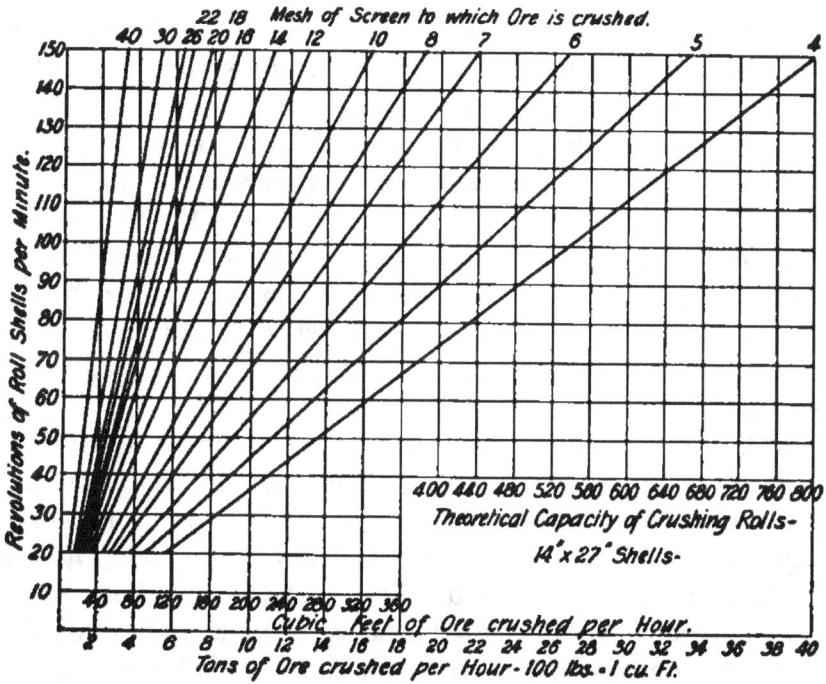

FIG. 10

through *90*, out to the point of intersection with the diagonal line marked *10* in the row of figures marked "Mesh of screen to which ore is crushed." This point of intersection lies nearest the vertical line marked *200* in the line for "Cubic feet of ore crushed per hour." Therefore the *theoretical* capacity for 90 revolutions and crushing to 10 mesh in 14 × 27 rolls is 200 cubic feet per hour.

A second row of figures at the bottom of the diagram gives the "Tons of ore crushed per hour," assuming

100 pounds to equal 1 cubic foot. The actual capacity for several sizes of rolls will be about 60 per cent. of the amounts shown in the diagram.

THE GRAVITY STAMP BATTERY

19. Stamps are under certain circumstances the best of the fine-crushing machines. On account of their simplicity of construction and operation, they are peculiarly adapted to enterprises in which the seat of operations is remote from railway facilities, while their comparative cheapness recommends them for operations of a temporary nature or where capital is limited. As a means of saving the mineral values of ore by amalgamation and concentration, they give a fair efficiency, varying with the character of the ore, the completeness of the plant, and the skill of the operator. The relation between their efficiency and that of a smelting or leaching process, in conjunction with the freight rates, frequently determines whether the ore is to be treated on the spot or shipped to the nearest smelter or reduction works.

In the case of gold-bearing sulphides, the combination treatment is commonly employed. The ore, if it contains any free gold, is crushed under the stamps, and any gold which may be freed from the pyrite is caught and held by the amalgamated plates of the battery. The rest of the crushed ore is passed over suitable apparatus, by means of which the light gangue materials are washed away and the heavier sulphides left, thus greatly reducing the ore in bulk, while retaining practically all the values. The sulphides, or "concentrates," are then shipped to the nearest smelter or reduction works for the final treatment.

20. Stamp Shoes.—The stamp consists of a wrought-iron or steel "stem," 10 to 14 feet long, a cast-iron "head" or "boss," and a chilled-iron or steel "shoe." The two ends of the stem are interchangeable, being slightly tapered to form blunt conical wedges, one of which fits tightly into a hole in the upper end of the boss. In the bottom of the boss is another hole, similar to that in the top, but larger, into which the conical shank of the shoe fits loosely, being wedged in by strips of wood for wet crushing, or by iron for dry

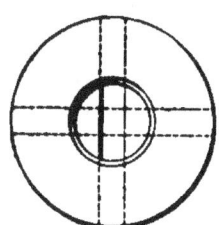

work. In the latter case, the head is strengthened by a ring shrunk around the bottom. The construction of the shoe is shown in Fig. 12 and constitutes the striking surface of the stamp. The

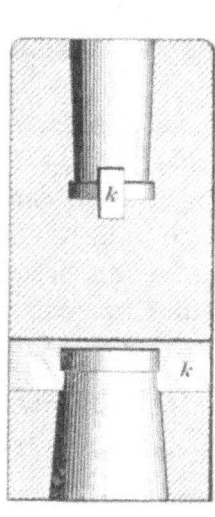

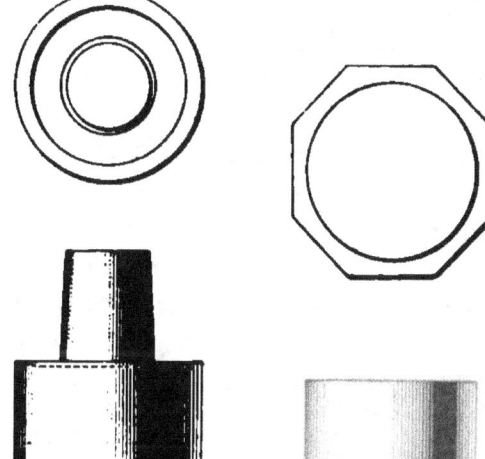

FIG. 11   FIG. 12   FIG. 13

shoe and stem may be released from the head by means of drift keys driven in through the slots *k*, Fig. 11.

21. Dies.—The stationary die upon which the stamp falls is of chilled iron or steel, with a striking surface whose diameter is equal to that of the shoe and a rectangular or octagonal base, Fig. 13. The latter is preferable and is now almost universally used, as it saves iron and renders the removal of the dies easier when "cleaning up" the battery or making re-

FIG. 14

pairs. For dry crushing, round dies are used—preferably with lugs cast on the bottom, as shown in Fig. 14. By giving these dies a quarter turn in the recess in which they rest, the lugs catch and bind them firmly in position.   Dies are also made with wedge-shaped bases, dovetailing into a socket in the mortar.   There seems to be considerable difference concerning the merits of steel and cast iron for stamp shoes and dies.   Some millmen use chrome or manganese steel; others prefer cast-iron dies and steel shoes; still others condemn this practice and state that iron should not be used in conjunction with steel.   In competitive trials cast-iron dies and shoes crushed 1,680 tons, while chrome-steel shoes and dies crushed 16,800 tons.   The wear on iron shoes and dies is between 2 and 3 pounds per ton of ore crushed in California, where in Colorado it is about 1 pound. When the boss is worn down to within 1 inch of the foot-plate, it is to be replaced.   Dies wear more slowly than shoes because protected by a layer of 2 or 3 inches of pulp, but they are replaced more frequently, because of their small height.   A small loss of metal in a die renders it unfit for use. Dies should all be renewed at the same time; if one breaks, one worn as much as the others should take its place in preference to a new die.

22. Cams.—The stamp is raised by means of a double-armed cam keyed to a revolving shaft and catching under a

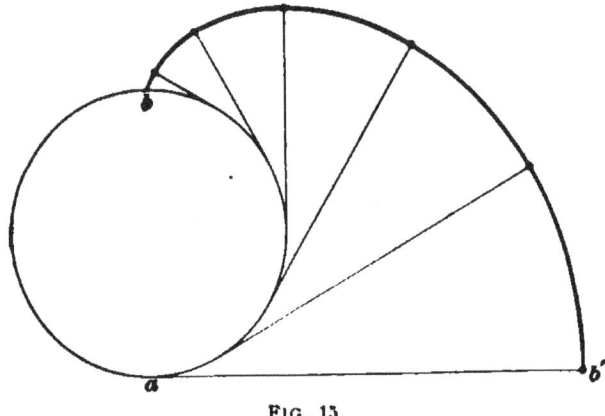

FIG. 13

tappet fastened to the stamp stem.   The curve of the cam is laid off as the involute of a circle, the radius of which is equal to the horizontal distance between the centers of the cam shaft and the stamp stem. The curve may be obtained by attaching to any point *a*, Fig. 15, of a

circular disk of the proper diameter one end of a piece of string, which is then carried around the circumference of the disk to a point *b* diametrically opposite the first point. By holding the disk still and unwinding the string, at the same time keeping it taut, the point *b* will trace the line *b b'*, which represents the lifting curve of the cam. In practice, however, the curve is slightly flattened at the end. Fig. 16 shows right- and left-hand cams, the upper cam being constructed to revolve to the right, the lower one to the left.

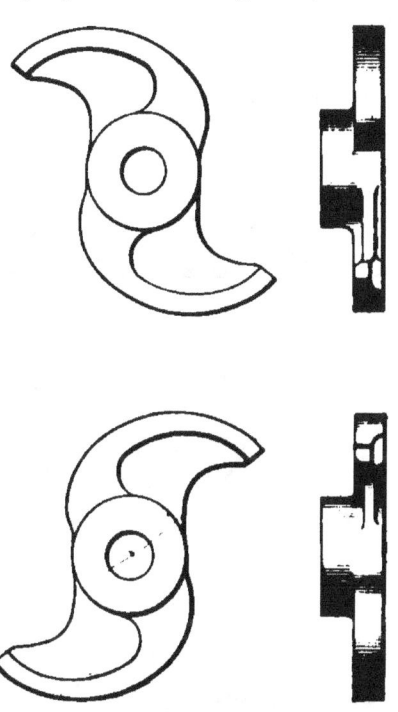

FIG. 16

Cams are usually fastened to the cam shaft by carefully fitted steel keys. There are usually two key seats in each shaft, one-third of the circumference of the shaft apart, though formerly one seat sufficed for all the cams of a set. A later form of fastening is a taper bushing fastened to the shaft by means of pins, as in the Blanton cam, Fig. 17. The shaft has corresponding holes bored in it to receive the pins from

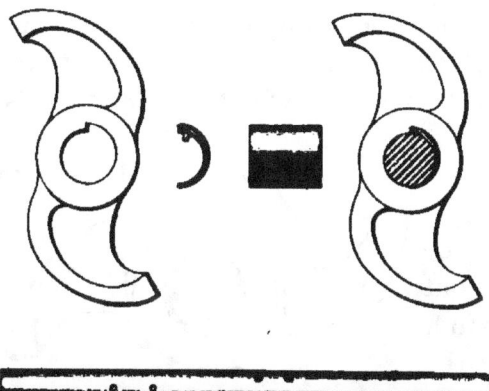

FIG. 17

the bushings. Another device is shown in Fig. 18. The bushings tighten automatically and give a more even bearing than the ordinary key. They possess the further advantage of being much easier to adjust and remove than the old form; in fact, an entire set of cams with this fastening can be

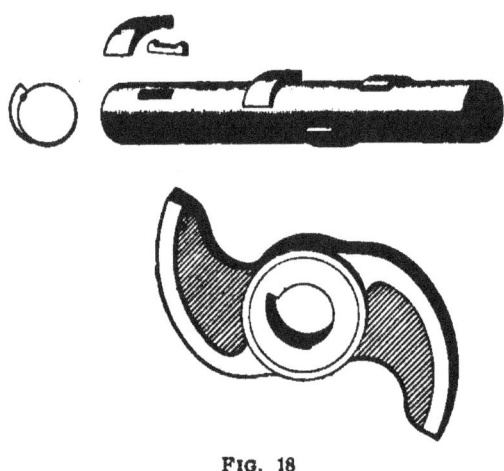

removed and replaced in the same length of time that is frequently required for the removal of a single cam with the old fastening.

The cam shaft, usually of turned wrought iron or mild steel, is driven by a belt from a countershaft. The cam-shaft pulley is of wood, on an iron hub, while the countershaft

FIG. 18

pulley is iron. The countershaft pulley is either fixed or has a friction clutch. The belt is kept taut by ordinary tightening pulleys.

As the cam shaft revolves, the cam lifts the stamp gradually, and at the same time gives the tappet a slight rotary movement, so that the stamp does not fall in the same place on the die twice in succession. When the stamp is raised to the end of the stroke, or "drop," the cam slips from under the tappet, and the stamp, weighing from 600 to 1,000 pounds, drops.

The new Blanton cam is shown in Fig. 19 and the cam shaft in Fig. 20. This arrangement permits the cam to be secured to the shaft without keys or loose parts of any kind. The cam is notched to fit the shaft, which has ten taper

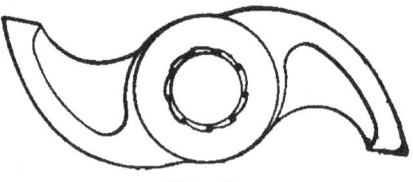

FIG. 19

faces made with mathematical precision, thus insuring a proper fit and division of the cams on the shaft.

It is only necessary to place the cams on the shaft in their respective positions, and tighten on the taper faces sufficiently to hold them in position until put in operation, when

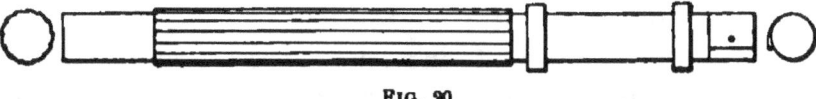

FIG. 20

they will tighten themselves further on the shaft in proportion to the amount of work they have to do.

23. Tappet.—The tappet, shown in section in Fig. 21, is made of cast iron, bored true inside, and is fixed to the stem by means of a gib *g*, which is pressed firmly against

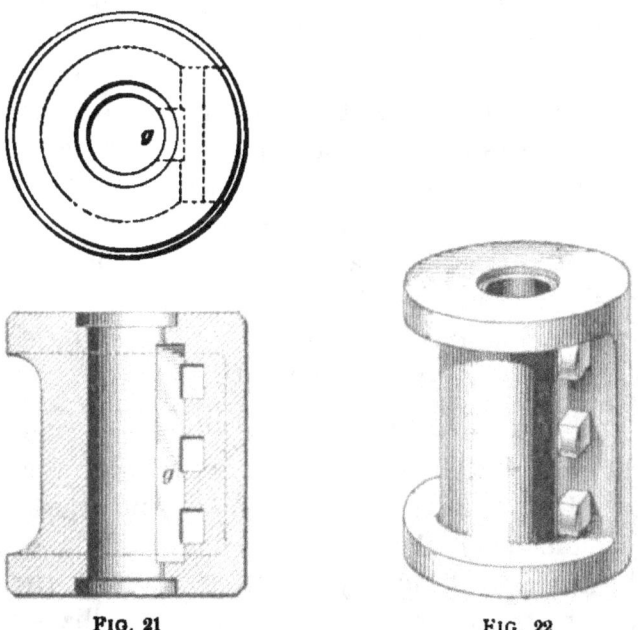

FIG. 21   FIG. 22

the stem by the keys shown in Fig. 22. The wear on the stamp and die is met by raising the tappet so that the drop is kept constant. Each time the cam raises the tappet it imparts two motions—namely, a lifting or vertical and a rotary or horizontal. This rotary motion is due to the cam's curve sliding upon the plane surface of the tappet. It is very important to have this slight horizontal turn given to the stamp.

When a battery is "**hung up**" for cleaning up or repairs, the stamps are supported by hardwood "**fingers,**" shown in Fig. 24, tipped with iron, which are fixed in sockets on a jack-shaft, usually at the back of the battery. There is one jack-shaft to each battery of stamps. The fingers are moved independently of one another by a handle in the back. They hold the tappet high enough to clear the cam in its revolution. To hang a stamp, a tapering stick faced with iron is inserted between the cam and the tappet, raising the latter so that the finger can be slipped under.

24. Arrangement of Stamps.—Stamps are usually arranged in batteries of five each, working in a heavy cast-iron mortar, in the bottom of which are placed the dies. Two- and three-stamp batteries are made for prospecting purposes; while steam and pneumatic stamps have only one stamp to the mortar. There are few permanent stamp mills of less than 10 stamps. The number of stamps at mills is usually increased in multiples of 10, the batteries being set in pairs, with the middle battery post common to the frames of both. The cams of both batteries are fixed on the same shaft, which is supported in the middle by a bearing box on the common post. Instead of the double shaft, two short shafts are sometimes used with one common bearing.

25. The Frame.—The frame in which the stamps move is usually made of well-seasoned heavy timbers bolted tightly together. Iron frames are also made, but they are not to be recommended where timber is available and the battery is intended to stand any length of time, as the jarring works the bolts and rivets loose and crystallizes the iron of the frames, rendering it brittle and apt to break. They are very convenient, however, when a portable battery is desired.

Wood frames are of two general types: the A *frame* and the *knee frame*. The old A frame is, however, being rapidly superseded by the knee frame. In the A frame, shown in Fig. 23, the countershaft is usually placed on the battery sills, almost directly under the cam shaft, and the belt runs

nearly vertical.   With the knee frame shown in Fig. 24 the
countershaft may be set level with the cam shaft, if desired,
the tightener dispensed with, and each battery driven independently by a friction-clutch pulley on the countershaft,

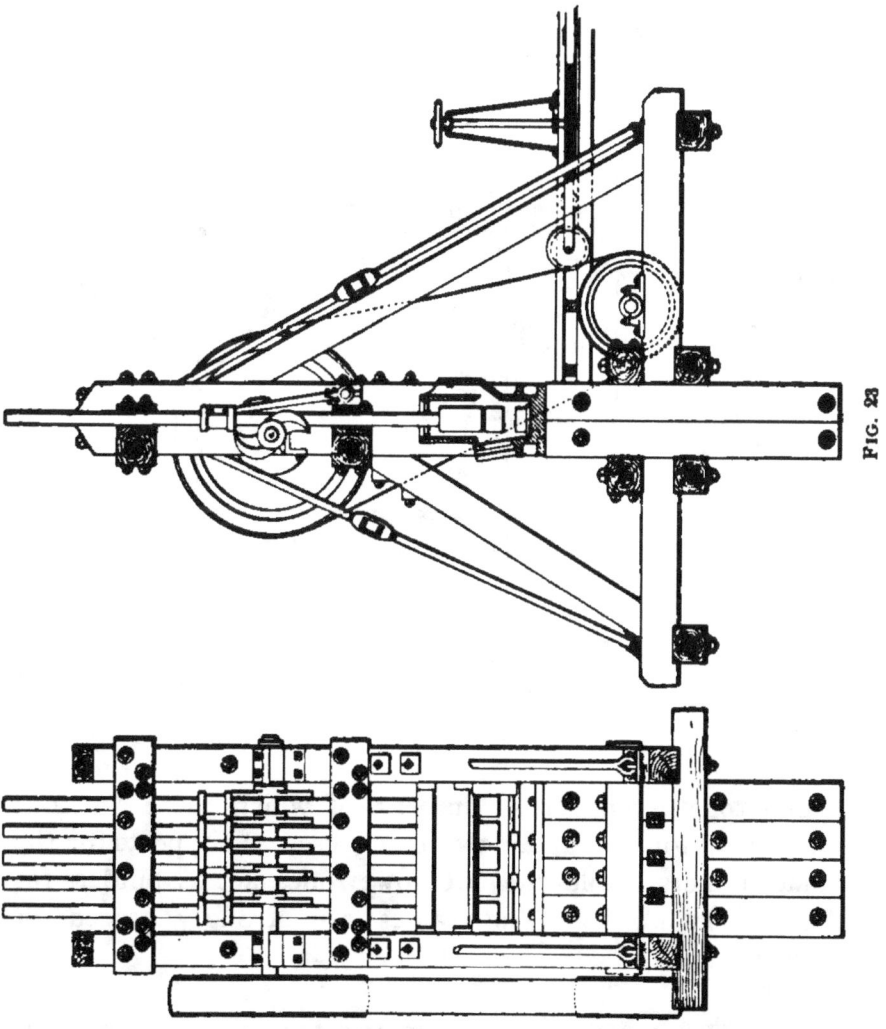

FIG. 23

so that it may be stopped and started without affecting the
rest.   The knees support a floor convenient in hanging up
stamps and in inspecting and repairing the batteries.   The
frames are braced either on the back or front, according as
the countershaft is set behind or in front of the battery.

Fig. 25 shows a back-knee frame with the countershaft on the battery sills.

The battery posts are usually of 12″ × 24″ timbers with their lower ends framed into the 12″ × 12″ sills. They are

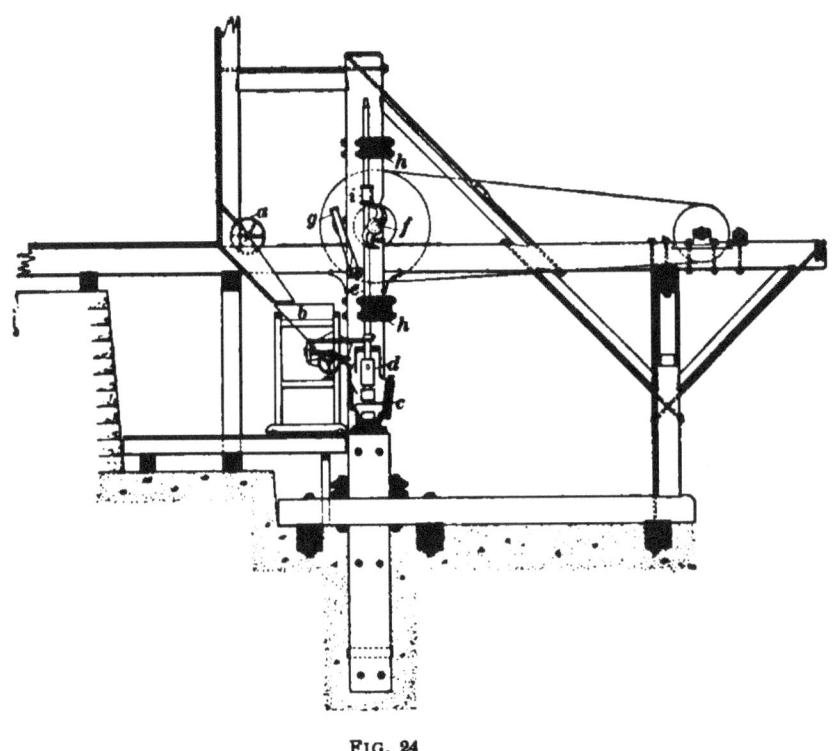

FIG. 24

bound together by cross timbers, of which there are four sets, as shown, the upper two being used as supports for the guide timbers, while the two lower ones act as binders for the battery blocks. The frames are strongly braced for both tension and compression.

26. The Guides.—The stamps are directed in their fall by two sets of guides bolted to the cross timbers of the battery frame. The lower set is generally about 18 inches above the top of the mortar, and the upper set about 7 feet higher. These guides are usually made entirely of wood, Fig. 26, though combination guides of

wood and iron shown in Figs. 27 and 28 present some
advantages in the way
of quick repairs and
economy of timber, as
scrap timber can be used
for guide blocks, and the
iron frame is practically
imperishable. They also
may have the friction
along instead of across
the grain, which increases
the life of the block.
Iron guides have also
been tried, but are not as
satisfactory as wooden
and combination guides.

Wooden guides for an
ordinary five-stamp bat-
tery are usually made of
4″ × 12″ hardwood plank,
and consist of two tim-
bers bolted face to face,

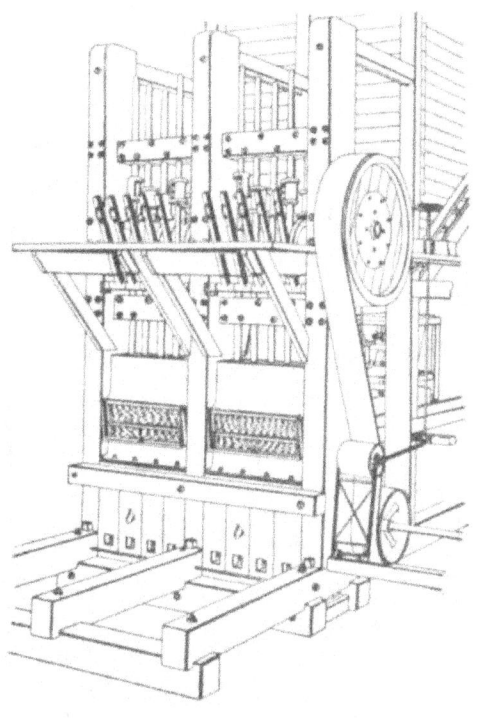

FIG. 25

with grooves on the inner faces for the passage of the
stamp stems. The grooves are not cut hemispherical, but
are left slightly shallow,
with thin strips of wood
placed between the two
timbers, so that as the
grooves wear deeper lost

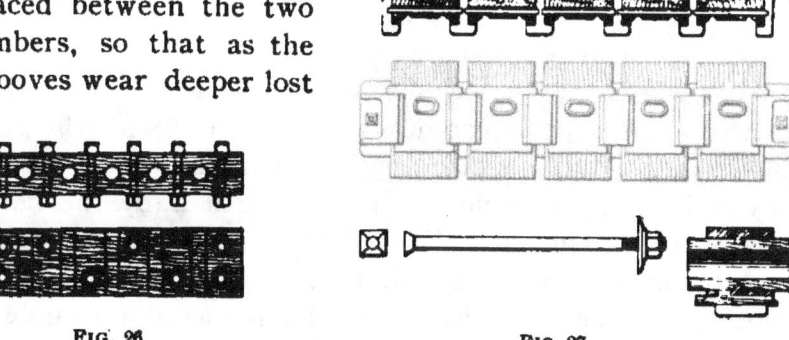

FIG. 26   FIG. 27

motion may be taken up by removal of the strips and by
then drawing the timbers towards each other. The wear

may be still further taken up by planing off the inner faces of the blocks.   The guide blocks fit in between the battery posts and are held firmly against the back of the cross timber by bolts passing through all three timbers.   In case repairs are needed, the back block can be removed by simply taking off the nuts and drawing it off.

Guides should be kept well lubricated with a paste of graphite and linseed oil, care being taken that none of the lubricant gets into the mortar, as it would interfere seriously with the recovery of the gold.

27.   Sectional guides of both wood and iron are also made, as shown in Fig. 28.   This construction permits of

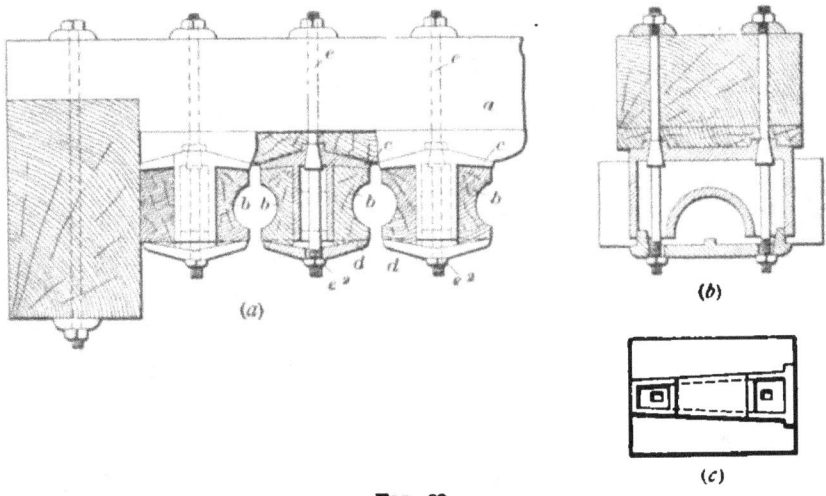

FIG. 28

their adjustment, so that lost motion caused by the stamp stem may be taken up as desired.   In Fig. 28 (*a*) the guide blocks *b* are held in place by two clamping plates *c* and *d*, which in turn are held against the girth *a* by the bolts *c* and nuts *c*[1].   The back plates *c* are dipped rearwardly to fit snugly into the girth beam, so that there will be no lost motion.   When it is desired to change a stamp guide or take up lost motion, the plate *c* is loosened and the block *b* is either taken out or moved forwards with shimmers.   If the guide is badly worn, it is discarded and a new one put in its place; if only slightly worn, it is shimmered to fit snugly.

28. Single-Discharge Mortar.—In Fig. 29 there are two styles of mortars shown in cross-section. In shape they resemble cast-iron troughs. The ore is fed into a slot f at the back, and the crushed ore finds its exit through the screen in the front of the mortar, on to an apron plate or into collecting troughs called launders. The left-hand

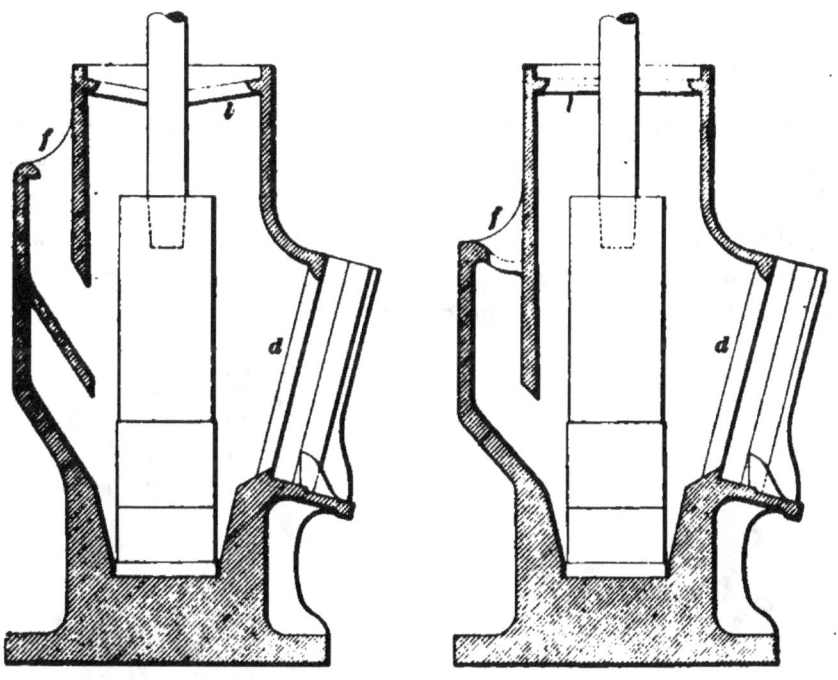

FIG. 29

mortar is arranged for inside amalgamation, both at the front and back, while the right-hand mortar in the figure is arranged for amalgamation on the screen side only. The chief difference between them is in the feeding arrangement. The back amalgamating plate, being put in a recess, is protected from the falling rock as it is fed into the mortar. There are many styles of mortars, some high and some low discharge, some with straight-screen fronts and others leaning outwards, as shown. Each style has an advocate, because that style may be particularly adapted to the work in hand; but the best mortars are built to meet the circumstances of high and low speed, inside and outside amalgamation, high and low drop, also convenience, and all these particulars cannot be embodied in one mortar.

29. Double-Discharge Mortars.—In Fig. 30 (*a*) and (*b*) two double-discharge mortars are shown. Such mortars are used where dry crushing is practiced—that is, no water enters the mortar to assist in washing the ore through the screen. For wet crushing they might increase the product where the ore was coarse, but it would be done at the expense of screens.

30. Feeding Ore to the Battery.—To obtain the maximum crushing duty of a battery, feeding should he carefully regulated and kept just short of the point where the striking of metal on metal is noticeable. If it is allowed to get below this, breakage is apt to follow the uncushioned fall of the stamp upon the die, and even if no break occurs,

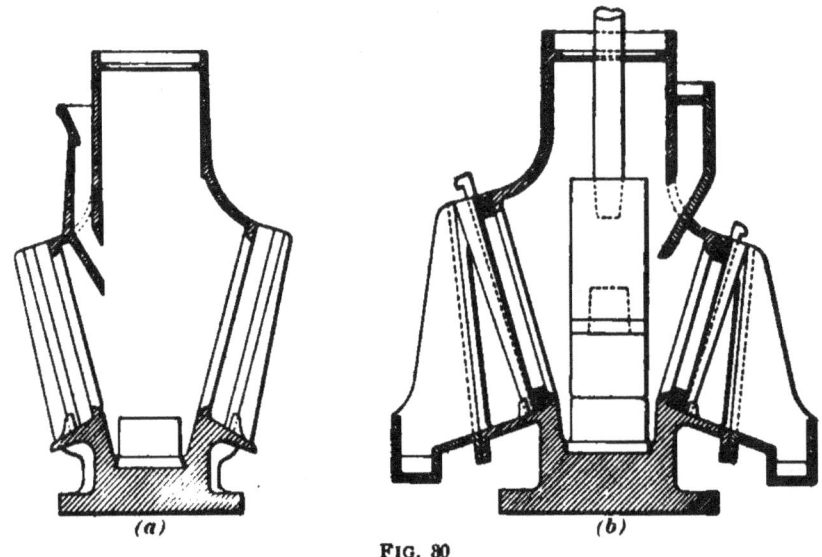

(*a*)   (*b*)

FIG. 30

energy will be wasted to no useful purpose. On the other hand, if too much ore is kept on the dies, the cushioning is so great as to lessen the crushing efficiency of the stamp, the drop is shortened, and the packing of ore and pulp around the stamp between strokes causes an appreciable suction at the beginning of the up stroke. Usually in wet crushing, 2 to 2½ inches of ore on the dies is sufficient.

31. Splash Boards.—The top of the mortar is covered tightly by "splash boards," which are two boards resting on narrow ledges *l*, *l*, Fig. 29, cast on the inside of the mortar, their edges meeting in a close joint down the middle of the mortar, with semicircular notches in each board for the passage of the stamp stems. These splash boards are best set slanting slightly towards the center, as in Fig. 31, with

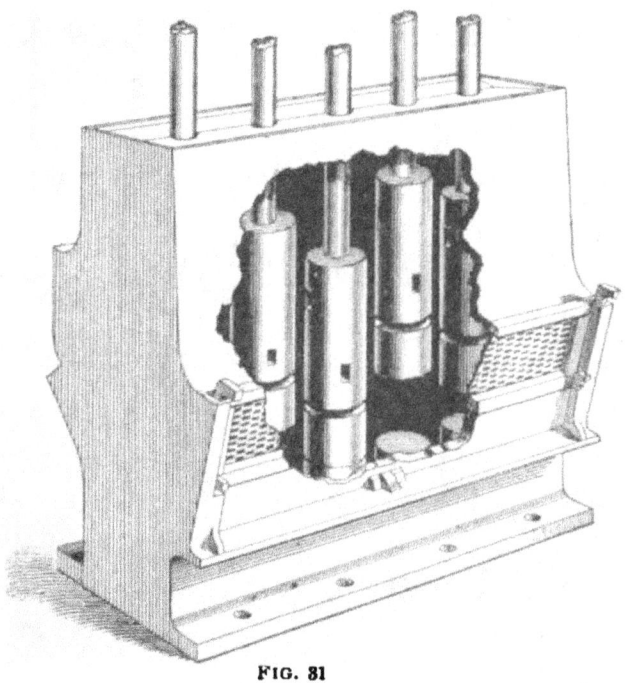

FIG. 31

their edges beveled to make tight joints. The feed opening has a piece of canvas hung inside of it, which prevents the splash from coming in there, and another strip of canvas is hung in the upper part of the discharge opening *d*, above the screen frame, which prevents the pulp from splashing out there and at the same time permits the insertion of the hand to examine the condition of the inside amalgam and to clean the screens when they become clogged. A board is sometimes used instead of canvas over the discharge opening.

32. The height of discharge of a mortar is the perpendicular distance between the tops of the dies and the bottom of the screen. It may vary from 4 inches to 16 inches in the different types of mortars. To regulate the height of discharge, "chuck blocks" *d*, Fig. 32, usually of wood, are placed between the bottom of the discharge opening of the mortar and the bottom of the screen frame. As the dies wear away, these blocks may be replaced by lower ones, thus

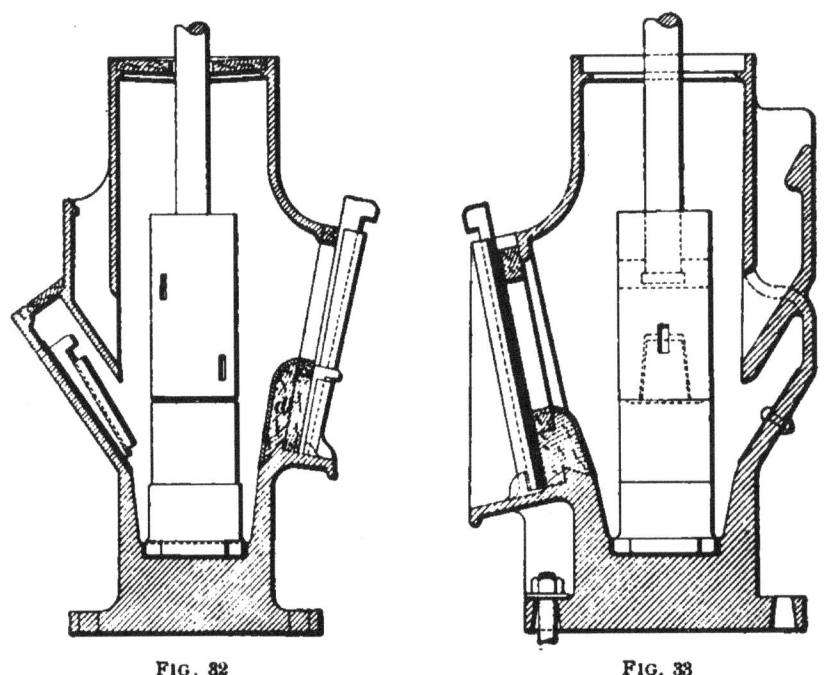

FIG. 82   FIG. 33

keeping the height of discharge nearly constant. In amalgamating mortars, the front amalgamated plates are fastened to the chuck blocks shown in Figs. 32 and 33, while the back copper plates are under the lip mentioned. An iron chuck block made of ¼-inch iron plate is considered by some to be better than wood, as the copper plate is then 1¾ inches farther away from the shoes, and the scouring action is consequently less, allowing the retention of more amalgam on the plates. The copper plates are riveted to the iron with copper rivets, which will not rust out, and the iron itself will last as long as the battery.

33. Sectional Mortars.—In some localities where transportation even by wagon is impracticable, mortars are cast in sections, no part weighing over 300 pounds. The sections are carefully planed and bolted together, as shown in Fig. 34. The upper part of the mortar is built up of wrought-iron or steel plates and angle irons.

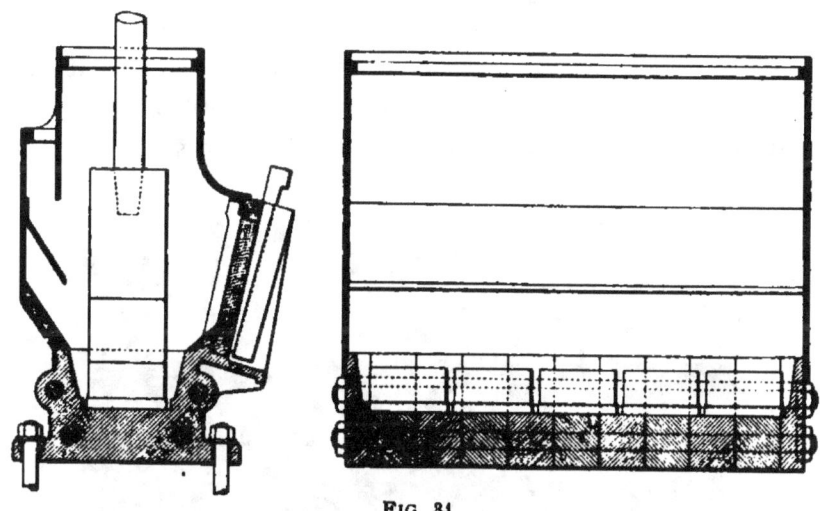

FIG. 34

34. Lined Mortars.—Mortars are subjected to a splash as each stamp falls and are gradually scoured by this action until in time they become too thin or too wide to work well. For this reason they are sometimes fitted with steel linings, which when worn may be replaced without heavy expense.

35. Screens.—The maximum size of the crushed product of a stamp battery is regulated by the size of the mesh of the discharge screens, and the efficiency of the battery depends to a considerable extent upon the care exercised in the selection of the screens. The size to be used on any ore should be determined by experimental "mill runs," with different mesh screens.

Screens made from steel or brass wire cloth are suited only to wide mortars with a high discharge; in such cases, they allow a freer passage of the pulp, which would be apt to clog in punched screens. In low-discharge, narrow mortars, punched screens are used almost entirely, as the increased force of the splash overcomes the tendency of the screens to clog, and they wear much longer than would wire screens, which, even in the wide, high-discharge mortars, last on an average only about one week. A Russia-iron screen is supposed to last at least two weeks before it cracks. Five per cent. of aluminum in copper makes an alloy termed aluminum bronze that lasts several times as long as Russia iron and does not crack, and when the holes are worn large, the screen can be remelted and made over.

FIG. 35

Screens are punched either with *slots* or *round holes*, as shown in Figs. 35 and 36 The slot screens are generally considered the better form, with the slots running diagonally. Screens are also made with the slots in vertical or horizontal rows, parallel and alternating, and in "burred" or "burr-slot" screens; the rough edge is left on the inside of the slot, and can be closed slightly as the slot wears by striking with a mallet. The slots are generally made about ¼ inch long and are punched as small as 60-mesh needle, which is equivalent to .014 of an inch.

36. Screen Frames.—Screens are fastened to rectangular wooden frames, which fit into the discharge opening of the mortars. For wire screens, the frame is divided by inch strips into three or four panels to prevent bulging, and diagonal-slot screens are strengthened by a mid rib. A patented screen is also manufactured, with the screen fitting into the frame in sections which are independently removable. This effects a considerable saving in screen cloth, as the bottom sections, which wear out faster than the top, may be renewed independently of the top panels, whereas in the single screen, the life of the bottom is the life of the

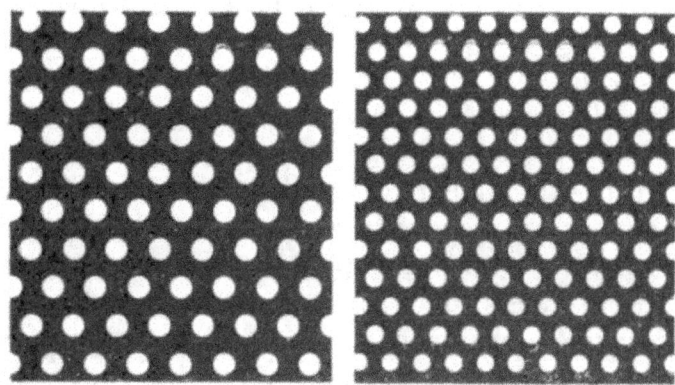

FIG. 36

screen. The frame is placed in the mortar with the screen on the inside, and is usually held in place by one horizontal and two or four vertical keys, Fig. 31, the frame being protected from the keys by strips of sheet iron screwed to it.

37. Order of Drop.—While the cams of a battery are all on one shaft, they are set so that the stamps fall one after another, usually in the following order: 1–4–2–5–3, supposing the stamps to be numbered consecutively from left to right, looking at the battery from the front. This order is preserved with a twofold object: first, to have the "swash" from side to side, so that each stamp as it falls will drive the ore and pulp under the one which drops next in order; and second, to distribute the pressure as evenly as possible and avoid any tendency towards rocking.

With a double battery (10 stamps), the scheme of dropping is extended to the whole ten stamps. Thus, starting at the left as before and numbering the stamps consecutively from 1 to 10, the order of drop would be as follows: 1–8–4–10–2–7–5–9–3–6. This order is sometimes varied according to the fancy of the designer or superintendent, but the idea is always to balance the strains throughout the battery, and have the pulp flow from the mortar in a series of waves.

38. Battery Blocks.—On account of the continuous and heavy jar to which the mortar is subjected, the timbers on which it rests must be very strong and have a very solid foundation. The usual arrangement of the foundation timbers or "battery block" is as follows : A trench is excavated down to solid strata, if such can be found at a reasonable depth, but in any case at least 7 feet deep and about 3 feet larger both ways than the battery block. In the bottom is laid a horizontal timber, or mudsill, of the same surface dimensions as the horizontal section of the battery block; this is carefully leveled and tamped and upon it are stood the vertical timbers which make up the battery block. The contact surfaces should be carefully planed (not sawed) to a perfect bearing. In very loose ground, where a good natural foundation cannot be had, it is customary either to sink piles or to put in an artificial foundation of concrete.

The battery block proper is made up of planed timbers bound firmly together by bolts. Modern practice favors the use of smaller timbers, or even 2-inch plank, spiked together, with joints lapping, and held in place by horizontal binding timbers, which are fastened by transverse bolts. There are usually two sets of these binders, the upper set—made of 8″ × 12″ timber—being usually just above the battery sills and framed into them, and the lower set—usually made of 12″ × 12″ timber—3 or four feet lower down. When plank is used in building up the block, the two upper binders are put flush with the top of the block. When large timbers are used for the block, the lower binders are sometimes omitted.

Plank blocks were first built with the faces of the plank parallel to the short dimension of the block, but this arrangement has given way to that in which the width of the plank runs lengthwise of the block (see *b*, *b*, Fig. 25), as this form is more convenient for repairs, it being only necessary, in replacing a block, to open up the front of the block and raise the mortar, when the plank can be torn out with picks. Plank blocks are much cheaper than the old form of block, made up of two or three huge timbers, which were clumsy to handle, awkward to work, and, moreover, expensive and difficult to obtain in clear lumber.

39. Anchor and Tie-Bolts.—The battery block being in position, the tie-bolts are screwed up until there can be absolutely no play between the various members. Dirt is then thrown in around the bottom and tamped lightly to hold the block in position and the top is planed perfectly level. The holes for the anchor bolts are then bored and the pockets for the nuts chiseled out. A sheet of rubber $\frac{1}{4}$ inch thick is usually next laid on the block, with holes for the passage of the anchor bolts; the latter are put in and the mortar is set in place and bolted firmly down. The pit is then filled and firmly tamped.

The anchor bolts, $3\frac{1}{2}$ to 4 feet long and $1\frac{1}{2}$ inches in diameter, are usually threaded at both ends, pockets being cut in the block for the lower nuts. The lower ends are sometimes turned into rings instead of being threaded, and 2-inch iron bars running horizontally through the block from side to side pass through these eyes, each bar serving for two anchor bolts on opposite sides.

40. In erecting a battery, all bearing surfaces should be planed perfectly true and level, and the centers of gravity of both mortars and stamps should be as nearly as possible in the same vertical plane as that of the battery block, to avoid rocking, which would tend to throw the battery out of line and cause the parts to wear unevenly, greatly shortening their life.

STEAM STAMPS

41. Construction and Capacity.—Besides the ordinary " gravity-stamp " battery just described, there are in use

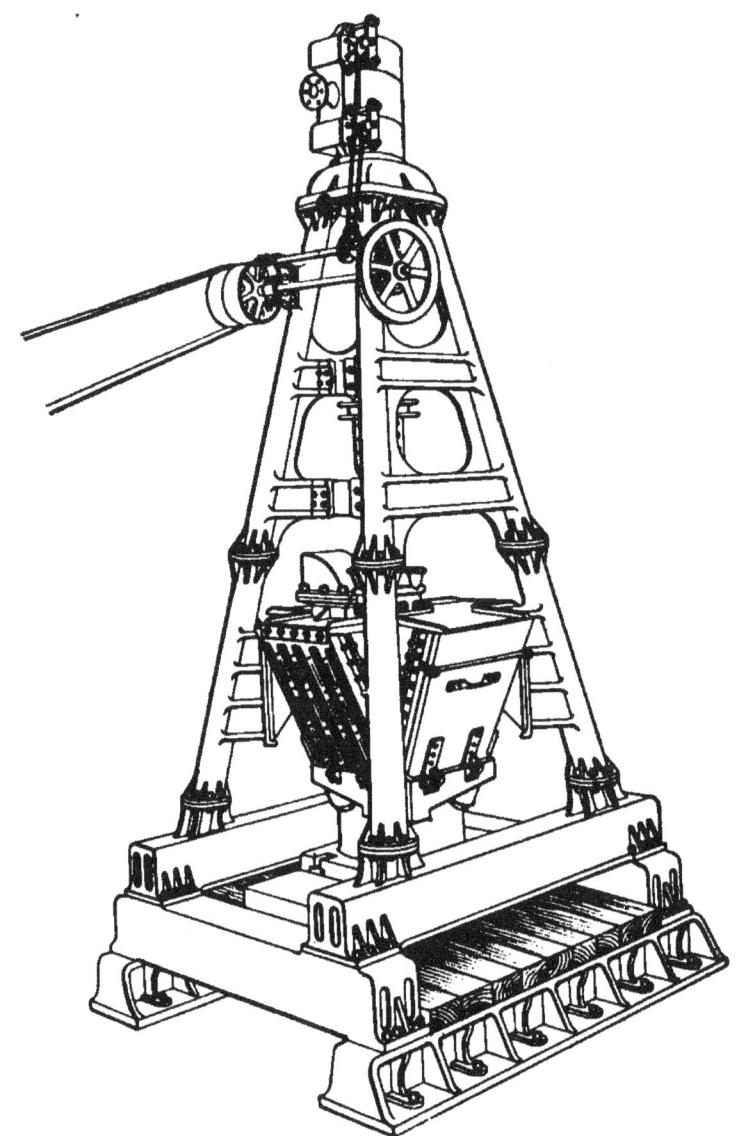

FIG. 37

steam stamps.   In the steam stamp, the stamp stem is the piston rod of a piston moving in a vertical steam cylinder.   The stamp is raised by the steam entering through

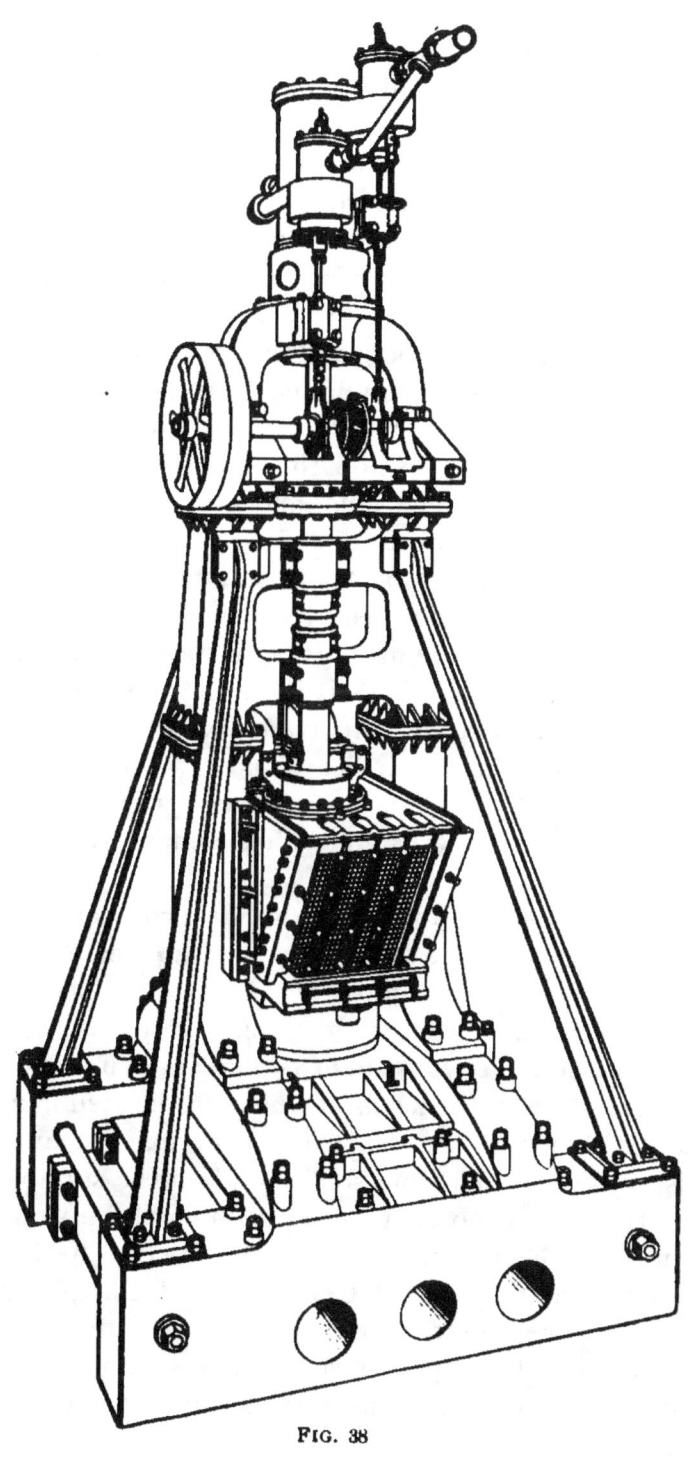

FIG. 38

the lower ports, and when it reaches the top of the stroke the bottom exhaust opens suddenly and at the same instant live steam is admitted through the upper ports, adding impetus to the fall of the stamp. The valve mechanism is such that the piston descends under full boiler pressure of steam, while the pressure on the up stroke may be regulated at will. Steam stamps dispense with the use of a mortar block (battery block), the anvil on which the mortar is built being set on spring timbers, as shown in Fig. 37, with 1 inch of rubber between the timbers and the anvil; or the anvil may be set on a solid cast-iron anvil block, Fig. 38. It is claimed by the manufacturers of the latter style that it increases the crushing capacity; but while this is probably so, it also increases the jar on the frame, and consequently the tendency to crystallize and break.

The crushing capacity of these stamps is enormous, one steam stamp with a 15-inch piston and a 30-inch stroke doing the work of 50 or 60 ordinary gravity stamps. At present the chief use of steam stamps is in milling free copper ores, but they bid fair to be an important factor in gold milling, as these stamps with their heavy blow, strange as it may seem, make less slimes than the ordinary gravity stamps—a great consideration in crushing before concentration.

42. Tremain Steam Stamp.—For prospecting or for small mines where an extensive stamp battery is not required, or where it is simply desired to develop the mine and test its value, the Tremain steam stamp has come into very general use. It consists of two steam cylinders, which operate two stamps, the heads of which are located in a single mortar, as shown in Fig. 39. In the Tremain steam stamp the pistons are turned from solid metal on the stamp stems, so that the area exposed to the steam on the up stroke is smaller than the area exposed to the steam on the down stroke by an amount equal to the area of the piston rod. This area on the up stroke is simply a ring about ¾ of an inch wide around the piston rod. Live steam is employed in raising the pistons, it being introduced on the under side, and, acting on the small area, raises the pistons and stamps, which weigh only about 300 pounds each. As one piston ascends, it moves a valve which admits live steam to the bottom of the other piston and vice versa, so that the valves of one stamp are operated by the piston of the other in such a manner that the stamps always alternate, one ascending as the other descends. After the steam has raised the stamp, a passage is opened which allows it to expand around, and acting on the greater area of the top of the piston, it urges the stamp down very much more rapidly than it would fall by gravity, so that the blow actually struck is equal to that of an 800- or 1,000-pound stamp. After the steam has urged the piston down, an exhaust valve is opened and the steam exhausts into the atmosphere as the stamp ascends. On account of the fact that these stamps are operated positively by steam, they can be run at a very much more rapid rate than ordinary gravity stamps, it being possible to use a speed of 200 or more drops per minute for each stamp, and hence the crushing power is very much increased.

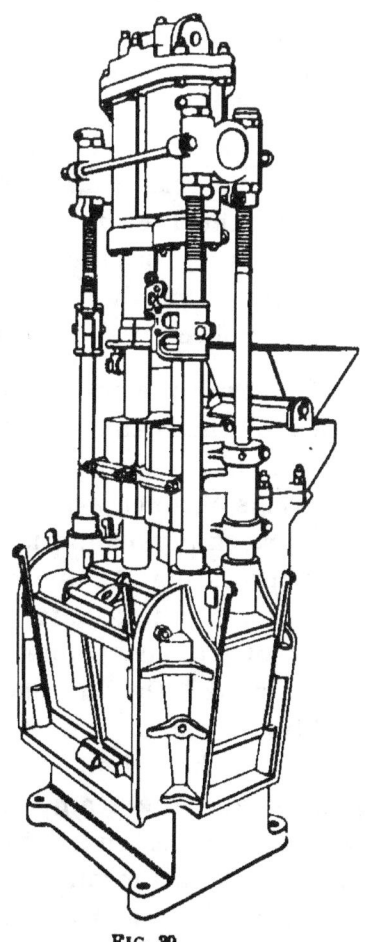

FIG. 39

Some of the advantages of this style of steam stamp are that it is comparatively light and portable and requires no extensive foundation. The stamp can be run by steam from the same boiler which runs the hoisting engine and mine pump, so that it does away with the necessity of an extra engine or separate motor for operating the stamp mill. The capacity of a two-stamp battery on average ore is greater than that of an ordinary five-stamp gravity battery, and in some cases mines have been operated for a considerable period with Tremain stamps only. In case it is found desirable to move the mill from one location to another, the battery can be taken up and transported with as great ease as either the boiler or engine, and this portable feature is one of the most important factors in the selection of this style of mill. Where the Tremain stamp is used without a crusher, it will be necessary to break up the coarse rock with sledges, but steam stamps can handle much coarser material than the gravity stamps. Owing to the fact that the Tremain stamp, like other steam stamps, produces less fine or slimed material, it is especially applicable for cases in which it is desirable to concentrate ores after they have been passed over the amalgamation plates.

43. Pneumatic Stamps.—The style of stamp usually called by this name is not driven by compressed air, as its name would indicate, but usually consists of two stamps, the stems of which are attached to pistons of small diameter which work in pneumatic cylinders. These cylinders have a reciprocating up-and-down motion given to them by means of a crank-shaft, with which they are connected by means of ordinary connecting-rods. When one of the cylinders is raised, the air beneath the piston is compressed and so lifts the stem and stamp; similarly, on the downward stroke of the cylinder, the air above the piston is compressed and the stamp driven down more rapidly than it would have descended by gravity alone. The stamps are provided with a means for rotating them, so as to give better or more even wear to the shoes or dies. This style of stamp has been used principally for working the tin ores in Cornwall, and their output is from 20 to 30 tons per head per day, using a 36-mesh screen. The power required is usually about 25 H. P. per head. Like the steam stamps, these pneumatic stamps produce less slime than ordinary gravity stamps.

44. Quick and Slow Drop Stamps.—Stamp batteries for the treatment of gold ores are divided into two general classes, according to the kind of work for which they are intended. For " free-milling " ores, in which the gold occurs native in comparatively good-sized grains, unassociated with pyrites, or "sulphurets," the *California* battery is used. The stamps in this type of machine are very heavy—800 to 1,100 pounds—the average weight being about 950 pounds. They fall through only a short distance, from 4 to 8 inches being the average, but they make from 80 to 100 drops per minute. The mortar is built with a low discharge from 4 inches to 8 inches, and in the original form no battery plates (amalgamated copper plates placed inside the mortar) were used, all the gold saved being caught on the apron plates. In most modern batteries of this type, however, an effort is made to save part of the gold in the mortar.

When the discoveries of rich veins of gold-bearing pyrites were first made in Colorado, batteries of the California type were put to work on the ore. They gave very unsatisfactory results, however, compared with the assay value of the ore, and in the attempt to increase the saving, the millmen evolved the so-called *Gilpin County* battery. The weight of the stamp was gradually cut down and the drop lengthened, and the width and the height of discharge of the mortar were at the same time increased, finally resulting in the typical Gilpin County battery, with stamps averaging about 600 pounds in weight and falling through a height of from 16 to 20 inches into a roomy mortar with a high discharge (12 to 16 inches) and battery plates. The stamp battery averages about 30 drops per minute.

The reason for the change lies in the fact that in gold-bearing pyrites the gold is in a very finely divided form, lying in the cleavage planes of the pyrite. In crushing this ore, a great deal of the gold is set free, but is so fine that it is apt to be floated away without coming in contact with the amalgamated plates. This fact was not known to the early millmen, and at first they put up mills designed to amalgamate all the gold on the apron plates, as they had been accustomed to do in milling free-gold ores, but clean-ups failed to give the proportion of the assay value of the ore that they had anticipated, so they concluded that the gold must be combined very closely with the pyrites, and turned their attention to the concentration of the latter, with little or no attempt at amalgamation. The results were still unsatisfactory; but the true condition of the gold at last became known to them, and to meet it they first put amalgamated plates into the mortar. This was an improvement, but it was found that with quick drop and low discharge, the battery plates were not given time enough to satisfactorily perform their work; so, step by step, the weight of the stamp was diminished and the depth and width of the mortar increased, until the extreme type of high-drop battery was obtained. As is usual in such cases, the changes were carried rather to extremes, and modern batteries are usually a compromise between the two old forms. About 75 per cent. of the total gold saved by amalgamation is now caught in the battery. The pyrites are concentrated and shipped to the nearest smelter or reduction works for treatment, and at some large mills the tailings are leached by potassium cyanide solution.

ROLLER MILLS

45. Chilian Mills.—The term *roller mills*, as here used, comprises all the modern forms of crushing machinery in which the crushing is done by rollers moving in a horizontal pan. These machines are constructed on the same general plan as the ancient *Chilian mill* which crushed ore by heavy stone rollers in a stone pan.

The modern machine of this type is essentially a steel or iron pan with a vertical spindle passing up through the center and bearing two or more radial arms, at the extremities of which are the rollers of chilled iron or steel. The pan in the path of the rollers is also lined with chilled iron or steel, furnishing a durable crushing face. Crushing is done either wet or dry, the crushed pulp being discharged through screens in the sides of the pan.

Roller mills are of two general types: those in which the axis of the roller is a continuation of the radial arm, and which may, for convenience, be called *radial roller mills;* and those in which the roller revolves on an independent vertical axis, which may be called *centrifugal roller mills.*

46. Radial Roller Mills.—In some machines of this type, the central shaft is stationary and the pan revolves, in others the pan is revolved in a direction opposite that in which the rollers are moving.

The rollers are usually cylindrical. Moving as they do in a circular path, the outer end of each roller has to traverse a longer arc in the same time than points farther in towards the central spindle, the distance any point traverses being proportional to its distance from the center of revolution. The various points on the surface of a cylindrical roller all revolve at the same velocity about the axis of the roller; and the outer end of a horizontal cylindrical roller revolving about a vertical axis is obliged to slide in order to make up the difference between its linear velocity around its own axis and its velocity around the central vertical spindle. The inside end, on the other hand, probably drags somewhat, and at only one point along the whole line of contact between the roller and the pan is the motion purely *rolling.* As the idea of the machine is to crush entirely by rolling contact, this is theoretically very bad, but practically it answers very well for all ordinary work, the only serious objection being the uneven wear on the crushing faces. But when it comes to crushing ore to a very fine, uniform size, as for leaching, the difficulty assumes a practical aspect, the amount of sliming due to the slip of the rollers being considerable.

This difficulty is somewhat overcome by making the rollers and pan bottom conical, and having the apex of the cone angle of the rollers correspond with the apex of the cone of the bottom. In this form every point on the roller along the line of contact with the pan moves with the same velocity as the corresponding point on the pan bottom, and consequently without slip. In the Schranz mill the central spindle is stationary and the pan bottom revolves, driving the rollers by friction. Other radial roller mills use cylindrical rolls with a comparatively narrow face, so that the difference in velocity between the two ends of the roll is very small, and there is consequently little slip.

47. Modern Chilian Mill.—As has been explained in the previous article, the radial roller mills do not crush the ore primarily, but both crush and triturate, or grind, it.   There are certain cases in which this triturating action is very desirable, as, for instance, in the preparation of clay material, and hence most, if not all, of the clay mills are of the Chilian mill type.

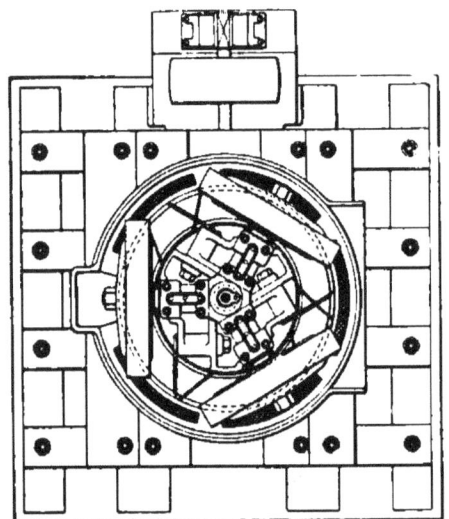

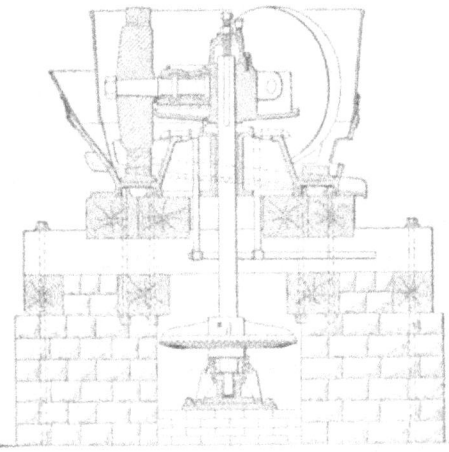

FIG. 40

In Old Mexico and other locations where rich gold and silver ores occur in regions containing practically no water, it is often necessary to reduce the ore by the patio process, and for this purpose the material is ground in Chilian mills driven by mule power. The mills were originally of the old crude type, but at present the greater number of them are of improved construction, and many of them are similar to that illustrated in Fig. 40, which shows a plan and section of a Chilian mill provided with three rollers. Silver ores intended for pan amalgamation may also be ground or pulverized in Chilian mills, and a large number of millmen prefer this style of machine, even for use with gold ores which are to be amalgamated on plates outside of the machine. They have also been employed for the regrinding of middlings in concentrating works, but are not as well adapted for this purpose as the Huntington mill. The Chilian mill usually has but two crushing rolls or wheels.

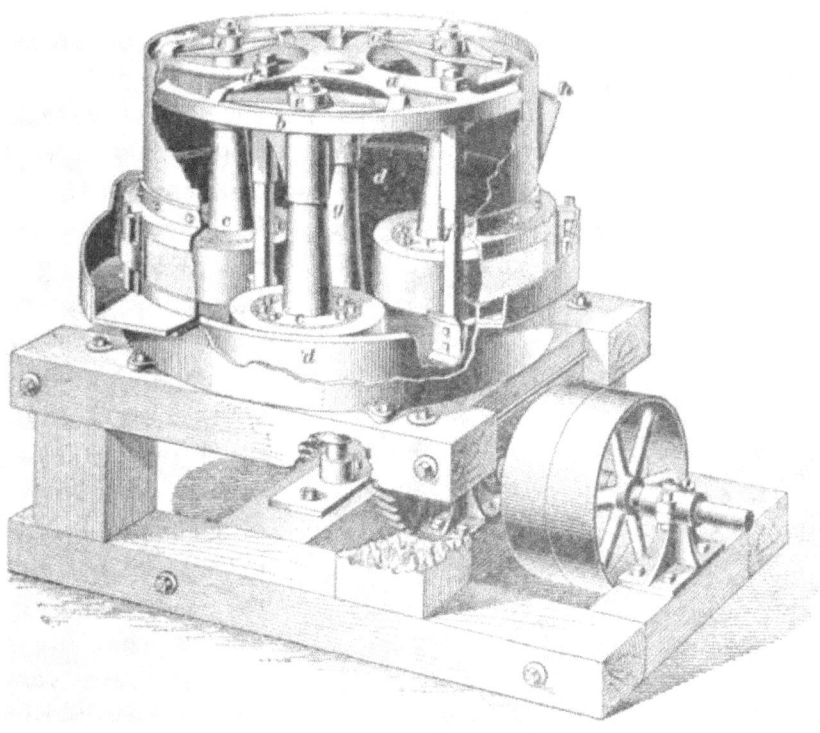

FIG. 41

48. Centrifugal Roller Mills.—The second type of roller mill is termed *centrifugal*, because the crushing action is largely due to the centrifugal action of the rollers, which are hung so that they are free to swing outwards in a radial direction, the rapid revolution of the spindle causing the rollers to press against a hardened ring die in the side of the pan, immediately below the discharge screens. These mills are used for wet crushing and amalgamation, also for regrinding the intermediate products in concentration mills. The type is represented by the *Huntington roller mill*, which will be described in detail, as its success and extensive adoption in gold metallurgy have placed it upon a high level as a crushing and amalgamating machine.

49. Huntington Mill.—The Huntington mill, shown in Fig. 41, consists of an iron pan, through the center of which passes the vertical shaft *g*. The arms *a* extend horizontally and terminate in a ring *b*, from which the rollers *c* are hung so as to swing freely in a radial direction on the yoke *e*, which is shown in detail in Fig. 42. The rollers are also free to revolve on their own axes. In front of each roller is a scraper *f*, which keeps the ore from packing. The rollers are hung so that they clear the bottom of the pan by about an inch. The central shaft, revolving at a rate of from 45 to 75 revolutions per minute, causes the rollers to swing outwards against the ring *d*, the pressure varying as the square of the speed of the mill. The crushing power of the mill, being equal to the product of the pressure and the velocity, consequently varies as the cube of the velocity; that is, twice the speed gives eight times the crushing power.

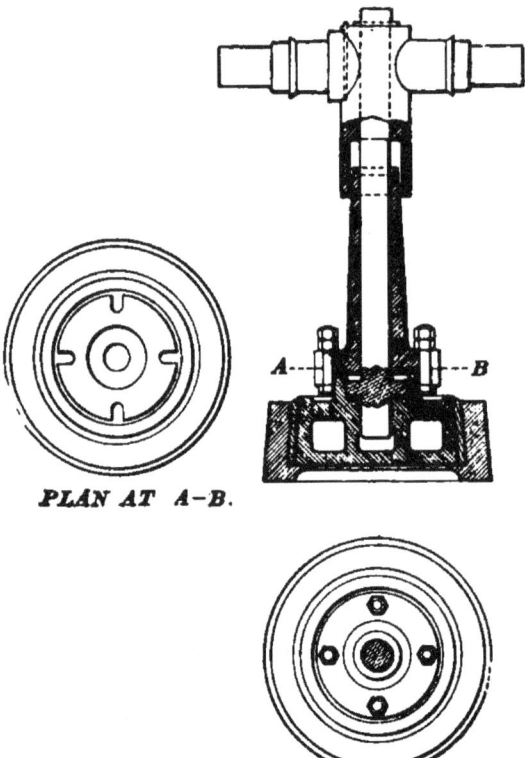

PLAN AT A–B.

PLAN.

Fig. 42

The ore, which has been previously broken to ⅛ inch in diameter or smaller, is fed into the hopper *h* on the side of the pan. The rollers and scrapers throw it out to the rim of the pan, and as fast as it is crushed it passes through the discharge screens and trough shown to the left of the figure on to the amalgamated plates, not shown in the figure. The greater part of the gold, on being liberated from the gangue, sinks to the bottom of the pan, and is caught there by the quicksilver, of which there are from 17 to 25 pounds. The clearance of the rollers prevents their "flouring" the mer-cury, and at the same time they are close enough to keep the surface of the mercury agitated and in the best possible condition for amalgamating.

This mill is particularly adapted for the treatment of brittle sulphide ores, which under the stamps are apt to slime.

It is made in three sizes, $3\frac{1}{2}$, 5, and 6 feet in diameter, the second being the most commonly used. A 5-foot mill will crush from 10 to 20 tons of rock in 24 hours, through a 30-mesh screen, at an expenditure of from 10 to 12 horse-power. The first cost of the mill is considerably less than for a stamp mill of the same capacity, and the parts are fairly cheap and easily replaceable. The power per ton of ore crushed is much less than for the stamp mill, the pulp is in better condition for concentration, and the loss of mercury is minimized. Its disadvantages are that the wear on the parts is great—particularly on screens, dies, and shoes— and that the corroding action of the acid in some mine water and in decomposing pyritic ores soon renders the machine unfit for use. This latter source of trouble is also present in the stamp battery, but the great thickness of its permanent parts makes it of less moment than in the Huntington mill.

Huntington mills have an especial field in the regrinding of intermediate products in concentrating mills.

50. The Kinkead Mill.—This mill is simply a large automatic mortar-and-pestle arrangement. The mill is shown in Fig. 43. The upper portion a of the spindle is vertical and revolves in fixed bearings directly over the center of the pan, which is shaped as shown in the figure.

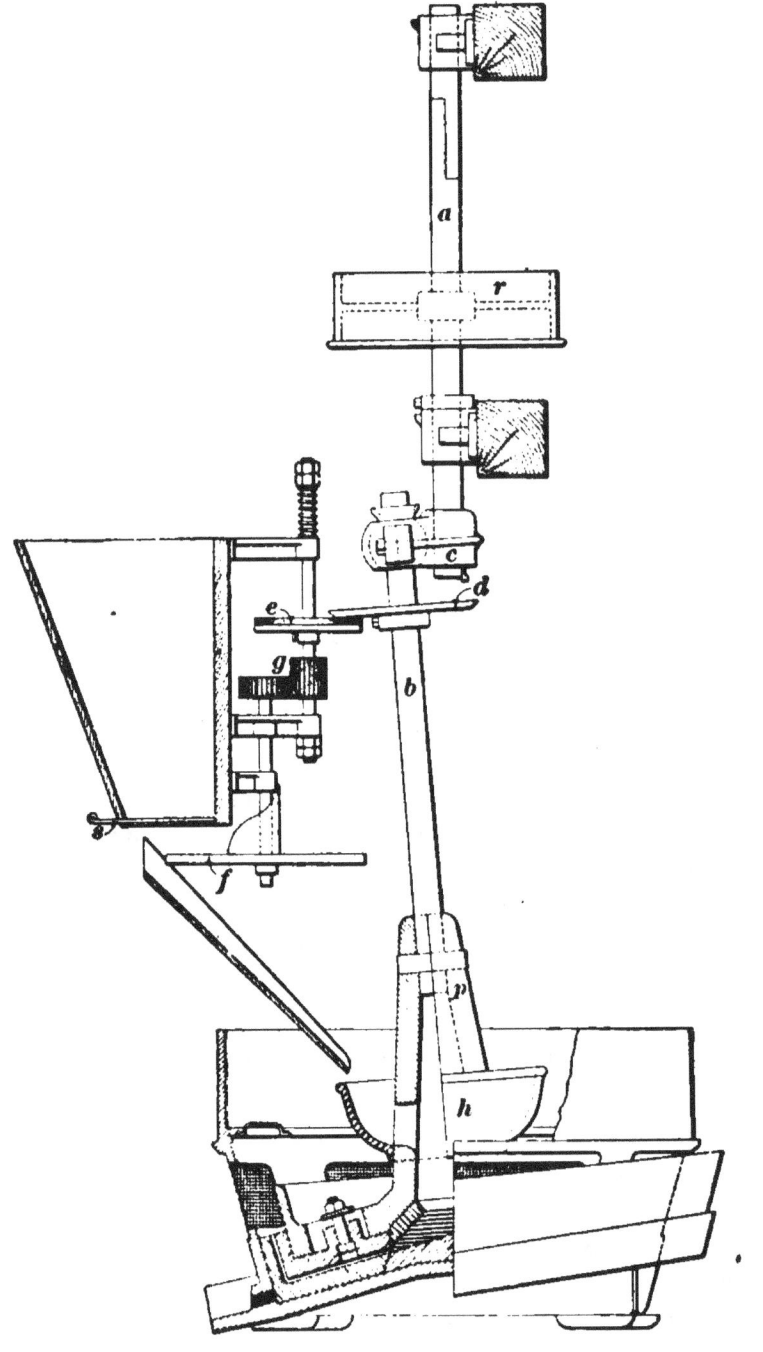

Fig. 43

The portion *b* of the spindle is connected to the lower end of *a* by the offset crank *c*, and slants slightly inwards towards the bottom, where it connects with the shoe or pestle *p*, the offset giving the shoe a gyratory motion. The machine is driven by a belt on pulley *r*. As the ore comes from the crusher it is fed into the hopper *h*. The corrugations shown on the under side of this hopper cause the rock to be rapidly crushed to a small size, which can conveniently work down to the rim, where the final pulverization is accomplished. The large screening area allows the pulp to escape as soon as it is sufficiently reduced, and this, together with the fact that the action of the machine is almost entirely a pinching action, reduces the sliming to a minimum. The pulp passes into a discharge launder, which extends all round the mortar and carries all the pulp to an amalgamated plate in front. All crushing parts of the machine are made of crucible steel and are replaceable.

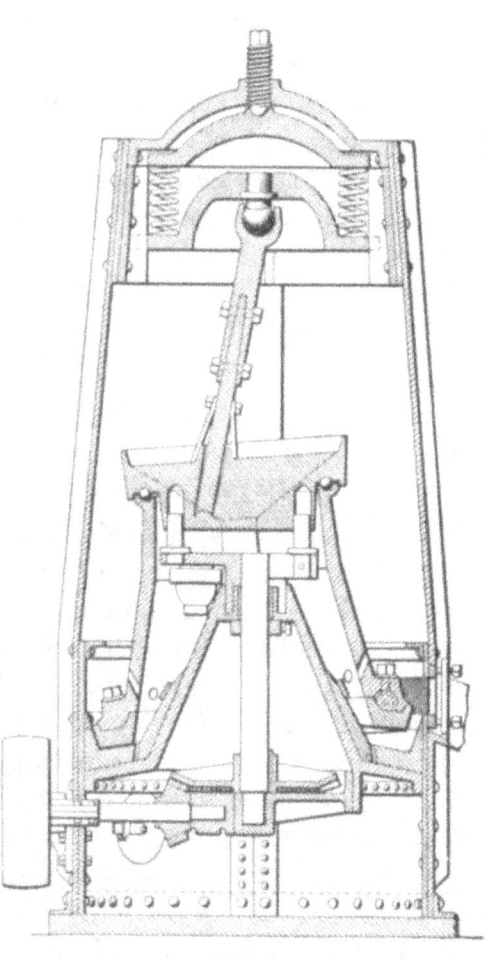

FIG. 44

The feeding of the machine is entirely automatic. The hopper slide *s* is pulled out to allow the desired amount of ore to fall on to the feed plate *f*. This plate is connected through the gears *g* to the friction plate *c*. For a short time, in every revolution of the main spindle, the disk *d* presses on *e* and revolves it slightly around; this motion is transmitted to *f*, which also turns for a short distance, the ore on the plate being pushed against a sheet-iron guide and a portion of it forced off the plate each time the disk moves. The shaft on which the disk *e* is fastened is hung from a spring, to prevent any excessive pressure between *e* and *d*, which would be apt to break one or the other.

51. Oscillating Mill.—A somewhat similar mill, shown in Fig. 44, has the power applied to the central shaft from below by means of bevel gears. The motion is transferred to a crushing shoe by an iron disk set eccentrically on the shaft and having on its periphery two loose rollers, which as the shaft revolves are carried along with it, pressing against a ring on the inside of the shoe and causing it to oscillate without revolving.

BALL PULVERIZERS

52. Krupp Ball Mill.—Ball pulverizers comprise all machines in which the crushing is done by means of balls rolling in a cylinder. The Krupp ball mill is of the multiple-ball type and is used for dry grinding. The machine consists of a revolving drum made up of hard-steel plates, inside of which are a large number of chrome-steel balls of various sizes. Outside of this drum are first a perforated sheet-steel cylinder and then a cylindrical sieve, concentric with the crushing drum and revolving with it. The whole is housed carefully with sheet iron, with an offtake for dust at the top and a discharge funnel at the bottom. The crushing drum is made up of segmental steel plates *a* and *b*, shown in Fig. 45. The plates *a* are perforated over the front half of their area, and the back portion is thickened and bent spirally, as shown in the cross-section, causing a slight step between the adjacent ends of segments. The ore is fed into the drum through the hopper *h*. Attached to the shaft of the drum at the feed opening are a set of helical spokes *s*, which act as a screw conveyer, feeding the ore gradually into the drum as it revolves, and at the same time rendering it impossible for the balls or the ore to escape back into the hopper. As the ore is broken up by the tumbling balls, it passes through the holes in the plates on to the sheet-iron screen c, the holes in which are considerably smaller than those in the drum. Through these

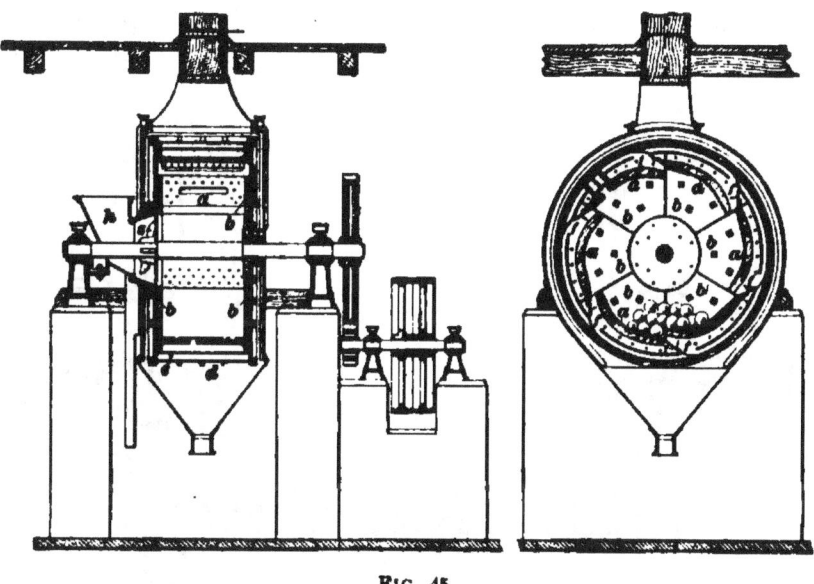

FIG. 45

holes the finer stuff passes on to the outside wire screen d, and the material too coarse to pass the screen rolls back as the mill revolves on to the plates b, and thence into the drum to be recrushed. The plates b are strips of sheet iron extending the full width of the screen from the front end of each crushing plate, through a slot in the first screen, to the outer or battery screen d. This slot also allows the return of the coarse material from the battery screen to the drum for further reduction. The ore which passes d finds its exit through the discharge funnel. The mill is run at a speed of from 20 to 45 revolutions per minute, according to the size of the machine, which varies ordinarily from 53 to 86 inches in diameter.

53. Alsing Pulverizer.—This is another form of multiple-ball machine used for very fine crushing. It is, in fact, a pulverizer in the true sense of the word.   The material is crushed to 10 mesh or finer before being introduced into the pulverizer and is discharged from it as an impalpable powder.

The machine, as shown in Fig. 46, is automatic and has a continuous discharge.   The material is fed in at one end by a screw feeder *s* in a hollow trunnion and is discharged through the trunnion at the opposite end, the size of the

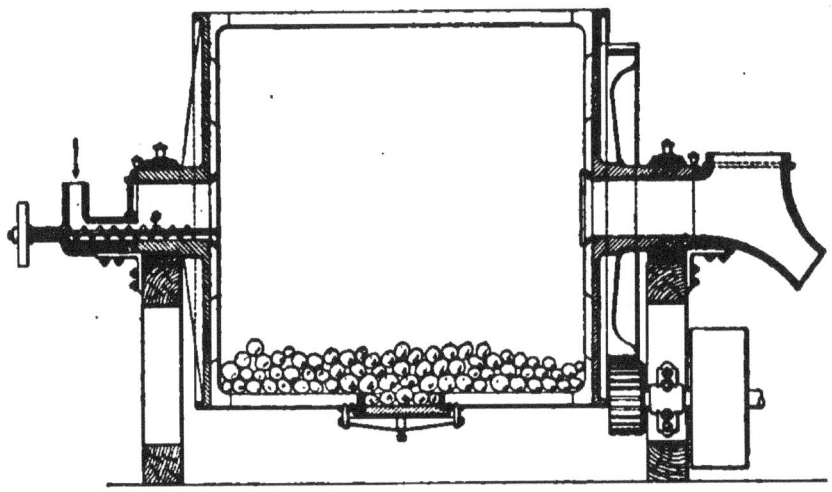

FIG. 46

product being regulated by the rapidity of the feed—the slower the feed, the slower the discharge, and consequently the more complete the pulverization.   The cylinder is 8 feet long and 3 to 6 feet in diameter, lined with hard, vitreous porcelain, and makes from 25 to 35 revolutions per minute. A large number of spherical flint pebbles constitute the crushing apparatus.   It would seem at first thought that the size of the product could not be uniform, but in practice this mill has ground 15 tons of ore in 24 hours from 10 mesh, and 99 per cent. of the product passed a 125-mesh screen. The rolling of the machine works the larger particles towards the bottom, so that by the time the ore has traveled the length of the cylinder it is pretty thoroughly crushed. The machine may be used for either wet or dry work.   There

are a number of patent mills which work on the same plan
of the Alsing, the more recent being improvements upon
that machine. They are used chiefly for grinding cement
and pigments.

54. Allis-Chalmers Ball Mill.—Fig. 47 illustrates a
ball pulverizing mill which has been quite extensively

FIG. 47

employed in preparing ore for cyanide and chlorination
treatment when they require that the ore be pulverized very
fine. The ore is first reduced to about four-mesh size, or
smaller, after which it is fed to the pulverizing machine,
which consists of a sheet-steel shell having cast-iron heads
on which are cast trunnions that work in bearings so as to
support the cylinder. One of the heads is provided with a
gear for driving or rotating the cylinder. The heads are
lined with plates in the form of sectors, bolted to the head
near the center as shown and secured at their outer ends
by the linings of the cylindrical portion. The cylindrical
portion is lined with curved plates, held in place by means

of wooden strips or steel wedges driven between them after they are set in position. The shell linings are so arranged that they form a series of ridges which have a slightly spiral course in such a direction as to feed the ore and crushing balls towards the feed end of the cylinder. The material to be crushed is fed into one end of the apparatus through the hollow trunnion and crowds its way to the other end, where the discharge takes place through another hollow trunnion. The rate of discharge is governed by the rate of feeding, on account of the fact that the ore naturally tends to move towards the feeding end, and travels in the opposite direction only when forced to do so. The balls which accomplish the crushing are about $2\frac{1}{2}$ inches in diameter, are made of chilled white iron, and weigh approximately $2\frac{1}{2}$ pounds each. It requires about three tons of these balls to each crushing barrel, and in order to make up for wear, it is necessary to occasionally charge new balls in with the ore. The feed end of the machine is provided with spiral arms which prevent any material inside from backing out through this end. A series of experiments to test some of these machines gave the following results: The material fed to the crusher consisted of 60 per cent. ore broken to screen four mesh; 30 per cent. to eight mesh; and 10 per cent. finer than eight mesh. The product from the barrel which passed through 100-mesh screen amounted to thirty tons in twenty-four hours. In pulverizing ore this fine, about 50 per cent. of it was reduced as fine as 150 mesh, and the remaining 50 per cent. varied between 150 mesh and 100 mesh. The wear and tear on the balls amounted to only three pounds per ton of material pulverized. The crushing was done wet and the wear on the balls made up by charging new balls into the machine without stopping. The speed of the pulverizing barrel was from 12 to 15 revolutions per minute, requiring about 17 H. P.

One great advantage that this style of pulverizer has over other crushing machines is that no sizing or screening is necessary. The material fed to the crushing barrel is passed through finishing rolls first and discharged sufficiently uniform in size for its desired purpose. When crushing wet in the above experiments, from five to seven gallons of water per minute were required for each barrel. The crushing may be accomplished dry, and when this is the case, the wear on the balls is increased.

In the illustration the discharge arrangement has been removed from the front end of the machine and a portion of the casing broken away to show the inside of the barrel.

55. Globe Mill.—The mill shown in Fig. 48 differs essentially from those described, as only one ball *b* is employed

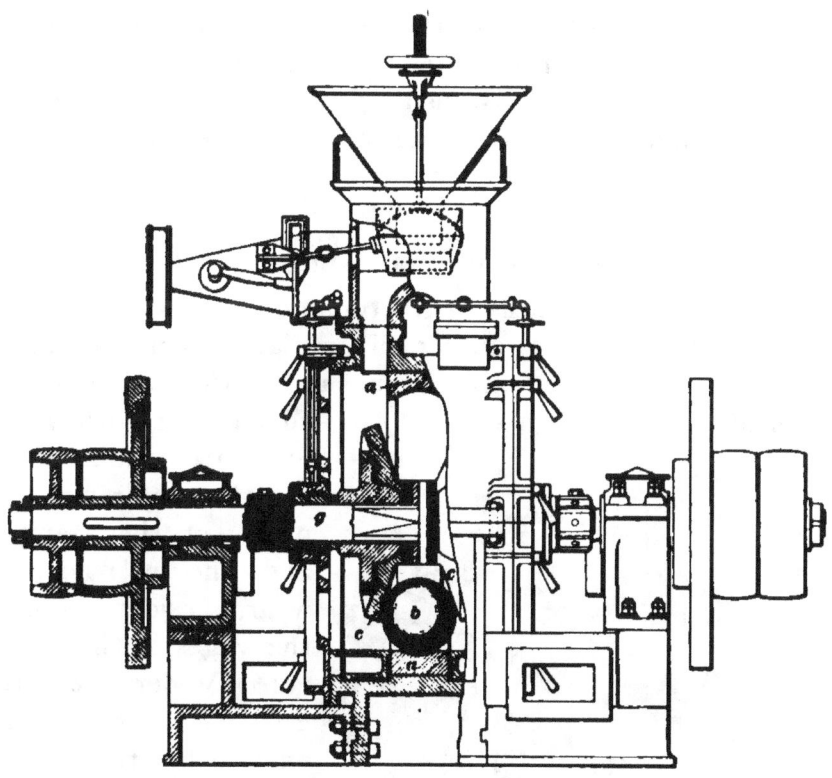

FIG. 48

to do the crushing, and the cylinder *a* is stationary. The ball moves about in the cylinder, motion being imparted to it by the frictional contact of two steel disks *c* fastened to the main shaft *g*, and flaring slightly away from each other, but pressing lightly against the ball. The machine revolves at the rate of about 300 revolutions per minute, giving the ball in a 5-foot mill a velocity of about 75 feet per second, so that after it once gets well in motion, very little friction from the disks is necessary to keep it going, and the wear on the disks is, consequently, surprisingly small. The ore is fed into the machine by an automatic feeder at the top and falls into the grooved path of the ball *b*, which is pressed strongly outwards by the centrifugal force due to its rapid motion. The crushing action occurs along the whole path of the ball, as its rapid motion draws the ore around with it. The mill is adapted for either wet or dry crushing. The pulp is discharged through screens in the ends of the drum, against sheet-iron splash plates, whence it falls into the discharge boxes. The wet-crushing mill can be fitted with inside amalgamated plates if desired, and the coarser gold saved inside the mill.

AUTOMATIC ORE FEEDERS

56. Objects Attained.—Automatic feeders for stamp batteries have almost universally replaced hand feeding. When running with a uniform ore, automatic feeders, once carefully adjusted, will work day in and day out with very

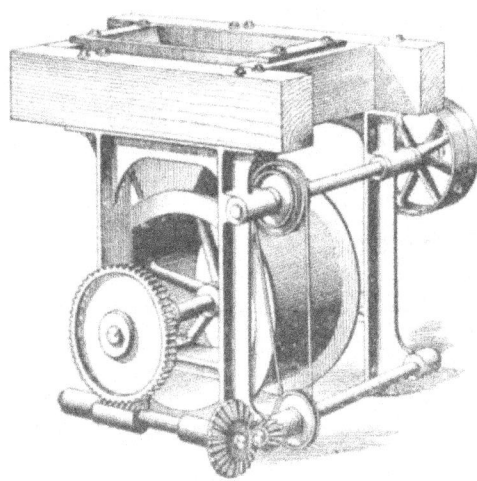

little attendance, giving the maximum capacity of the stamps and the minimum wear, besides saving the wages of the feeders. Modern feeders cut out a certain amount of ore at each stroke of a certain stamp and push it off into the motor, making the feed regular.

FIG. 49

57. The Roller Feeder.—Another and more modern form of roller feeder

is shown in Fig. 49.   The roller, like that in the previous
machine, revolves a little with each drop of the driving
stamp, and carries out with it a little ore each time.   The
feed from this machine is apt to be irregular, as the roller
may slip without bringing out any ore, and large pieces may
clog the machine.

58. Tulloch Feeder.—The Tulloch feeder, shown in
Fig. 50, feeds perfectly, is cheap, light, and so very simple

FIG. 50

in construction and operation that any blacksmith can make
whatever repairs may be necessary.   The feeder is of the
shaking-tray type.   The hopper *a* holds about 1,500 pounds
of ore, which runs into it directly from the crusher or is
dumped in from a car.

The tray *b* is hung from the frame timbers in such a
manner that it swings forwards of its own weight until lugs
beneath strike the jar rod.   It is swung backwards by an
arm on the rocker-shaft *c*, connected to the under side of
the tray.   The crank-arm *d* of the rocker-shaft is connected
with the short arm of the lever *e*, the long arm of which is
connected with the tappet rod *f*.   This tappet rod of 1-inch

steel is opposite and parallel to the middle stamp stem and
passes through a hole bored through the lower stem guides.
The head of the rod, upon which is set a rubber buffer g, is
struck by the stamp tappet towards the end of its fall, and
pressing down the lever c, throws the tray back. Some of
the ore in the tray in front of the door h is pushed off, the
amount corresponding to the length of the swing of the
table, while the tray, falling back into position again as
soon as the tappet rod is released by raising the stamp
stem, carries forwards an equal amount of ore from the
inside of the hopper. A spring is sometimes used at
the back of the tray to assist the forward motion. The
frame of the feeder is mounted on rollers, so that the
machine can be readily moved about. The feed regulates
itself automatically; if the bed of ore on the dies becomes
too thick, the lowest position of the stamp is raised in consequence, shortening the length of the stroke of the tappet
rod, and consequently diminishing the feed of the machine,
until the bed has worked down again. If the bed gets too
thin, on the other hand, the feed of the machine increases
correspondingly.

59. Challenge Feeder.—The Challenge feeder, shown
in Fig. 51, is particularly adapted for very wet and sticky
ore, on account of the fact that the ore is scraped off in
place of being shaken off. It is heavier and more expensive
than the Tulloch and much more intricate in construction,
but is very strong and durable and feeds well. The hopper
feeds on to the inclined cast-iron plate or disk a, which is
revolved by a bevel gear beneath, driven by a friction
disk f—or, in a modification of the Challenge, by a ratchet-
and-pawl arrangement—which is connected by a system of
levers to a tappet rod, as in the Tulloch feeder, and turns a
short distance with each blow on the rod. The friction disk
(or pawls) and the levers and tappet rod are brought back
into position after each blow by a flat steel spring s and link l.
With each partial revolution of the plate, the wing b scrapes
off a little ore into the machine being fed. The highest

position of the tappet rod is controlled by the hand wheel c, while the length of its stroke varies with the length of the

FIG. 51

drop of the stamp, which in turn depends on the thickness of the bed of ore on the dies, so that the machine is self-regulating.

60. Belt-Driven Feeders.—Automatic feeders were formerly driven by the blow of one of the outside stamps, either No. 1 or No. 5, but the ore was found to be distributed better if the blow was given by the middle stamp; therefore, all feeders are now constructed with the tappet rod opposite the middle of the machine, under the center stamp tappet, unless otherwise ordered.

Both the Tulloch and the Challenge feeders can be adapted to feeding Huntington or other roller mills, the levers being operated by cams on belt-driven shafts instead of by tappet rods. Belt-driven feeders are not self-regulating and must be very carefully adjusted.

ORE DRESSING AND MILLING

(PART 2)

CLASSIFYING MACHINERY

1. The ore as it comes from the crushing machines ranges in size from impalpable dust to the largest pieces which can pass the machine. In all processes of treatment, uniformity in the size of the product is sought, as such ore makes an open bed, through which the leaching gas or solution and the wash water can readily permeate, and thereby a more uniform extraction be obtained. For this reason, screens or other sizing appliances are placed between the crushing machines and the apparatus for the further treatment of the ore. The ore which passes the screens goes on to the leaching vats or tanks, while the coarse ore passes through another crushing machine or else goes back into the original machine to be further reduced. If, on the other hand, the entire product of a crushing machine, as a set of rolls, for instance, were conveyed directly to a leaching vat, the bed would clog in some places, from excess of fines, and at others be too open. The leaching solution would then merely travel up and down through these open spaces and never fairly permeate into the clogged spots, so that their values would be practically untouched. And even if the ore bedded well and percolated evenly, nevertheless, since the time of extraction is the time required for the solution to thoroughly penetrate the coarsest particles, the operation would be unduly prolonged. By screening the ore and recrushing the coarse particles, the maximum size of the product can be reduced as much as desired without

§ 26

For notice of copyright, see page immediately following the title page.

affecting the minimum—the practical limit of this recrushing being the point where the cost of recrushing balances the gain in time and extraction. This point varies with different ores, and can be determined only by experiment.

In preparing ores for concentration, also, a certain amount of sizing is necessary; for, while the fundamental principle of all concentration is that the minerals of higher specific gravity will sink to the bottom and those of lower specific gravity will range themselves above in inverse order of their specific gravities, yet there are complications which enter into and modify this hypothesis. Of two minerals of different specific gravities that are broken up into particles of a uniform size, the heavier will readily arrange itself in a layer below the lighter, nevertheless, if they be of varying sizes, they will have a tendency to arrange themselves not only according to their specific gravities but also according to the *law of equal falling particles*. This law is *that bodies falling free, in a fluid, fall at a speed proportional to the weight divided by the resistance*. Now, the weight of a body is proportional to the volume, and hence increases much more rapidly than the resistance, which is made up of three separate forces—namely, the frictional resistance of the fluid, which is proportional to the total lateral surface exposed; the vertical reaction (floating force) of the fluid, which is equal to the weight of the fluid displaced; and the cohesive force of the water, which must be overcome in order that the body shall sink. Therefore, if a lot of unsized ore is thrown into water, the particles will sink with a speed proportional to their weight divided by the resistance of the water. This will result in a bed of mixed ore at the bottom, the lower portion being composed mostly of the coarser particles of heavier mineral, but mingled with these will be found many coarse fragments of gangue rock and mineral not wholly freed from gangue. These latter particles will be larger than the pure mineral particles, but the proportion between their weight and the resistance they met in falling will be the same as in the case of the pure mineral lying at the same level. This arrangement will continue up through the bed, the proportion of gangue, however, becoming greater towards the top layer, which will consist almost entirely of fine gangue. The separation is known as **hydraulic classification.**

In most apparatus the fall is too short to allow of complete separation of the equal-falling classes, and the fall is further retarded by the friction of the particles upon one another, but the law, nevertheless, enters indisputably into the action of all concentrators. Indeed, these unfavorable conditions for concentration make it doubly desirable that the sizing should be uniform, in order that the separation shall be as complete as possible; for, incomplete as is the concentration according to this law, it is still a step in the right direction, and the more uniform the ore, the longer the step. Were perfect uniformity of product possible in crushing, a practically perfect separation of the minerals by water alone might be expected, the only drawback being the impossibility of entirely disengaging the mineral from the gangue. In actual practice, this incomplete separation gives rise to a product known as "**middlings,**" or unseparated ore and gangue, which in such a bed as the one mentioned above will occupy a position between the pure mineral "**headings**" and the pure gangue "**tailings.**"

SIZING MACHINERY

2. Drum Screen or Sizing Trommel.—The sizing apparatus most commonly used, in fact, the only *purely* sizing apparatus, is the screen. In the case of jaw crushers, rolls, and gyratory crushers, separated screens must be used to properly size the material crushed. The most common way of doing this is by winding screen cloth on a

FIG. 1

cylindrical or polygonal frame, as shown in Fig. 1, which revolves slowly on an axis slightly inclined from the horizontal, and running the ore from the crusher into this drum. The inclination of the drum allows the coarse material, which will not pass the meshes of the screen, to run down to the lower end, where it runs into a chute *a*, and is either returned by a belt or chain elevator to the original crushing machine, or, as in large mills, is carried to a finer crusher. The screened product is conducted by means of a hopper and a trough *b* to another apparatus for further treatment, or, as is frequently the case, to a second, and sometimes third and fourth screen, as shown in Fig. 4, each of somewhat smaller mesh than its predecessor. This limits the variation in the size of the product from any particular drum to the difference between the diameter of its meshes

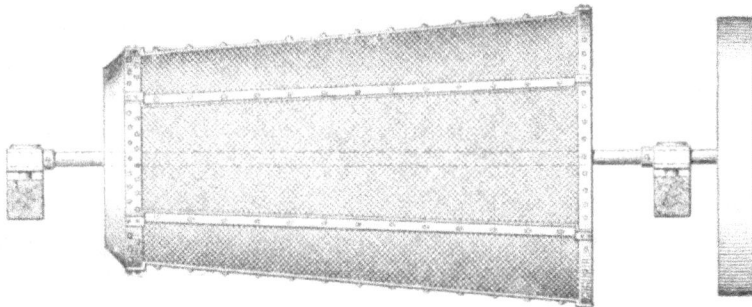

Fig. 2

and of those of the preceding drum. Such multiple screening is resorted to only when several distinct sizes of product are desired, or, in fine crushing by successive stages, to lessen the duty and increase the effectiveness of each crushing machine.

3. Conical Revolving Screens.—Drum screens are sometimes set with their axes horizontal, doing away with end thrust in the bearings, the slope necessary for the discharge of screenings being obtained by making the frame conical or pyramidal with the discharge at the wide end. In Fig. 2 is shown a conical revolving screen with longitudinal straps for stiffening the construction and holding the wire cloth in place. In case one section of screen cloth needs replacing, it can be accomplished without as much delay or expense as when a screen cloth is one entire piece.

4. Variable-Mesh Screens.—To save the room and expense of a multiple-screen system, various schemes have been devised for crowding the entire series of screens on to a single frame. The simplest of these is to divide the drum into sections, each of which is covered by a screen of different mesh, making practically a series of separate screens

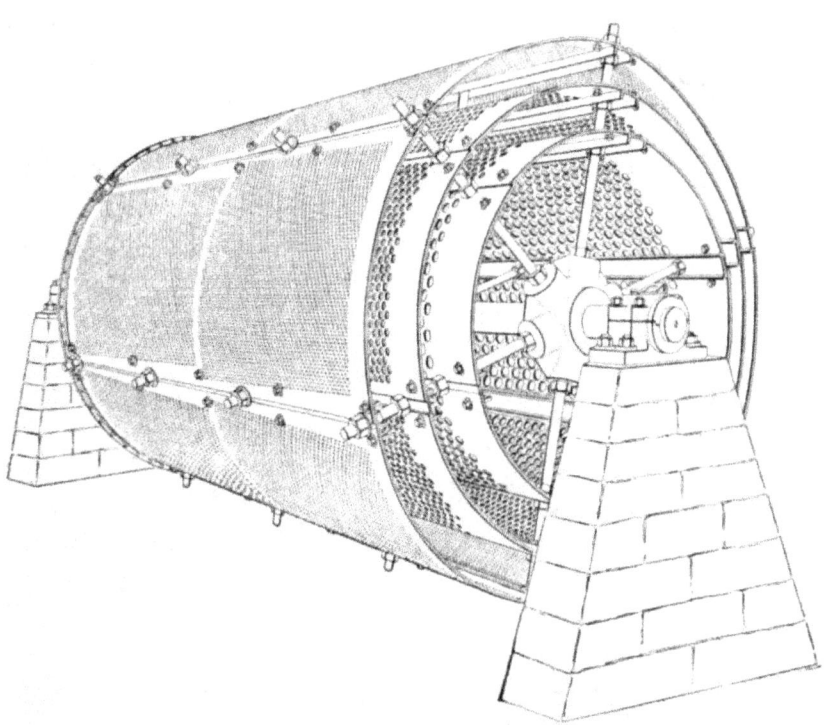

Fig. 3

upon a common axis, the fine screen being at the head and the coarse screen at the mouth of the drum. The large receiving hopper is divided into sections corresponding to the divisions of the drum, with delivery pipes from each section. This scheme is open to the objections that if more than two sizes of screen are desired, the drum must be made inconveniently long or the screens will not have time to do their work fully, and that the most wear falls on the finest screen.

A better scheme is to arrange the screens in a series of concentric drums, as shown in Fig. 3, each with a separate discharge trough for its screenings. In this arrangement the coarse screen is in the center, next to the shaft, where the material is coarsest and the meshes of each consecutive screen become smaller towards the outside. Conical screens of this type are sometimes made with every other cone reversed, so that adjacent screens discharge at opposite ends of the drum. In this case, as in simple conical screens, the axis is horizontal.

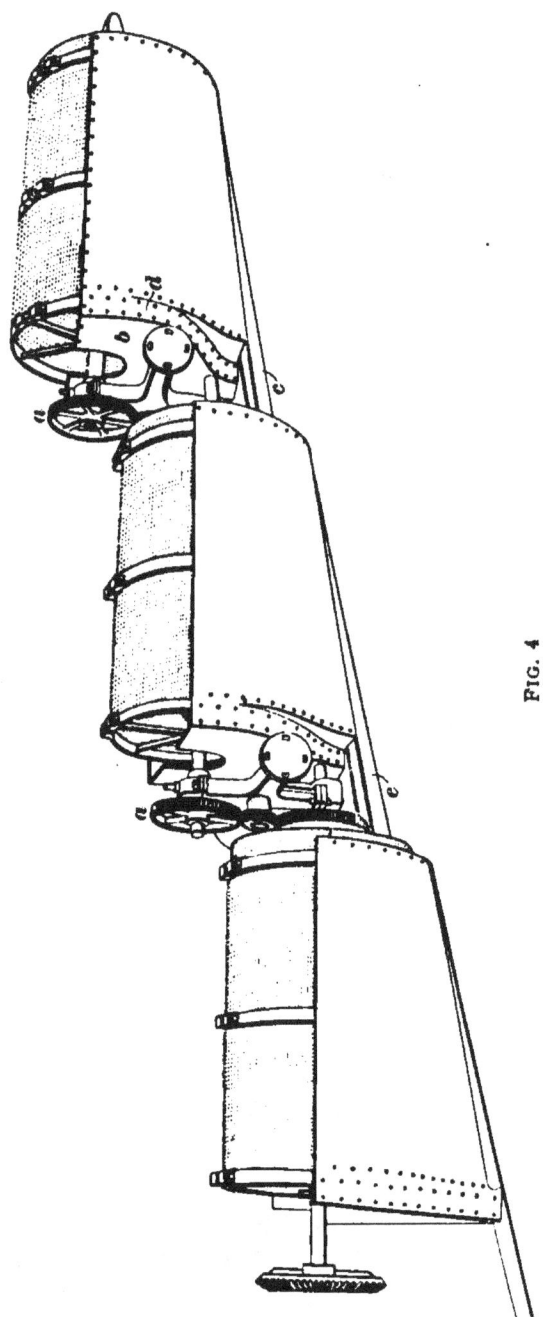

FIG. 4

5. Wet Sizing and Washing Screens. — Revolving screens are sometimes made with hollow shafts having small holes bored at right angles to the length of the shaft and connecting its inner and outer circumferences. One end of the shaft is closed, so that when water is forced into it through the other end it spurts through the holes into the screen, washing the material inside. These trommels may be revolved in a trough containing water, besides using the jets. Such trommels are used in gold dredging and in iron and phosphate mining to remove clay that adheres to the rock. The revolving screens shown in Fig. 4 are for wet sizing where coarse concentration is practiced. The treatment depends upon even sizing, which the wet sizing screens are quite effective in producing. The screens shown are connected by gearing a. The first screen removes all the pieces too large for treatment, by passing them out into the hopper b, from which they are returned to the crushing machinery. The particles which go through the first screen are washed down the trough c into the next screen, where they are sized again, the coarse particles passing out of the hopper corresponding to d, and those that go through the screen meshes travel into the next finer screen by way of the trough c.

6. Shaking Screens or Box Screens.—Shaking screens are but little used in ore dressing, the simple construction of drum screens, together with their economy of space and admirable working qualities, having led them to supersede all other forms for automatic sizing. Shaking screens are practical, however, and are used to some extent in coal screening. A reciprocating motion is given the screen by means of an eccentric or by a cam and springs. The mechanism in either case is much more intricate than that of a drum screen to accomplish the same work, and the shaking screen occupies considerable more space. Moreover, drum screens are frequently hung from overhead timbers, leaving the floor beneath clear for any purpose desired, while shaking-screen frames are usually set on the floor, taking up room at the expense of other machinery or the convenience of the workmen. The greatest objection to shaking screens in concentrating works is that they have such a jarring effect upon the building, and the average concentrating mill has enough jarring without introducing any more. The capacity of shaking screens is greater for equal areas of surface than the capacity of revolving screens.

7. Wet and Dry Screening.—Screening, like all other work about a mill, is preferably done wet whenever consistent with the further treatment of the ore, as the water facilitates the operation, lessens the wear on the screen, and renders housing unnecessary. Jets of water are sometimes played on the screen to prevent clogging, when there is not sufficient water in the pulp as it comes from the crushing machine. In case the ore is damp or wet, it cannot be screened satisfactorily without first being passed through a drier. Drying, while it makes more dust, renders the ore friable and permits the rolls or stamps to crush it quicker.

8. Screen Cloths.—The materials used for covering screen frames are punched sheet metal and woven-wire cloth. Both materials may be used for all sizes of product, but as a rule the sheet-metal screen is better adapted for coarse material, while wire screens work best on the fine stuff, as their open area per square inch is greater for the same size of mesh. For wet crushing, the screens can be made of brass wire, which does not rust. The relative merits of punched-metal and wire-cloth screens are a subject of dispute, but the round-hole screen is the commonest form for punched sheet-metal screens, and after it the longitudinal, oval, or rectangular slot. Square holes and diagonal slots are also common. Wire screens are almost invariably made with the meshes set square with the frame.

9. Grizzlies.—As the ore comes into the mill, it is dumped on to stationary inclined gratings, termed "grizzlies," and shown in Fig. 5. These are usually made of flat steel bars, set edgewise, about 1½ or 2 inches apart, and running lengthwise down the grating. As the ore slides down the grizzly, which is inclined at an angle of from 45 to 55 degrees in the direction of its length, the ore which is small enough falls between the bars into an ore bin below, while the coarse stuff slides down on to the feed floor.

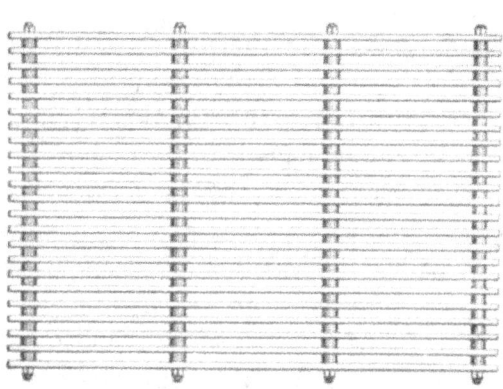

FIG. 5

The rock breaker located here crushes the ore to the proper size and then discharges it into the bins, where it mixes with the ore that passed through the grizzly.

The grizzlies vary from 3 to 6 feet in width and from 8 to 12 feet in length, 4 ft. × 10 ft. being the usual size. The bars are held in position by round-iron rods, usually three in number, one in the middle and one near each end; they are spaced by cast-iron washers, through which the rods pass. The bars are sometimes made with the lower edge thinner than the upper, the idea being to have the openings slightly wider at the bottom than at the top, and thus prevent ore sticking in the grizzly.

CLASSIFYING MACHINERY

10. Object of Sorting.—Screens of very fine mesh cannot very well be used for automatic work around a mill, as they are expensive, delicate, and altogether too slow in their action. Now, in the operation of any crushing machine, however coarse the maximum or average size of the product may be, there are produced a considerable proportion of "fines," the proportion increasing rapidly as the crushing faces of the machine are brought closer together. These "fines" are made up of particles varying in size from fine sand to an impalpable dust or slime. In character they are, like the original ore, a mixture, more or less intimate, of gangue and mineral; the only difference is that the more brittle portion of the ore (usually the mineral portion) is present in larger proportion, as its brittleness tends to make it break up fine. This makes it doubly desirable that these fines should be saved, and to that end many machines have been devised. Most of these machines depend for their operation upon the specific gravity of minerals and the tenacity with which the various minerals of an ore cling to a smooth surface against the force of a current of water. The machines were at first fed with the screen-sized material direct from the stamps, but this resulted in too wide a range of product from the concentrator, while the saving was not what it should have been; so that at present a classifying apparatus is usually interposed between the battery screen and the concentrators, *assorting the material into equal falling classes*, and each class is carried to a separate concentrator. The duty of each concentrator is thus lightened and the separation made much cleaner; for not only is the range of size of material to be treated by any one machine thus decreased, but the heavy, pure mineral in each class is confined entirely to the smaller particles and the pure gangue to the larger, with the combined mineral and gangue ranged in between in sizes relative to the proportions of mineral and gangue present. The smaller particles present much less surface to the water, in proportion to their weight, than the larger particles, and consequently tend to cling more tenaciously to the surface of the concentrator, and by regulating the current any proportion of the material desired can be kept from washing away.

11. Spitzkasten.—Spitzkasten, shown in Fig. 6, are troughs with pointed bottoms, arranged in a series, with settling pits at intervals, in which the various classes of material settle. These pits are allowed to fill up with sediment and are then cleaned. Each box discharges into one somewhat larger than itself. As the pulp stream flows through the series, each class of material settles out as it comes to a certain box, according to the strength of the current at that point.

In all modern machines of this type, the separation is made by introducing a rising current of water at the bottom of the box, as shown in Fig. 6.   The material settles against

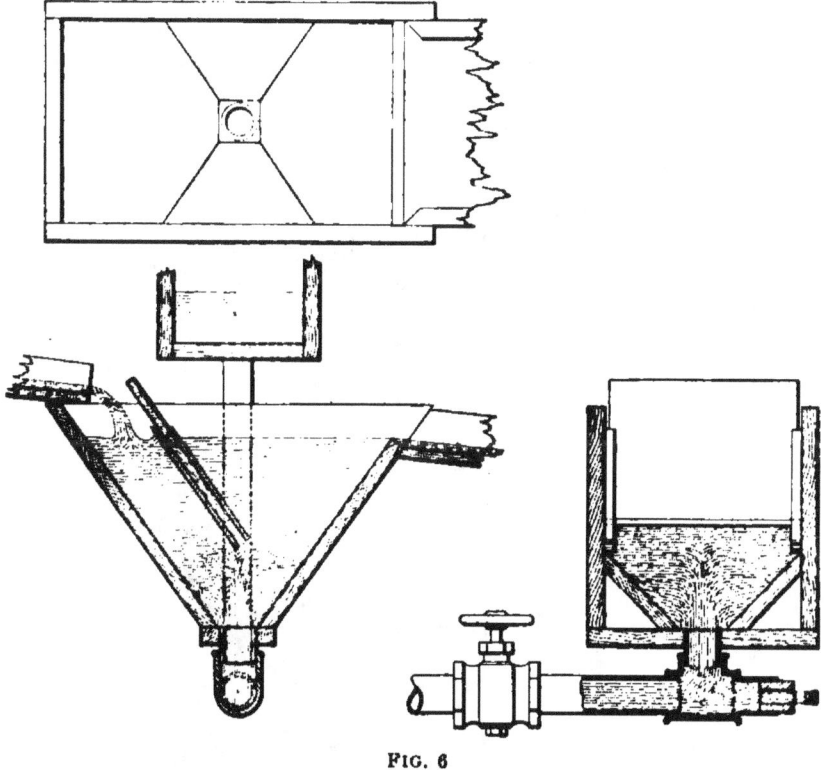

FIG. 6

this current into the tee below and is washed out at the orifice.   In this way all slimes are washed out, and the concentrates are very clean.   A partition or "diving board" is set in a box to divide it into a downward and upward current and prevent surface currents from traveling directly across the box.

The level of the water in the trough from which the wash water is taken is somewhat higher than that in the boxes, giving the desired pressure for an upward current, and the force and amount can be regulated by valves in the pipes, as shown. The first boxes of the series are usually quite small, and the current of the pulp stream as it passes through is correspondingly swift; so that only the heaviest ore particles settle out in the first box. As the boxes increase in size the force of the current diminishes and the finer pulp settles out, the size of the material settling in each box becoming successively smaller down the series.

12. Allis Classifier.—The Allis classifier shown in perspective and section in Figs. 7 and 8, respectively, is a

FIG. 7

modification of the spitzlutte. Referring to the section, Fig. 8, the pulp flows through the trough and screen as indicated by the arrows. The partition *a* divides the machine into two main divisions, corresponding to the down-flowing and up-flowing arms of the spitzlutte. The up-flowing arm, which extends under the main box like the tail of a *y*, is divided by partitions *b*, *c*, and *d* parallel to *a*—there being as many of these partitions as there are to be classes of products formed. The wash water enters through a pipe at the side of the box into a compartment behind the classifying trough proper and passes down and under the partition p, through a space left for this purpose, and rises on the other side. The pulp flows through the trough and screen down one arm of the apparatus and up the other. The lighter

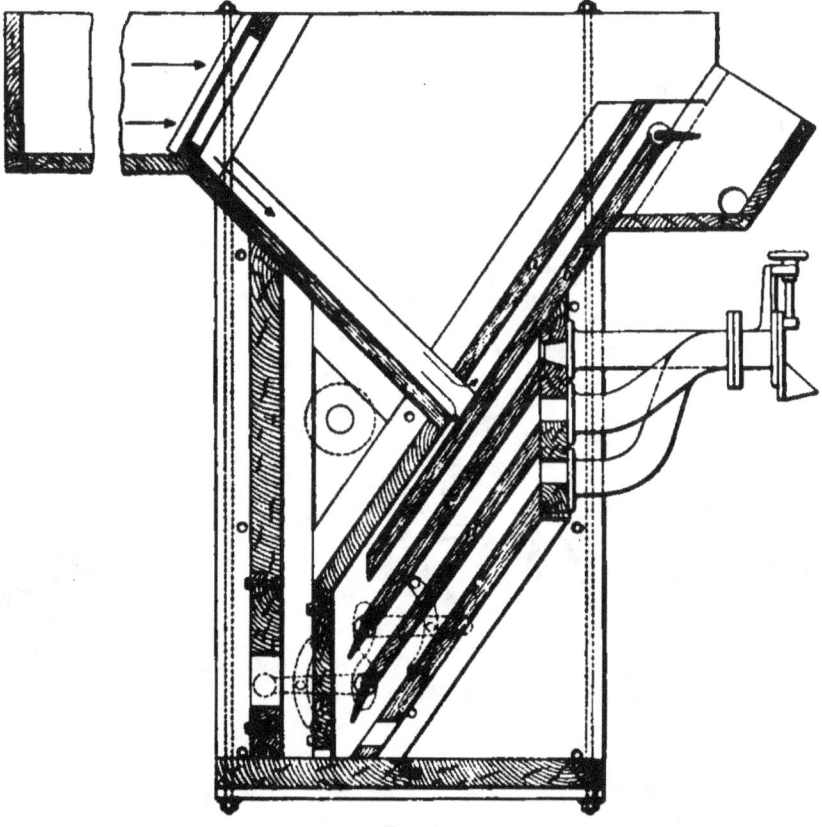

FIG. 8

particles of gangue are carried up with the ascending stream through the space between a and b into the tailings box. A metal lip at the top of a, extending over the top of the tailings slot, prevents the overflow from the receiving trough or down-flowing arm from running back down the tailings slot in case the machine gets too full. The heavier mineral particles settle to different depths, according to their densities, before they are carried up into their respective spouts, the wash current growing stronger as they descend. The tops of the discharge spouts are all nearly on the same level, so that all classes are discharged with the same force. The tongues shown at the ends of the partitions regulate the size and amount of each class, or the partitions may be made with slides which are adjustable to different depths by rods with thumbscrews attached, running up through the tailings trough.

13. Cone Classifier.—The cone classifier is the apparatus shown in Fig. 9, and is made entirely of iron. The

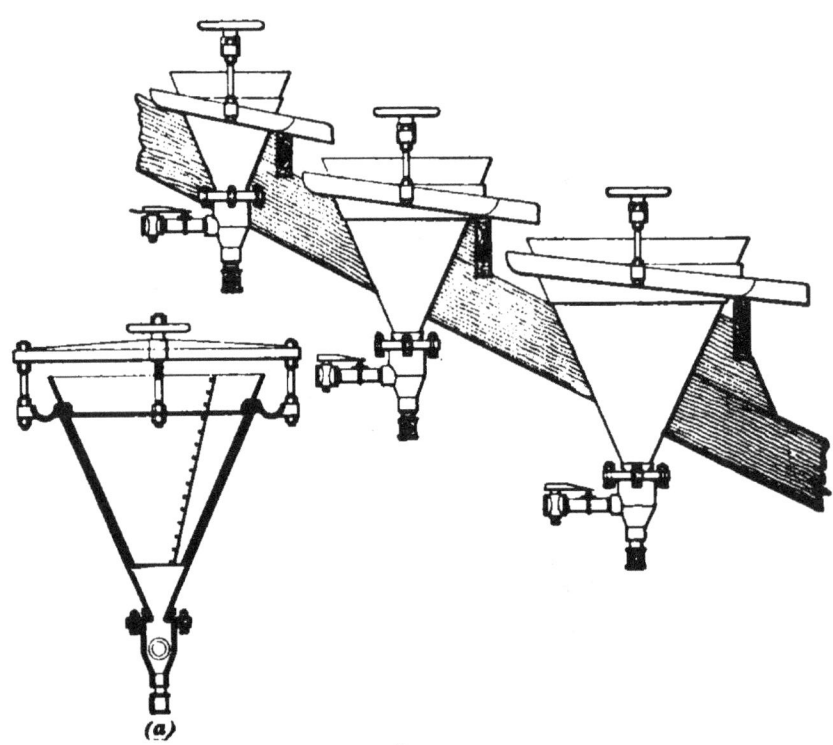

(a)

FIG. 9

outside cone is of cast iron; the inside cone, usually made of boiler iron, is open at the bottom and can be adjusted by means of a hand wheel and screw. The construction is shown in Fig. 9 (a). The pulp flows into the inner cone and down through the open bottom; here it meets a rising current of wash water and flows up through the space between the two cones. The particles of minerals have to settle against the combined upward force of the wash water and the rising pulp stream. Those particles which have weight enough to do this settle to the bottom and are discharged through pipes on to their respective concentrators, while the lighter particles are carried over with the main pulp stream into the launder and thence into the next larger classifier, where the whole operation is repeated, but with a slower current, on account of the greater size of the apparatus, the average size of the product being proportionately smaller. By varying the width of the space between the cones in each classifier and the amount of wash water, the separation may be made as close as desired and may be carried through a number of classifiers—more than three, however, is unusual. The old-style wooden classifiers in American mills have been largely replaced by these cone classifiers, as they are much more convenient and compact. Cone classifiers range in sizes from 12 to 40 inches in diameter and weigh from 100 to 625 pounds.

14. Trough Classifiers.—In the Lake Superior copper regions, classifiers of another form, known as trough

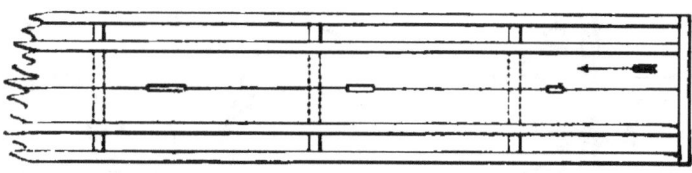

FIG. 10

separators, are used for a rough classification of the native copper ores, preliminary to jigging. The *Lake Superior trough separator*, shown in Fig. 10, is a double **V** trough, the space between the inner and outer troughs being divided at intervals, so as to make the apparatus in effect a series of double **V** boxes. Towards the lower end of each box a slot is cut in the bottom of the inner trough to allow communication between the two. The pulp is run in through the inner trough in a continuous stream, while the outer trough contains the wash water, which is kept at a level somewhat higher than that of the pulp in the inner trough, in order to maintain a steady upward current of water through the slots, against which current the mineral must settle.   The coarser particles naturally settle to the bottom first and sink through the slots into the outer trough, and thence through the discharge openings to the jigs, the finer stuff being carried farther along before settling.   The slots of the inner trough are made longer towards the lower end of the trough, so that the smaller particles shall have more time in which to find their way through into the lower trough.

15.   The *Calumet classifier*, shown in Fig. 11, is another type of trough classifier.   It consists of a trough which

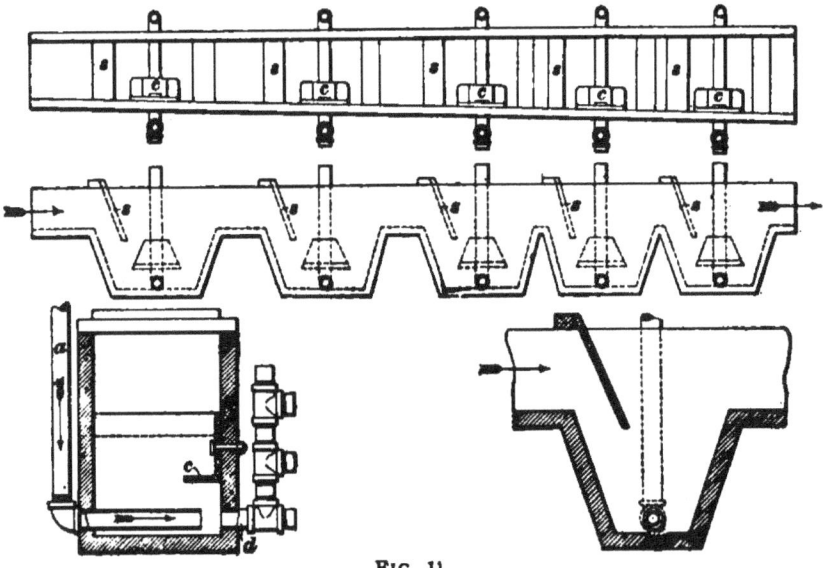

FIG. 11

widens slightly towards its lower end and has a series of boxes or pockets in the bottom.   The pulp flows through the entire series, the stops *s* deflecting the stream downwards into the bottoms of the boxes, so that all the material is subjected to the action of the wash water which enters through the pipe *a* and discharges directly against the discharge spigot *d*.   The spigot is not large enough to carry off all the pulp directly, so it swirls and eddies in the bottom of the box, keeping the sand stirred up, and allowing only those heavier particles which have weight enough to settle in this commotion to be washed out through the discharge.   The shield *c* deflects the currents set up by the wash water and prevents them from rising to the surface, thus confining the agitation to the bottom.   The force of the wash water can be regulated at will, the classes issuing from the spigot responding readily to any change in the force of the wash water.   The force of the wash water in the lower boxes is less than in the upper, and the average size of the product proportionately smaller. By using for the discharge spout a vertical pipe with three or four openings at different levels, the amount of water discharged may be regulated—the discharge being more rapid when the lowest hole is used and slowest when only the top hole is open.

16. Hydraulic classifiers are, as a matter of fact, concentrators, but are treated under a separate head because they are invariably used in preparing ores for further concentration.   The jig, on the other hand, is just as essentially a classifier in its action as the hydraulic classifiers just described, but the classes which it yields are treated as final concentrates; hence the machine is always classed with concentrators.

SETTLING BOXES

17. Object of Settlers.—It is always desirable, if possible, to crush and size ore with a large excess of water, as this greatly facilitates both operations.   This excess of water is, however, frequently undesirable, or at least unnecessary, in the subsequent treatment of the ore.   If a too thin pulp interferes with the subsequent operations—as is the case with many concentrators of the vanner type or in the pan amalgamation of silver ores—the battery pulp is run into settling boxes, where the suspended material settles out and is withdrawn at the bottom with just sufficient water to give it the proper consistency for concentration or amalgamation, as the case may be. The concentrates from some machines are delivered with a great deal of water, and are usually settled out and drawn off very thick, to be dried for further treatment, storage, or shipment. Tailings from concentrating and amalgamating mills are likewise sometimes settled out and dried and then treated by the cyanide process for what gold remains in them.

In case of a scanty water supply—quite a common drawback to milling operations—the necessity for dry crushing may be avoided by drawing off the superfluous water from the pulp and tailings and using it over and over again. The adoption of this scheme has allowed the working of many large deposits which would otherwise have remained untouched.

18. Settling Ponds.—Pulp settling is a very old practice. It was formerly done in large settling pits sunk in the ground, through which the pulp stream flowed. The sediment was allowed to accumulate until it came so near the surface of the water that the surface currents commenced to cut channels in the deposit, when the pulp stream was deflected into another pit and the sediment in the first pit shoveled out. The pits were usually rectangular, with steep, sloping sides. The large settling tank or vat which has replaced the settling pit is a huge rectangular pointed box, set on a framework above the ground. These tanks are sometimes set in series, beginning with tanks only a few feet long and wide, and ending in tanks of enormous dimensions, somewhat on the plan of spitzkasten, but on a much larger scale. The larger the tank, the greater the diminution in the velocity of the pulp stream on flowing into it; consequently, the boxes arranged in series in this way would be in effect classifiers as well as settlers, the heavier and coarser equal-falling particles settling out in the smaller boxes, while the finer and lighter particles would remain in suspension until they reached the almost motionless bodies of water in the larger tanks.

19. Automatic Settling Boxes.—Besides the trouble of hand cleaning, which necessitates either the construction of two tanks or sets of tanks, to be worked alternately, or else the shutting down of the mill while the tanks are being cleaned, the old settling tank, as well as the more primitive pit, presents several other disadvantages. Their size and consequent cost is one great drawback. In addition to this, a great deal of fine material is floated across

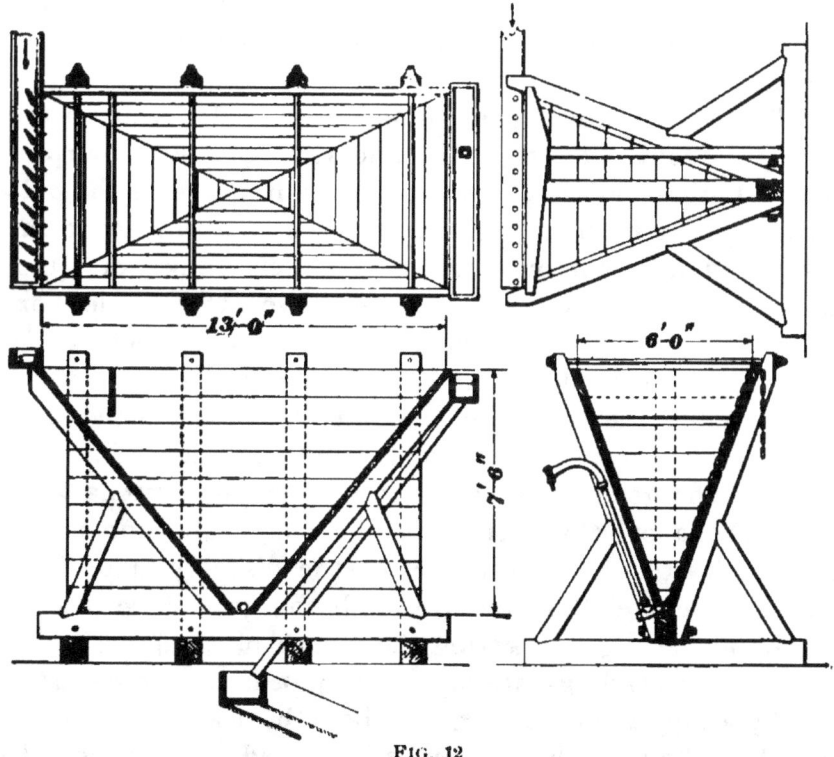

Fig. 12

by surface currents, and when the deposit of sediment approaches the surface, more or less of it is washed over the lower edge of the box. The continuous, automatic-discharge settling box removes all these objectionable features at a single stroke. The size of the box is reduced to reasonable proportions, surface currents are prevented by the use of a diving board, and the operation of the apparatus is continuous, the sediment being removed as fast as formed and not being allowed to accumulate in the box.

The construction of the settling box is illustrated in Fig. 12. A box of the dimensions given in the figure will handle the pulp from five stamps, or even ten under favorable conditions. The sides should slope at least 50 degrees from the horizontal, or the sediment will stick to them instead of sinking to the bottom and discharging. The pulp is fed in through a distributor at the head, with holes and guide tongues, and the clean water discharges over the lower edge—which is cut 2 or 3 inches lower than the sides for this purpose—into a trough, and thence into the water launder.

All the pulp must pass under a vertical diving board across the tank near the head, and this serves the purpose of preventing surface currents; that is, it prevents the pulp stream from running right across the tank and over the other end in a narrow current between banks of quiet water instead of spreading equally over the whole box. This diving board also serves the purpose of completely submerging the particles of ore, thus preventing their being floated off on the surface of the water, supported by a film of air—the source of considerable loss in milling. The sediment discharges through a 1½- or 2-inch siphon discharge, a few inches above the bottom of the box. The upper portion, or, better still, the whole, of the discharge pipe is of rubber hose, and the pressure of discharge can be altered by simply raising or lowering the mouth of the hose, which is closed by a sliding-gate tap. The pipe is connected to the box by a nipple and tee, or, if a hose be used for the full length of the pipe, the tee can be dispensed with, as the only reason for using it is that an opening may be had at the bottom of the box so that it can be completely emptied if desired, and with a hose discharge-pipe this can be had by merely dropping the nose.

The form of the settling box has considerable influence on the size of the box required. If the location is such that the number of square feet is limited, it will be found better to employ a wide short box than to employ a long narrow box, on account of the fact that the relative percentage of the materials settling from the water depends upon the degree to which the velocity of the current is retarded, or, in other words, the nearer the flow is brought to rest, the more thorough will be the settling of the contents. If two boxes can be employed, it is usually better to divide the flow and send half of it through each box than to place the two boxes one after the other and depend on each of them to extract a portion of the material from the flow.

CONCENTRATING MACHINERY ·

20. Concentrator is the general name applied to all machines for reducing the mineral values of an ore into smaller bulk, in order to get rid of as much superfluous material as practicable. Among the Western smelters it is customary to vary the smelting charge with the character of the ore as regards fluxing, or, what is practically the same thing, to have a uniform charge for a neutral ore (one in which iron oxide and silica are present in proper proportions), and then pay a fixed premium for every additional unit (per cent.) of iron, or require a bonus or excess charge on every unit of silica beyond neutrality. For instance, if a mine at a considerable distance from a smelter is producing a quartz ore carrying 10 per cent. iron pyrites—the other 90 per cent. being quartz—and $12 per ton in gold, the owner would be apt to find, if he shipped the ore direct, that, after paying the freight and the smelting charges, including the bonus on the silica, there would be little left of his $12. But the quartz being much lighter than the pyrite, he finds that after crushing he can, by the use of suitable apparatus, wash away the greater part of the gangue rock, leaving behind the pyrites, in which all the values are contained. In this way he dispenses with a great portion of his freight charges, and if he carries the concentration far enough, he may, instead of paying a bonus on excess silica, receive a premium on excess iron. There is always more or less loss of mineral in concentrating, but by the careful use of good apparatus this can be kept down to a nominal figure. The limit to which concentration may be profitably carried is the point beyond which the cost of concentration, together with the inevitable loss of values in the tailings, exceeds the saving in freight and the treatment charges.

21. Concentrating apparatus may be divided into two general classes: First, machines in which the separation is performed by means of an intermittent upward current of water, which tends to arrange the particles in layers, in the order of their specific gravities. This class comprises all jigging apparatus. Second, machines in which the separation is mainly due to the superior tenacity with which the particles of the heavier mineral cling to a smooth surface against the force of a stream of water. This class includes buddles, belt and table concentrators.

22. Jigs.—Jigs are almost universally used for concentrating the coarser sizes of ore, but are inefficient for ore which will pass through a screen having less than 30 openings to the linear inch, or 900 holes per square inch; ores below this size are usually concentrated on bumping tables, vanners, or buddles, or other slime concentrators, according to the fineness.

All hydraulic jigs work upon the same principle, that is, the tendency of ore particles in water, when approximately of the same size, to arrange themselves in layers, according to their specific gravity, when the bed of material is kept sufficiently open to allow the particles to move freely among themselves. This is accomplished in jigs by giving a column of water a pulsating motion or by giving the grating and screen upon which the ore lies a short reciprocating motion, the resistance of the water lifting up the ore on the down stroke of the piston or grating and the particles assorting themselves as they settle back. The pulsating motion of the water makes the operation of the machine continuous, as the particles of a certain density are never allowed to get below a certain level; for, so long as the bed is properly preserved, there will always be a layer of heavy mineral upon the screen, which it will be impossible for the lighter mineral to displace, so that the latter is confined entirely to a level above the bed of heavy mineral, though the particles of heavy mineral may work down through it, a little at each stroke, to the bed below. In jigging coarse material, the holes in the screen upon which the bed rests are made smaller than the ore to be jigged, and the latter forms its own bed as described, the different classes being discharged through various forms of pipe and slide discharges above the screen. Material has been successfully jigged in this way as coarse as 2¼ inches in diameter and as fine as 10 mesh. For jigging the smaller sizes, however, it is customary to have the meshes of the screen rather larger than the particles of ore to be jigged, so that the whole surface of the screen may be utilized for discharging the concentrates. A bed 1 to 4 inches thick, of coarse mineral, of the same or slightly greater specific gravity than the mineral to be concentrated, is arranged directly above the screen. This bed is usually made up of coarse pieces of the same mineral as that to be concentrated. The fragments composing the bed are all too large to pass through the screen. Through this bed the fragments of the mineral to be concentrated work their way, and passing through the screen, fall into the hutch below, where they accumulate and are discharged at intervals. This method is used very largely in American gold milling, where sizes seldom run above ½ or ¾ inch.

23. Middlings.—Crushing can never completely disengage the ore from the gangue, nor can screening or even hydraulic sizing be made so close that there will not be a considerable variation in the size of the particles making up any one class, so that in jigging it is practically impossible to get a perfectly clean separation of ore and gangue. Even among the cleanest concentrates there is always some gangue and ore combined, and also in the cleanest tailings; and in all concentration, by jigging or otherwise, there is always an intermediate product between concentrates and tailings, known as "middlings," which is made up of combined gangue and ore.   In jigging, these middlings form a bed or stratum between the clean concentrates and the tailings—or, in the case of jigging through a bed of coarse material, just above this bed—which is discharged separately.   If

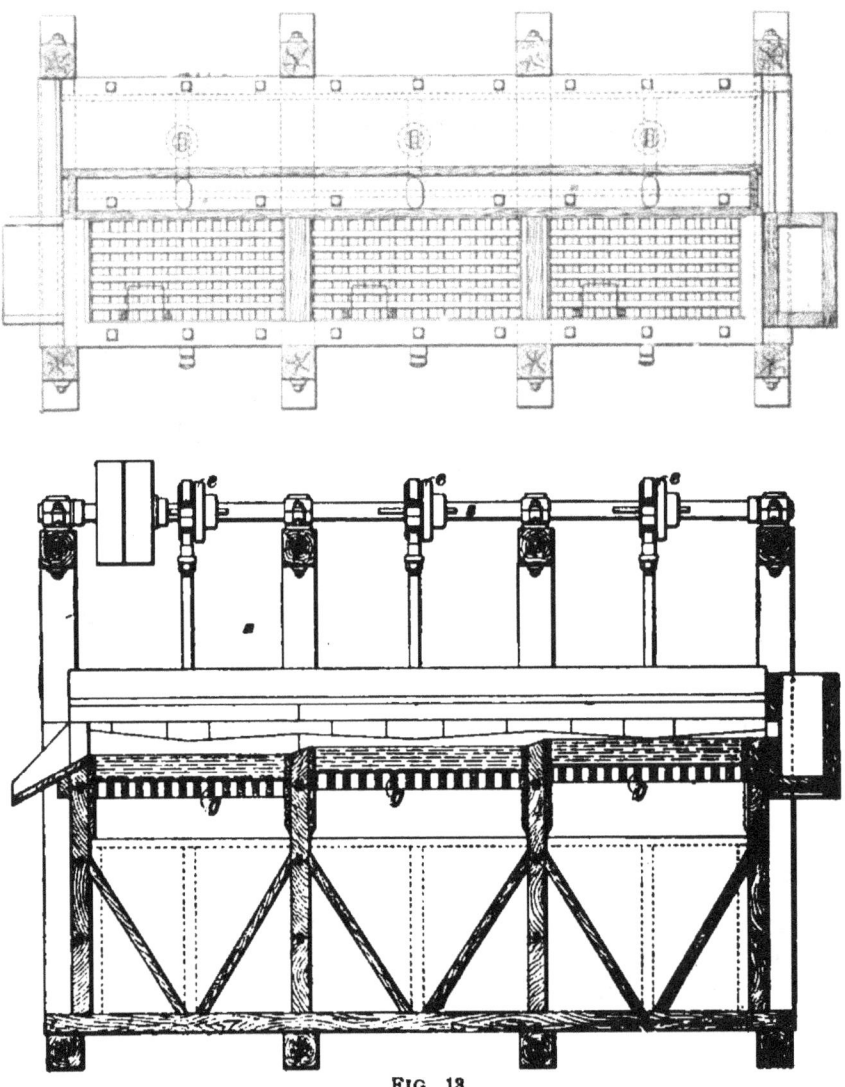

FIG. 13

practicable, the middlings are usually recrushed and returned to a finer jig to be further concentrated.

24. Stationary-Screen Jigs.—The ordinary type of jig belongs to the class in which the grating supporting the

screen is stationary.   In construction, all jigs of this type
are essentially the same.   The machine consists of a rectangular box or tank, divided, for the upper part of its
depth, into two compartments by a vertical partition.   A

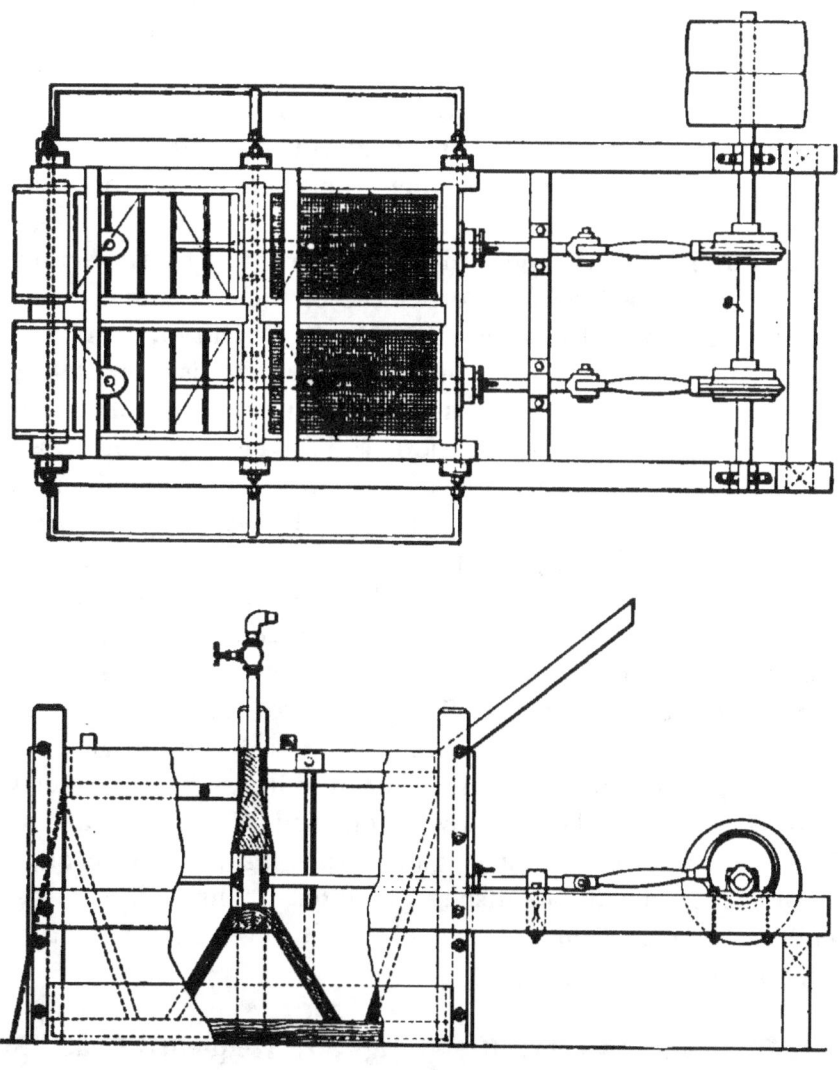

FIG. 14

space is left open below this partition to allow free passage
for the water between the two compartments.   In one compartment is a stationary grating *g*, Fig. 13, of wooden bars
supporting a wire screen upon which the ore is bedded; in

the other is a piston or plunger, which is moved usually up and down, as in Fig. 13, but sometimes horizontally, as in Fig. 14, by a crank, eccentric, or other reciprocating device on the shaft *s*. The horizontal-plunger jig is not very extensively used, as it presents no decided advantages over the vertical-plunger type, requires more floor space, and it is hard to keep the packing about the piston rod water-tight. Jig plungers are made to fit loosely, and in the vertical-plunger type of jig are sometimes perforated with auger holes, in order to reduce the suction on the back stroke.

In starting a jig, the bed is first arranged on the screen as nearly as possible in the order the particles would arrange themselves under the action of a jig in operation; water is then run into the tank until the ore bed is completely covered, when the piston is started up, giving the water column a quick, dancing motion, which keeps the bed open and assists in the separation of the classes. Ore and water are fed in, either together or separately, at a rate to keep pace with the discharge of the machine and make its operation continuous. If they are fed separately, the ore is fed in at the head of the machine on to the screen and the water is fed into the piston compartment; otherwise both are fed on to the screen. The bottom of the tank is made hopper-shaped, with a hole and plug or a discharge gate for removing the concentrates. The middlings were formerly allowed to accumulate, and were cleaned off at intervals by hand, but in most modern jigs they are discharged through automatic, continuous-discharge gates. The tailings discharge over one end of the box, left lower than the other for that purpose.

25. Compartment Jigs.—Jigs are frequently made in sets of two, three, or four, or what are known as two-, three-, or four-compartment jigs, one long tank being partitioned off into that many main compartments. Each one of the latter is further subdivided into screen and plunger compartments. The grating and the tailings dam of each successive main compartment are somewhat lower than those of the preceding compartment, so that the overflow and tailings from each compartment are carried on into the next and further concentrated. These multiple-screen jigs are used when several grades of product are desired or when the ore contains more than two minerals which it is desired to separate from one another more or less completely. For instance, if an ore contains galena and pyrite, with a quartz gangue, a three-compartment jig would be used. The concentrates from the first compartment would be galena, almost pure; from the second, mixed galena and pyrites; and from the last compartment, nearly pure pyrites. If the gangue is a heavy one, like baryta, or there is another mineral in the ore which it was desired to separate, as zinc-blende, another compartment would be added, the concentrates from which in the latter case would be mixed blende and pyrites, with some gangue, particularly if the latter is heavy. The separation of three minerals may also be accomplished in a two-sieve jig, the mixed galena and pyrite forming a middlings class in the first compartment, above the bed of heavy, coarse mineral, while the concentrates from the second compartment are nearly pure pyrites.

The force of the water column in the different compartments is regulated by varying the length of the stroke of the piston. The plungers, in jigs of less than four compartments, are all operated from one shaft; in four-compartment jigs two shafts are generally used, with two pistons on each, the shaft for the last two compartments revolving somewhat more rapidly than that for the first two. It is in the method of varying the length of the piston strokes on a common shaft, independently of each other, that the chief difference between the jigs of this type lies.

26. Hartz Jig.—The Hartz jig is the commonest form of jig, and is typical of the vertical-plunger class. Fig. 13 shows a three-compartment Hartz jig. The stroke of the piston is regulated by means of a double eccentric c, made up of two eccentrics, one within the other. Fig. 15 shows the construction and principle of the eccentric. The inner

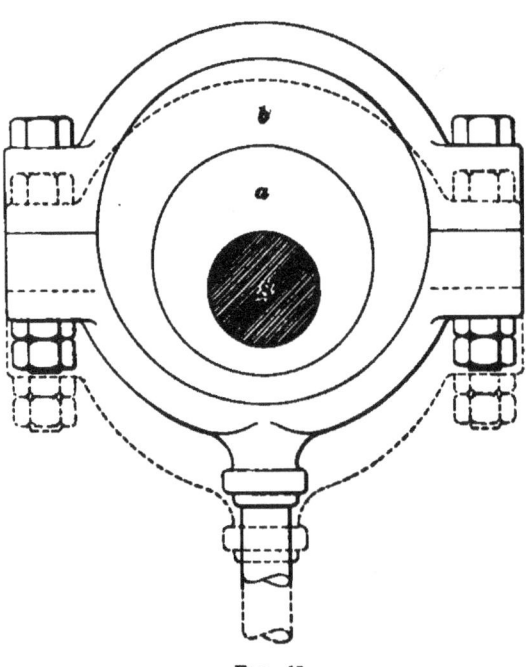

FIG. 15

eccentric *a* is fixed on the shaft *S*, usually ½ inch out of center. The outer eccentric *b* is set the same amount out of center with reference to *a*, about which it may be turned, being held in position while in operation by setscrews. When *b* is turned so as to carry its center on the outside of the center of *a*, opposite the shaft center, the total throw of the whole eccentric is equal to the sum of the throws of *a* and *b*. Thus, with *a* and *b* each ½ inch out of center, the entire eccentric would be 1 inch out of center, with a consequent throw of 2 inches. But if *b* be turned half way round *a* from this position, so that the center falls on the same side of the center of *a* as the shaft center, the center of *b* will coincide with the shaft center—each being the same distance from the center of *a* and on the same side—and the throw will consequently be reduced to zero, as shown by the broken lines in the figure. By turning *b* to any desired position between these two extremes, the throw can be varied from 0 to 2 inches.

27. Quick-Return Hartz Jig.—In the ordinary Hartz jig, the up and the down stroke require the same length of time, and consequently on high-speed jigs the suction on the back stroke is considerable. But by the use of a countershaft and the arrangement of crank, lever, and connecting-rod shown in Fig. 16, the down stroke of the piston is made to occupy only one-third of the time of the full double stroke—or one-third of one revolution of the countershaft—the other two-thirds being consumed on the up stroke. The diagram, Fig. 17, shows how this is accomplished. The small circle is the path of the crankpin p in the crank c, Fig. 16, and the large circle represents the path that would be described by the pin t in the end of the lever l in one complete revolution of the rocker-shaft e. The connecting-rod r, being fixed at one end to the crank and at the other

FIG. 16

to the lever, must necessarily, in any position it can possibly take, have one end somewhere on the circumference of each of these circles. The lengths of c, r, and l being known and the paths to which the connecting pins on c and l are confined being fixed, by assuming p at any point of the circumference of the smaller circle, and laying off the length r from this point to the circumference of the larger circle, we obtain the corresponding position of the connecting pin on the lever l. On the diagram, the corresponding points on the two circles are numbered alike. It will be

noted that the lever travels on the down stroke between its
two extreme positions, indicated by the points *1* and *3* on
the circumference of the larger circle, while the crank moves
between the corresponding positions on the smaller circle,
and that the latter distance is only one-third of the circumference of the circle.   The return stroke of the piston occupies the other two-thirds of the revolution of the crank-shaft.
(While *p* travels from *1* to *2* in this particular instance,

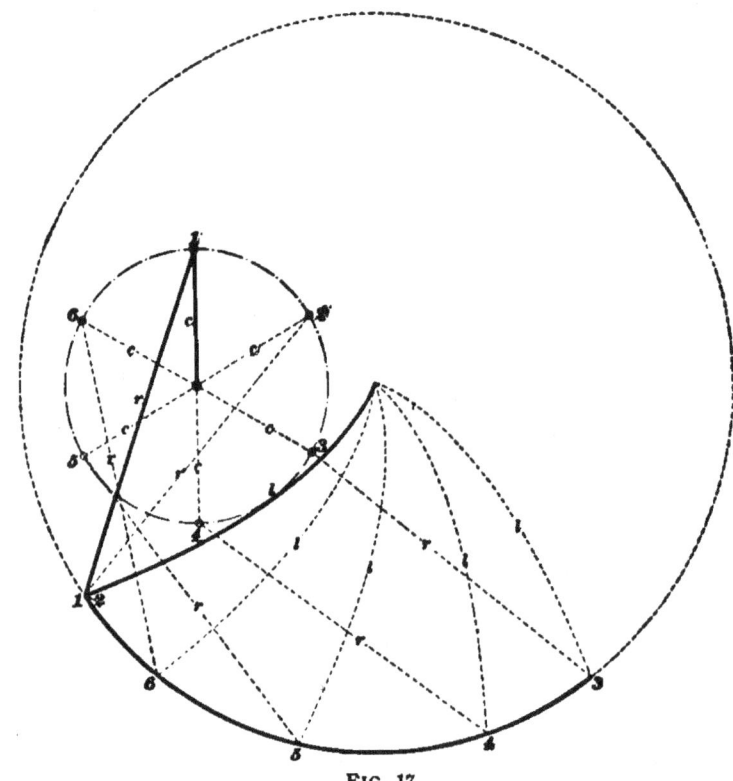

FIG. 17

t remains nearly motionless, and points *1* and *2* on the larger
circle exactly coincide, so that the down stroke of the piston
really occupies only one-sixth of a revolution of the crank
shaft.)   This device is used only on high-speed jigs with a
short stroke, as the eccentric moves only through a quarter
of a revolution, and in order to get a long stroke, the eccentric would have to be of quite large diameter and considerably out of center.   A similar scheme is also applied to

other types of jigs using cranks instead of eccentrics on the countershaft.

28. Slide Jigs.—The slide jig shown in Fig. 18 illustrates another adjustable reciprocating device.   In principle

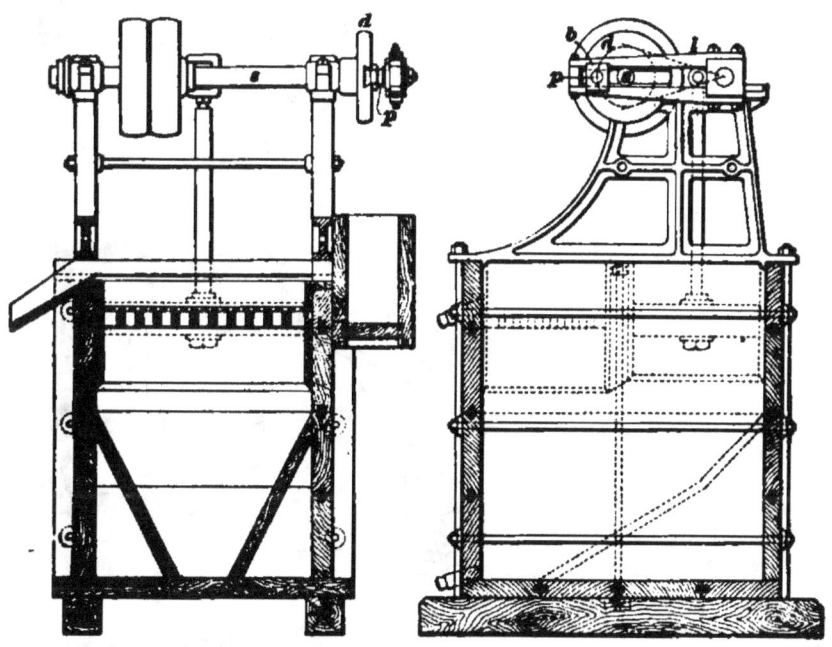

FIG. 18

it is somewhat similar to the quick-return Hartz jig, the down stroke occupying a shorter portion of the revolution of the shaft *s* than the up stroke.   The lever *l* is slotted and keyed on the rocker-shaft.   In the slot of *l* is a freely moving block *b*, which also serves as a bearing for the free end of a crankpin *p* extending out from the wristplate *d*.   As the shaft *s* revolves, the block *b* slides back and forth in the lever slot, and at the same time causes the lever to oscillate, this motion being transferred to the rocker-shaft, and from here through the crank to the piston.

The two extreme positions of the lever *l* are the points where the center line of the lever is tangent to the circle described by the center of the pin *p* in its revolution about the shaft *s*, as shown by the dotted lines in Fig. 18.   It is

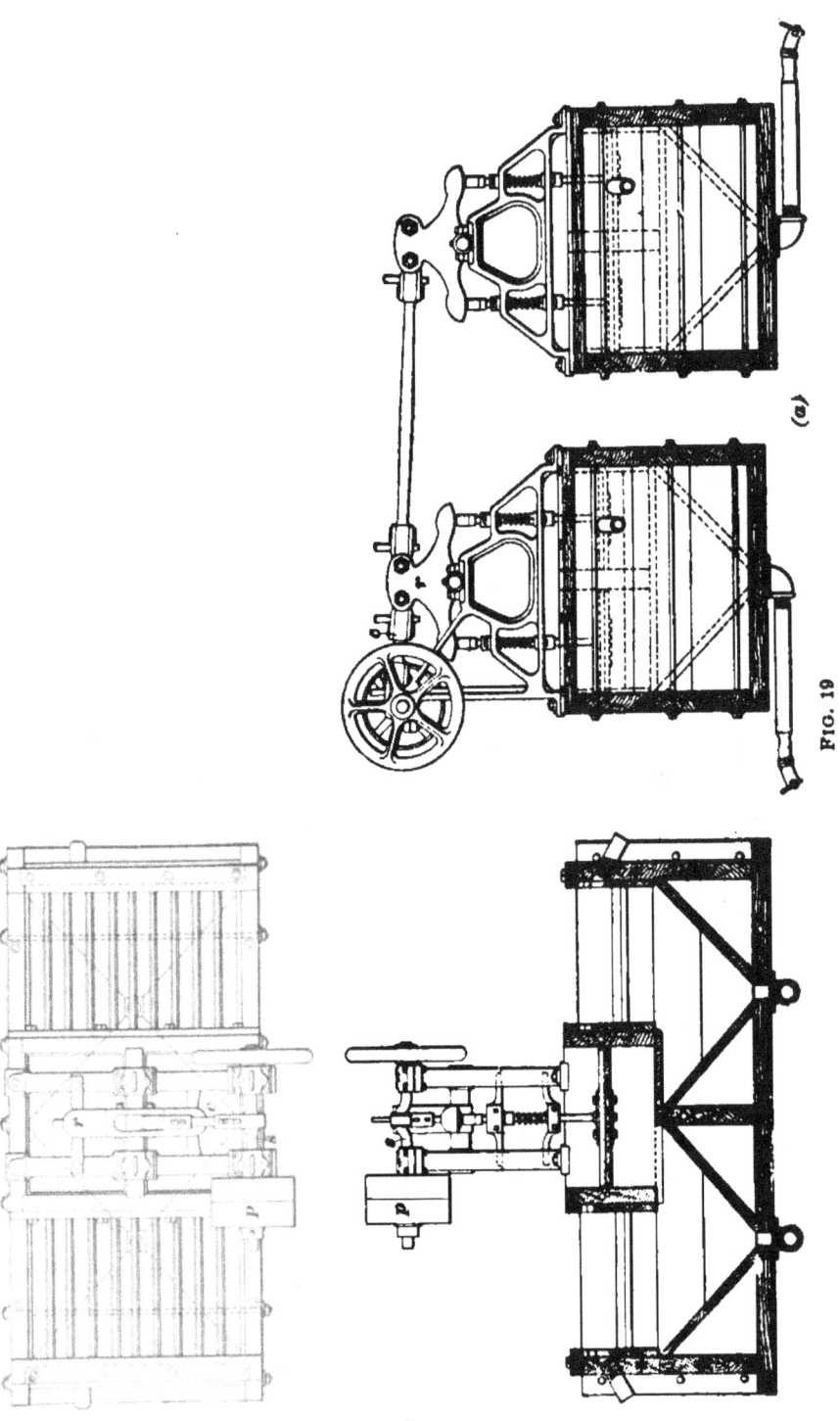

FIG. 19

apparent that the larger the circle described by the pin, the greater will be the difference between the duration of the up stroke and the down stroke. Consequently, in order to make this difference adjustable, as well as the length of the stroke, the pin p and wristplate d are so constructed that the distance between the centers of the pin p and the shaft s can be varied at will. The plate d is slotted diametrically across its face, and in this slot p slips and is held in place by a nut. If p is moved over to the center of the plate, the pin will merely revolve in the block b, without any up or down movement, and the lever and rocker-shaft will remain stationary. But if the pin is moved ever so little away from the center of the plate, it will have some throw and will start rocking the lever and rocker-shaft. As the distance of the pin from the center of the plate increases, the throw of the lever increases also, and consequently the length of the piston stroke, while the duration of the down stroke is decreased as the angle through which the lever moves increases. These jigs are suitable for coarse ores.

29. Collom Jig.—The Collom jig, shown in Fig. 19, illustrates another method of operating the pistons. In this method the pistons are not connected with the shaft, but are hung independently and held in position by springs and collars, as shown in the figure. The pulley p is set on a crank-shaft s, from which a connecting-rod c runs to the lever tappet r, giving the latter a rocking motion. As the tappet oscillates, the levers press alternately on the heads of the piston rods on either side of the shaft, the rods pressing down as the arms descend, while the springs return them to their normal position as the arms rise and release them. The length of the stroke may be varied by raising or lowering the set collar on the piston rod, against which the spring presses, thus lowering or raising, respectively, the normal position of the piston. As the lowest position of the piston is always the same, being the point to which it is depressed when the lever arm is in its lowest position, while the highest position is the normal position in which it is held by the spring alone, it is obvious that by raising or lowering the latter, the stroke is consequently increased or diminished, respectively, the motion of the lever tappet being entirely independent of that of the pistons. Two jigs of this type are sometimes operated from one driving shaft, as shown in Fig. 19 (*a*), by means of a connecting-rod between the two lever tappets; the motion is thus transmitted from the driving, or head, jig to the following, or tail, jig. The tail jig in this case is set at a slightly lower level than the head jig. Collom jigs are quite largely used for the concentration of copper ores.

30. Reciprocating-Screen Jigs.—Reciprocating-screen jigs are very little used, particularly in America. They get out of order more easily than piston jigs, the wear is greater, and the increased suction on the back stroke is also a disadvantage. They do away with the extra width required for piston compartments, but in all other respects are inferior to piston jigs. The best example of this type is **Green's jigger,** in which the screens are moved up and down by double eccentrics, like the plungers of stationary-screen jigs.

JIG DISCHARGES

31. Pipe Discharge.—The simplest form of pipe discharge is a pipe running up through the jig box and sieve. The lower end may be left open for continuous discharge, or kept closed and the concentrates discharged at intervals, as desired. The latter scheme presents two disadvantages. The first of these is that, the ore bed remaining on the screen for some time, the particles rub against one another and wear away a fine slime of rich ore, the greater part of which is lost. The second objection is that, to be sure that the concentrates are completely discharged, the discharge must be continued till the gangue commences to come through the discharge pipe, and the pipe will necessarily be left full of gangue, which will come out with the next clean-up.

On the other hand, when the continuous discharge is employed, the separation is not quite so clean, and the discharge must be very carefully regulated; but this can be done by means of slides in the lower ends of the pipes, by which the area of the discharge orifices may be adjusted. The use of continuous discharge saves the time that is lost, in discharging intermittently, by stopping the machine, and on the whole it may be considered the best practice under ordinary conditions.

32. Scope of Pipe Discharge.—The opening of the discharge pipes for the screen concentrates is flush with the top of the screen. If more than one class is to be concentrated in the same screen, as galena and pyrites, the pipe for the discharge of the lighter concentrates (pyrites) is carried up through the bed of galena. Three pipes are sometimes used in this way, two of them discharging clean minerals, and the third (the middle one) discharging a class composed of the two minerals mechanically combined. Middlings of gangue and ore may also be discharged through pipes, and the scheme may be employed for a middlings discharge in jigging through a bed by extending the pipe up through the bed of coarse, heavy material. Pipe discharges are largely used for fine and medium sized ore, but if the ore is very coarse, the pipes are apt to clog, and some form of discharge must be adopted that is less liable to pack and is more accessible for cleaning in case it does pack. Various devices have been invented with these objects in view, but the *Heberle gate* has superseded all others and is almost exclusively used in America for coarse jigs, and has to a great extent replaced the pipe discharge for the finer sizes.

33. Heberle Gate.—The beds of ore on the jig sieves, being kept loose and full of water, flow back and forth like heavy liquids; they can be run or siphoned off, and have, in fact, all the characteristics of fluids. The Heberle gate takes advantage of this fact. The gate, as illustrated in Fig. 20, consists of a rectangular opening *f* in the side of the jig box above the screen and an adjustable slide *c*, through which the concentrates discharge. Sometimes a double slide is used

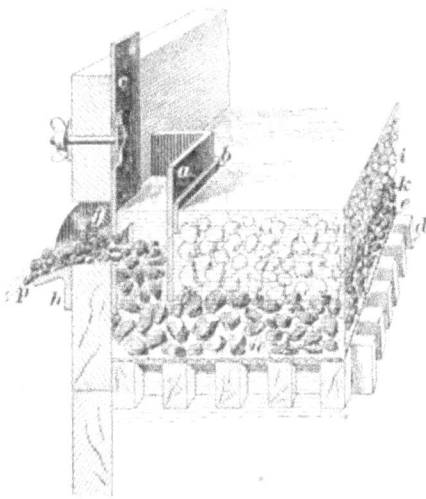

at *c*, so that both the top and bottom of the opening through which the concentrates discharge can be controlled. Behind the aperture there is a **U**-shaped piece *a*. This is secured against the side of the jig by means of the band *b*, which terminates in bolts that pass out through the side of the jig and are secured by nuts on the outside. By loosening these bolts, the shield can be moved up and down

FIG. 20

in such a manner as to regulate the distance between its lower edge and the face of the screen. Ordinarily for discharging concentrates, the shield *a* is placed so that the coarsest material can just pass under its lower edge without clogging between it and the screen *c*. The thickness of the bed of concentrates is controlled by raising and lowering the slide *c*, which regulates the height of the discharge opening. This can also be effected by regulating the supply of ore. The shield *a* prevents the tailings and middlings from flowing out through the opening in the slide *c*, but allows the concentrates to run under; and the weight of the ore and gangue on the bed forces them up inside of the shield, on the principle of the siphon, to a level considerably above the concentrates on the screen, but necessarily lower than the top of the material on the screen on account of the fact that the material on the jig is composed of heavy and light particles, while the concentrates are all heavy particles. A gate constructed on this principle can be used as a middlings discharge by simply raising the shield *a* to a sufficient height above the screen so that the middlings will flow over the concentrates and out through the siphon discharge. The lift being very short, the discharge is not apt to clog, and the top of the shield or gate a, being open, the jigman can tell at a glance just how the machine is working, and, if the gate clogs, can reach his hand in and clear it out. In the illustration, d represents the wooden grating which supports the screen e on which the jig bed rests; k represents the bed of concentrates, while i represents the gangue material above. The opening f in the side of the jig box is usually provided with a small spout h, over which the concentrates discharge as at p.

34. Tailings Discharge.—The jig tailings ordinarily discharge at the lower end of the box, over a tailings dam, the latter being one side, or a part of one side, cut lower than the rest of the box.

In the method known as the Hartz discharge, both the tailings and the concentrates from each compartment flow on to the screen of the next compartment—the tailings over the dam and the concentrates through a slit under it. There is a drop of about $2\frac{1}{2}$ inches between each screen, to prevent the material from backing up through the concentrates discharge, and the slit is protected by an apron which prevents the tailings from mixing with the concentrates as they flow over.

When water is scarce, and it is desirable to use as little as possible and still have automatic discharge, mechanical means are used to discharge the tailings. The most common mechanical discharge is a revolving paddle wheel which scoops the tailings over the dam as it revolves, or a scoop at the end of a suspended oscillating rod, working in the same way. The Archimedean helical screw has also been used for this purpose.

35. Stay Box.—Another water-saving device is a box or extra compartment built on at the lower end of the jig, called the stay box. The tailings discharge into the stay box through a slit which is 2 or 3 inches lower than the overflow of the stay box. This gives a constant head or back pressure against which the tailings must discharge. The box also acts as a settling box or hydraulic classifier, any heavy particles which may be in the tailings settling to the bottom, while the lighter tailings pass away with the overflow. Another form of stay box discharges entirely through a hole in the bottom, the opening being regulated automatically by a plug attached to a float which rises and falls with the water in the box. As the float rises or sinks, the discharge opening is altered to correspond. The tailings sink to the bottom and discharge, most of the water being retained.

36. Jigs are usually built of wood, held together by bolts and lagscrews, and if well constructed will last for 8 or 10 years—working a single shift. Sheet-iron jigs, though sometimes used, are not practicable for ore milling in general, as in the majority of mills the only water obtainable is pumped either directly from the mines or from streams into which the acid waters from the mines drain, and this corrodes the iron and soon renders it rotten and worthless. Moreover, the constant vibration shakes iron jigs to pieces in a very few years unless they are very strongly made.

Jig boxes are usually 3 feet to $3\frac{1}{2}$ feet square in the clear. The areas of the piston and screen compartments are usually equal, though for coarse jigging the piston compartment is sometimes reduced to $\frac{3}{4}$ or $\frac{2}{3}$ of the area of the screen. The space beneath the dividing partition should never be of less area than the piston compartment, to avoid contraction of the water column. The bottom of the jig box in fine jigs is sometimes built semicircular, as this form obstructs the water less than sharp angles and gives a more even flow. The speed of the jig varies from 75 strokes per minute for coarse ore to 200 and even 300 strokes for the separation of the very fine sizes. Strokes with a length of $5\frac{1}{2}$ inches are used in Europe for very coarse jigging, but in America these coarse sizes are seldom jigged, except at anthracite collieries, and 2 inches is usually the maximum stroke. The plunger compartment is usually covered by splash boards.

If chips of wood in the ore cause trouble by clogging the screens of the fine jigs, they may be collected by placing a strip of screen just back of the overflow of the coarse jigs, with its edge dipping slightly beneath the surface of the water.

BUDDLES

37. Stationary Buddles.—The buddle is one of the oldest forms of slime concentrators. Buddles are particularly efficient in treating slimes too fine to work well on belt or table concentrators, and are used chiefly for treating the tailings from other concentrators. The principle of the buddle is that of all concentrators—the settling of minerals in the order of their specific gravities.

From plane tables down which the pulp flowed to round buddles was a short but important step. Round buddles are merely circular tables, with the surfaces inclined either inwards or outwards at a slope of from 1 to 2 inches per foot. They are usually made of wood or sheet iron, with a wooden or cement working surface. The pulp is fed either at the center or at the rim, according as the buddle is an outward-flow or an inward-flow machine, and flows down the table, depositing its contents as it goes, in the order of their specific gravities. The outward-flow buddle is superior in operation to the inward-flow, as the pulp stream spreads out as it flows, consequently diminishing the force and allowing the material to settle out more completely, while the pulp stream on the inward-flow buddle contracts in its downward flow, giving a deeper and stronger current towards the tailing sluice and washing away considerable fine mineral with the tailings. Each form, however, has its advantages.

The feed of the inward-flow table being at the circumference, the area over which the headings are deposited is very much larger than in the case of the outward-flow table, where they are deposited near the apex of the cone. A great deal of gangue is also deposited at the head. The middlings product of the inward-flow table is smaller, coarser, and

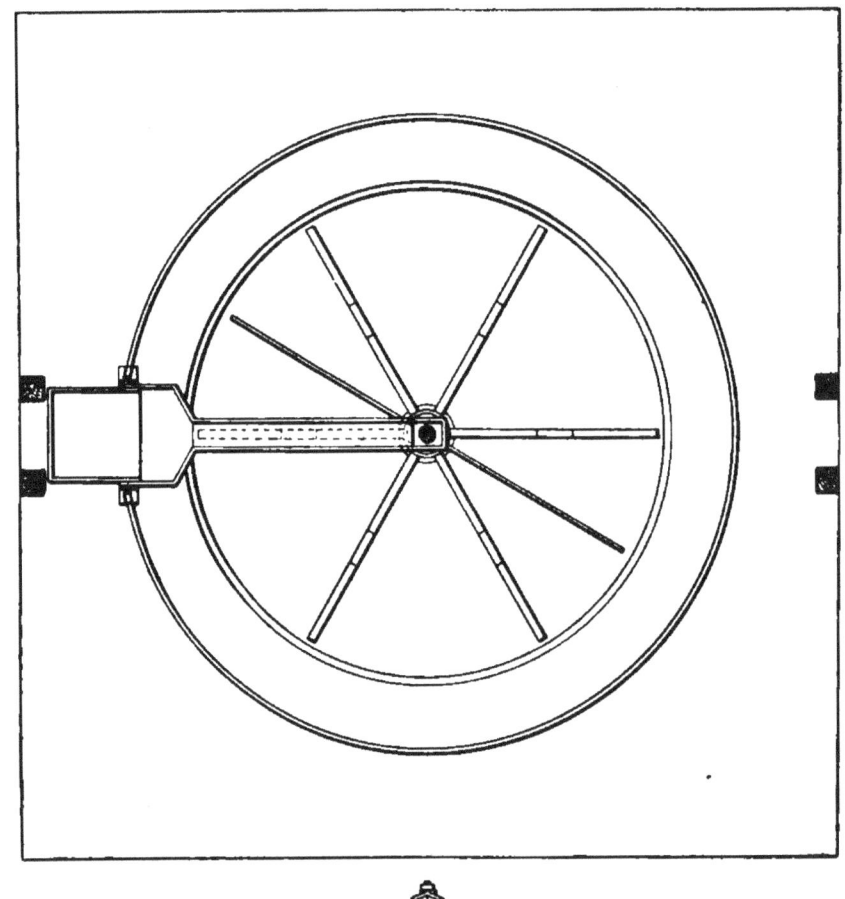

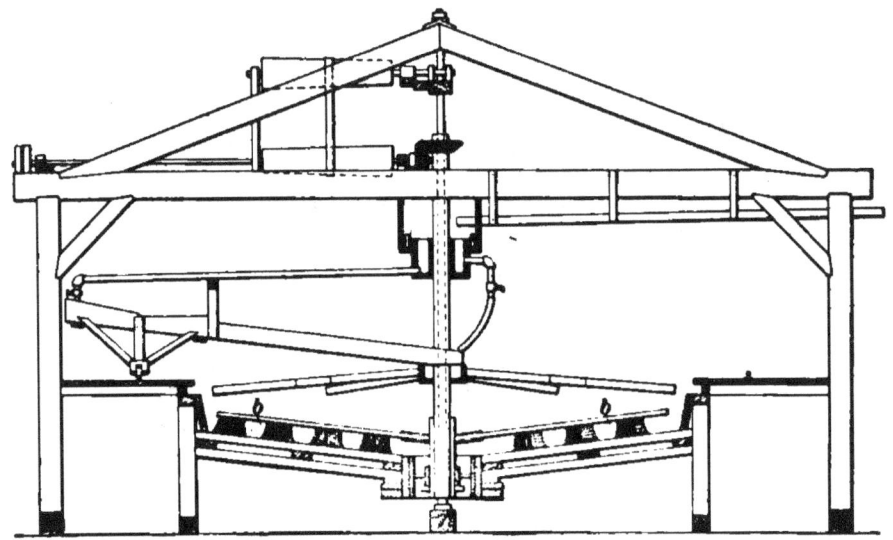

Fig. 21

richer than the corresponding class from the outward-flow table, and the tailings are apt to run rather high.

The outward-flow table, on the other hand, tends to produce a small, clean head class, shading off rapidly into a large, low-grade middle class, while the tailings are nearly barren.

38. Paine and Stephens Buddle.—The original form of round buddle was a stationary, convex, or concave conical table, 11 to 30 feet and upwards in diameter, upon which the pulp was fed, in the case of the inward-flow buddle, from slowly revolving feedpipes running from a main pipe in the center and discharging at the circumference, and, in the outward-flow buddle, from a slowly revolving central feed. The Paine and Stephens buddle, shown in Fig. 21, illustrates the best type of stationary inward-flow buddles. The brushes *b*, *b*, which are used on all intermittent-discharge buddles, are for the purpose of spreading the deposits evenly over the surface of the table. They radiate from the central shaft and revolve with it. The pulp on either inward- or outward-flow buddles flows down the inclined table, the heavy particles depositing first, and the light, or tailings, last. The buddle is usually arranged so that the feed and discharge gates and the brushes are raised automatically by a worm-gearing as the deposit accumulates.

When the buddle is full, which will occur in from 8 to 12 hours, according to the coarseness of the sand, the machine is stopped and the table cleaned. The bed is usually divided into three classes—heads, middlings, and tailings—the three rings are shoveled off separately, and both the heads and middlings, and sometimes the tailings, are retreated. The middlings are rewashed as before, and again divided into three classes, the heads going in with the first headings and the middlings mixed with other middlings and retreated. The headings are buddled again, the buddle discharge being raised about 3 inches, and the buddle covered with middlings from the tossing tubs. The heads are then charged. When this is completed, the deposit is again divided into three rings and cleared off, the heads being carried to the tossing tubs for further washing and the middles again buddled.

39. Revolving Buddles. — American milling practice has always been remarkable for an aversion towards intermittent-discharge machines, hand cleaning, and retreatment of products. It was this desire to avoid the formation and retreatment of middlings that gave rise to the invention of continuous discharge jigs, belt concentrators, etc., and it naturally showed itself in regard to the stationary-table buddle with its intermittent discharge and manifold retreatments, and the invention of the revolving continuous-discharge buddle was the result.

The sorting action of the revolving buddle is the same as that of the stationary type, but the operation is quite different. The table itself is given a revolving motion, the rate varying from one revolution in 5 minutes for very fine slimes to $2\frac{1}{2}$ revolutions per minute for pulp carrying 8 or 9 per cent. of fine sands. The higher the speed of the table, the greater is the capacity—the limit of the speed being determined by the grade of the tailings as compared with the gain in capacity. The feeding apparatus is stationary, and the pulp feed extends only about $\frac{1}{3}$ to $\frac{1}{2}$ of the way around the table, the remainder being fed by wash water. The inward-flow buddle is comparatively little used except in connection with the outward-flow machine. Revolving tables are made, like the stationary tables, of wood or sheet iron, with a working surface of either wood or cement. The frames are strongly braced. The tables range from 12 to 30 feet in diameter and are driven by bevel gears or worm-gearing, usually from above.

40. Inward-Flow Buddles. — The revolving buddle differs from the stationary buddle only in that the table is revolved, while the feed is stationary, and that the operation is made continuous by the use of wash water, which cleans the various classes off the table as fast as they are formed and carries each to its respective launder. The inward-flow revolving buddle is fed from a trough or launder at its circumference, which is divided so that pulp is fed about ⅓ of the way round and clear water the remainder. Flowing down the table, the pulp stream deposits its contents. The different classes are washed into their respective sections in the central pit by jets of water from stationary pipes running radially or diagonally across the zone on the table in which they are deposited. Thus, supposing there were three classes, headings, middlings, and tailings, a jet or series of jets about ⅓ of the way round from the forward end of the pulp-feed box, and extending diagonally up from the discharge pit as far as the layer of tailings is considered to reach, will wash this layer off continuously, for the table is revolving and constantly bringing fresh material under the jets. A second jet ⅔ of the way round, and extending across the middlings zone of the table, will carry the middlings to their launder; and a third jet, at the top of the table, washes the headings down just before the particular portion of the table upon which they are deposited comes again into the pulp stream. A revolving bristle brush is frequently used with the jets to clean the material off, as when water alone is used a slime forms on the surface of the table, which diminishes its efficiency.

41. Outward-Flow Buddles.—The outward-flow buddle is the type most generally used in America, as it produces barren tailings, and the product does not have to be put through so many operations and retreatments. Moreover, in American gold milling, buddles are used to treat only very fine slimes, such as the tailings from fine jigs and vanners, in which the greater part of the valuable mineral is extremely fine—much finer than the greater portion of the gangue—and the particles are carried some distance down the table before they have time to settle through the pulp stream. In this case, if the current gained velocity as it descended, it would sweep these tiny particles along with it and carry them over the tailings gate.

The feed of the outward-flow buddle is at the center, usually from a round iron box surrounding the shaft. This box feeds pulp through orifices in the bottom extending from ⅓ to ¼ way round its circumference and clear water the remainder. The pulp and water fall on to a fixed apron extending a short distance out over the table, and from here flow on to the slowly revolving table, where the mineral duly sorts itself. The table is cleaned by jets and brushes in the same manner as is the inward-flow table.

42. The **Collom buddle** is a rotary buddle, which is in form only a slight modification of the old stationary machine. The pulp is fed in through a trough instead of from a central distributor, and is spread by brushes, which are sometimes given a reciprocating motion to prevent the pulp from packing against them. The wash water is delivered from a central box or from an annular pipe at the head of the table. The table is cleared by jets, as in the other forms described. These tables are sometimes made in two or more annular sections, with a slight step between each; the sections sometimes have different slopes. They are also used for amalgamating tables, by cutting a series of annular grooves in the surface and filling them with mercury, which will amalgamate any gold coming in contact with it.

43. Evans Buddle.—It is sometimes desirable to protect the headings from the action of the wash water, and in such cases either a spiral apron is used, as in the Evans buddle, or the wash water is not fed over the apron, but issues from holes in a spiral pipe hung in such a manner that the best headings are not subjected to the action of the wash water.

The slime table, shown in Fig. 22, known as Evans buddle, protects the headings by means of a spiral apron placed at the center of the table. The pulp flows over half of the surface of the apron *a* through holes in the bottom of the distributor *d*, which is partitioned so as to feed pulp half way round and clear water the other half. From the apron the

pulp flows on to the revolving table *b*. Owing to the
spiral form of the apron, the top headings, as they sink,
pass at once under it, and are protected from any further
action of the pulp stream and wash water until they reap-
pear, at the end of the revolution, from under the widest
part of the apron, when the jet from *f* washes them down
into a division of the launder. The middlings are washed
off by the jets from the perforated pipe *e*, just ahead of the

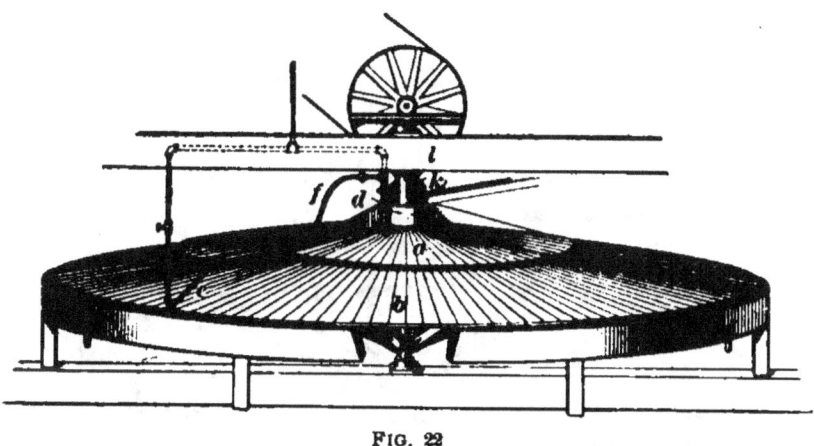

FIG. 22

headings jet, into another division, and the tailings flow off
through the remainder of the launder. The position of the
slime and water feeders on the apron is regulated by the division board *k*. The feed apron is hung from the frame *l* and
can be readily adjusted relatively to the table. The speed
of a 14-foot machine of this type is about 1 revolution in
80 seconds, and the capacity 25 to 30 tons per day of 24 hours.
The slope or pitch of the table is about $1\frac{1}{4}$ inches per foot
and of the apron $1\frac{3}{4}$ inches. The table revolves in the
direction *e* to *b*.

44. Multiple-Deck Buddles.—A frequent device to
save expense and economize in floor space is to place one
buddle above another, as shown in Fig. 23. Each table
may be fed separately, or the lower table may take the
tailings from the upper and further concentrate them. This
latter scheme is especially applicable when both inward and

outward flow buddles are used. The inward-flow table is
placed above and the rich tailings from this table discharge
directly on to the outward-flow table below, which catches
most of their values and delivers a barren tailing. In this
way the best points of each type are applied. This scheme
is employed with both stationary and revolving buddles.

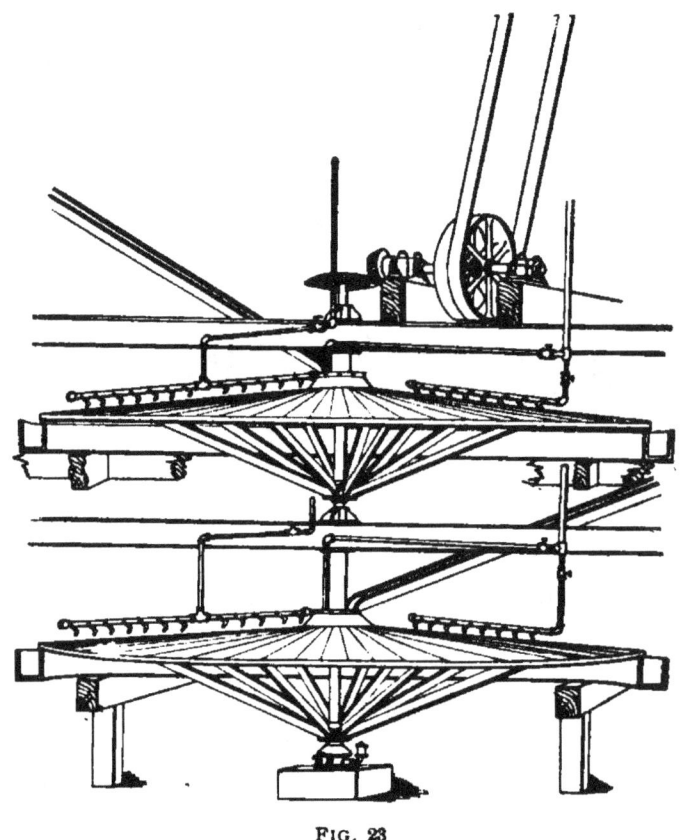

FIG. 23

Fig. 24 shows a sectional view of a triple-deck Linkenbach
buddle. This buddle is stationary, but differs from the
ordinary type in that the discharge is continuous, both the
feed and receiving launders revolving. Each table is about
1½ feet wider than the one above it. When two or three
grades of slimes are to be treated, the coarsest grade is treated
on the top table and the finer grades on the lower tables.

Referring to the figure, the pulp is fed on to the tables A
from pipes f. The wash water is fed through the hollow

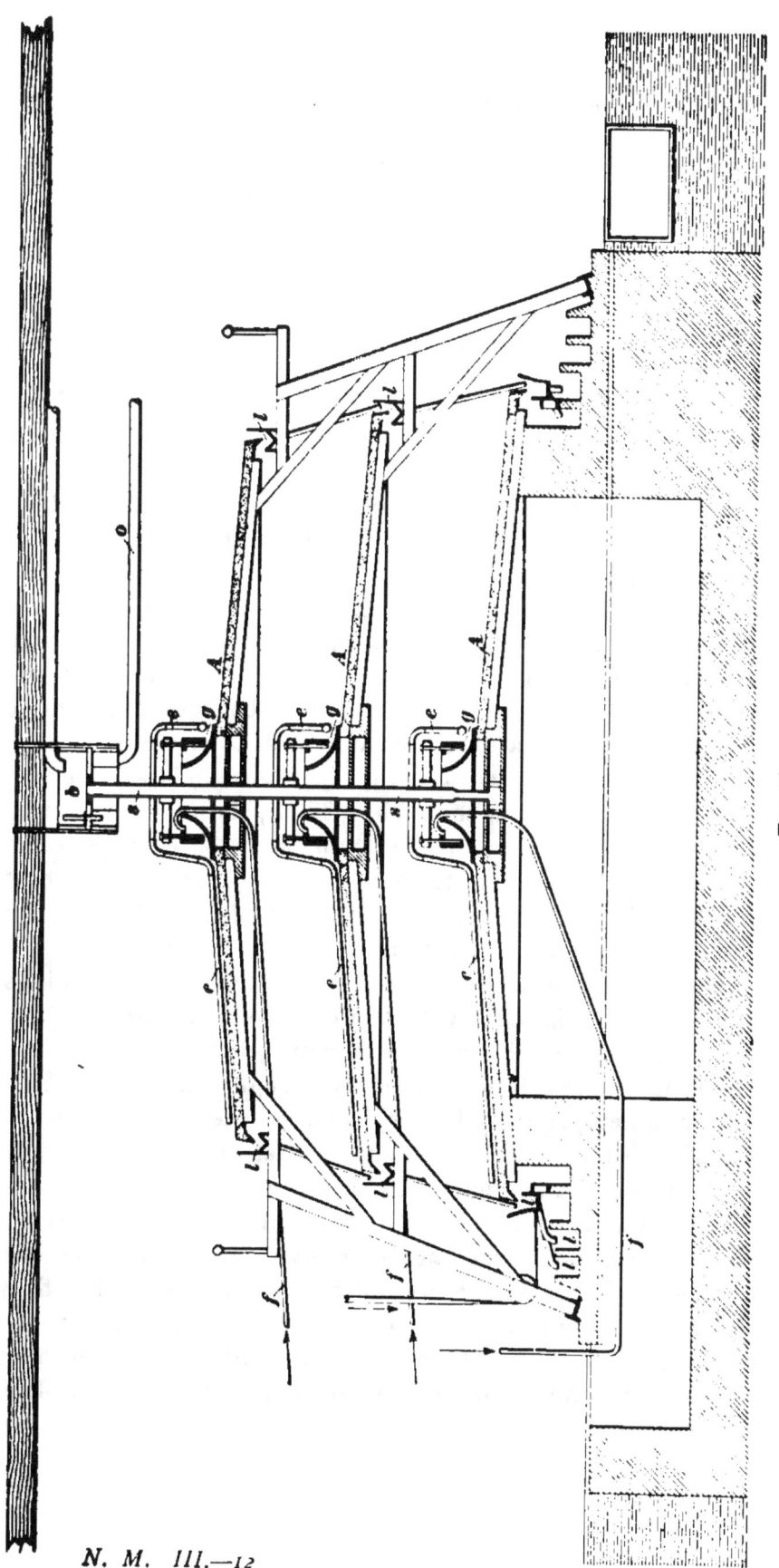

FIG. 24

N. M. III.—12

spindle *s* and the pipes *c*, which are attached to it and revolve with it. This spindle also carries the revolving cylindrical gates *g*, which deliver the pulp successively to all points around the circumference of the distributing aprons. Jets of water from the pipes *c* clean the classes off the table as fast as they are formed, carrying them down to the launders *l*, *l*. The launders for all but the bottom table are stationary and are made up of a ring of flat funnels, about 2 feet apart, delivering into pipes leading into the launder of the next table below. In this way all the tables discharge constantly into the lowest launder, which is supported on wheels and revolves at the same rate as the feed gate and cleaning jets. The revolving launder is divided by adjustable partitions into as many segments as it is desired to form classes of mineral. As this trough revolves at the same rate as the cleaning jets, any portion of it is always at the same position with reference to the discharge of the table and consequently catches the same class of material. Suppose, for instance, we are forming three classes—heads, middles, and tails—on the table. That portion of the trough closest behind the cleaning jets will catch the tailings which are at the lower edge of the table. By the time the highest of the tailings has washed down the table, the trough will have traversed some distance. At the point where the tailings end, we put a partition. The length of the tailings segment of the launder will be proportional to the width of the zone of material we desire to consider as tailings. The middlings will wash down into the trough behind this partition, another partition being placed between them and the headings which reach the bottom last, and consequently settle in the last portion of the launder. The use of open-bottom launders for the upper table is equivalent to having all the tables discharge into one revolving launder, as the classes, as fast as they are washed down the tables, flow right on through the funnels and pipes into their proper sections of the bottom launder as it revolves. Any overflow in the wash-water feed box *b* is taken care of by the overflow pipe *o*.

CONCENTRATING TUBS

45. Tossing Tubs.—The tossing tub, or keeve, is little used in America, but in Europe it is quite common in connection with buddles, particularly with those of the stationary type, the headings from which are always further concentrated and cleaned by tossing. It is simply a round

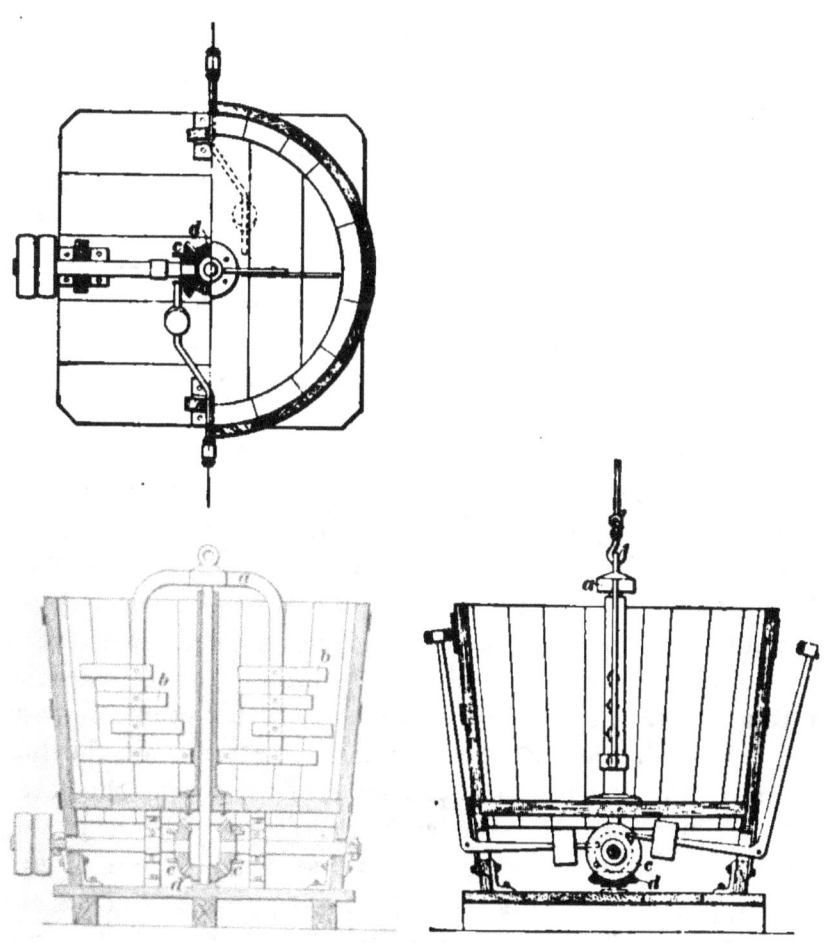

FIG. 25

wooden tub, shown in Fig. 25, about 4 feet in diameter and 2¼ feet deep in the clear, made usually of 2-inch material. This tub is stationary, but a vertical shaft or spindle passes up through a cast-iron sleeve or cone extending up in the center nearly to the top of the tub and is operated by a bevel gearing below. On this spindle and revolving with it is the yoke *a* bearing the stirring paddles *b*. This yoke may be lowered and raised into and out of gear by a light tackle over the tub. On opposite sides of the tub are two weighted bell-crank levers with light hammers attached to their upright arms. These bell-cranks are pivoted under the lower edge of the tub, and their horizontal arms extend nearly to the center, where they are engaged by pins set in the two vertical bevel wheels *c*, on opposite sides of the horizontal bevel wheel *d*. There are two pins in each wheel. As the wheels revolve, these pins raise the ends of the levers. As soon as the lever is released by the pin passing from under it, the weight causes it to drop back suddenly, and the hammer head on the other arm is brought up against the tub with a sharp blow. Each hammer strikes two blows at each revolution of the shaft, which is run at a speed of from 25 to 50 revolutions per minute—the usual speed being about 48 revolutions.

The tub before starting is filled about one-third full of water, the yoke is let down into gear, and the hammers are blocked back by wedges between the hammer arm and the side of the tub. The stirrer is then started up. The buddle headings are fed in till the tub is nearly full or the pulp reaches the proper consistency. The yoke is then lifted out by the block and tackle above it and the hammers thrown into gear by knocking out the wedges between them and the tub. The pulp is then allowed to settle while the hammers are tapping away on the sides of the tub at the rate of 50 to 100 blows a minute each. This rapid jarring keeps the water agitated, and the mineral settles in layers of equal falling particles. This settling operation requires from 15 to 20 minutes. The machine is then stopped and the water siphoned off. The upper 2 inches of the bed are usually thrown away as waste. The remainder of the bed is divided into two beds of equal thickness. The headings are set aside and the middles are retossed. The headings from this second tossing are combined with those from the first and sent to the roasters preparatory to some other treatment.

The second middles are returned to the buddles and are rebuddled along with the buddle middlings, and the upper layer of tailings is thrown away. The tossing tub is suitable only for the treatment of medium-grade slimes. Sand slimes coarser than 16 to 20 mesh are too coarse to separate properly, and very fine pulp slimes do not settle well.

46. Dolly Tub.—The Dolly tub is another form of slime-concentrating apparatus. It consists of a stationary

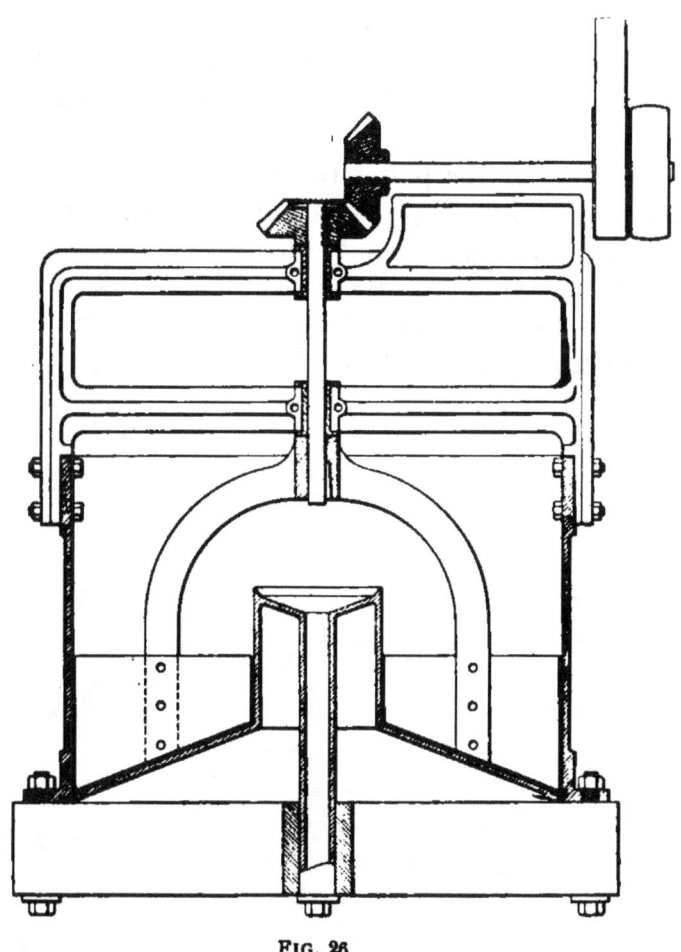

FIG. 26

wooden or iron tub, having preferably a conical bottom, slanting from the center to the sides, as in Fig. 26, with a raised funnel discharge at the center; a suspended vertical shaft driven by a bevel gearing above carries four arms on which are fixed paddles fitting loosely into the annular space between the sides of the tub and the central discharge cylinder. The feed and discharge of the machine may be either continuous or intermittent. The pulp in the continuous-discharge machine is fed into the tub near the side. The heavier particles sink to the bottom, and the centrifugal force set up by the revolving paddles, combined with the sloping bottom, carries them to the outside edge, where they discharge through slits or holes. The lighter particles, on the other hand, are kept in suspension by the motion of the water and are gradually drawn into the discharge funnel at the center. The paddles are sometimes dispensed with by delivering the pulp stream tangentially into the tub, the force of the stream thus delivered setting the water in the tub in motion in a circular direction. The whole of the water may be introduced along with the slimes, or part may be delivered through separate pipes, at different points in the circumference of the tub.

ORE DRESSING AND MILLING

(PART 3)

CONCENTRATION

TABLE CONCENTRATION

1. Theory of Concentration. — All concentration, whether wet, dry, or centrifugal, depends upon specific gravity primarily and upon size secondarily, the two together agreeing with the law of equally falling bodies. Assuming all particles to be of practically uniform size, the sorting power of water, for instance, will arrange particles falling through it in layers according to their varying specific gravities, the heaviest going to the bottom and the lightest arranging themselves on top with intermediate ones between in their order. The law of equally falling bodies, falling free in a fluid, as water, may be represented by the formula

$$S = \frac{w}{r} = \frac{w}{(f + c + b)},$$

where S = speed, w = weight, r = resistance, f = friction, c = cohesion, and b = buoyancy. Now, in unsized ores the separation will not be so satisfactory, for in any case the following is true, because f and c vary as the surface and b as the volume, the volume varying more rapidly than the surface, since

$$b = \tfrac{1}{3}\pi r^3 \text{ and surface} = 4\pi r^2.$$

§ 27

For notice of copyright, see page immediately following the title page.

Suppose $r = 1$, then $b = \frac{4}{3} \pi \times 1^3 = 1\frac{1}{3}\pi$, and the surface $= 4\pi \times 1^2 = 4\pi$.

But suppose $r = 10$, then $b = \frac{4}{3}\pi \times 10^3 = 1333\frac{1}{3}\pi$, and the surface $= 4\pi \times 10^2 = 400\pi$, so that $r = f + c + b$ varies less than the volume, while w varies directly as the volume, therefore varies faster than r, and the ratio $\frac{w}{r} = S$ increases;

and presently a point is reached in unsized ores where a comparatively large body of gangue, or gangue with ore attached, will fall as swiftly as a smaller body of greater specific gravity, or more swiftly than a body of ore smaller still. This results in some gangue in the lower layer and some fine ore in the superposed layer, the ore decreasing and the gangue increasing in quantity upwards. In each layer the particles of ore and gangue, though differing in size, have the same ratio of $\frac{w}{r}$. This law also holds in sized ores, but is unaffected by the complications stated above, and the concentrates should be, therefore, comparatively clean.

2. Bumping Tables.—Bumping, or percussion, tables are largely used for concentrating ores, in which the gold is so intimately associated with iron pyrites that it is not freed by stamping. The bumping table is essentially a suspended table which is capable of a limited movement, and is subjected to a series of blows or shocks in the plane of its motion. The shocks are delivered by drawing the whole table back by means of cams, and releasing it, when strong springs will throw it forwards suddenly, and the end of a beam, which is a part of the table, strikes against a fixed block or buffer, bringing the table up with a jar. The table is slanted away from the head or bumping end, and the pulp is fed on at this end, running down in a thin sheet. The heavier particles in the pulp settle to the bottom and cling to the surface. When the table strikes the buffer, the sudden jar causes the mineral to creep a little farther up towards the head of the table. Such particles as are heavy enough, in proportion to the amount of surface they expose, will resist the down-flowing sheet of water, until finally they reach the head. The shock also serves to separate the particles from one another and to completely submerge all particles, and then very little mineral is floated away by a coating of air.

The surface of the table should be absolutely true and even, so that all the pulp is subjected to the same action, and the top should be as thin as possible without danger of buckling under the jar, then the shock of the bump will send a violent tremor through the whole sheet. If a table be watched while clean water is being run over it, at each blow myriads of small drops of water will be observed to jump up from the plate at right angles. This action, in the regular operation of the table, keeps the pump agitated and exercises a certain amount of classifying action on the particles. Bumping tables have a large capacity, make a fairly clean separation, and are simple in construction and operation; but as the loss of slimes is heavy, they are not to be used for very close concentration.

3. Gilpin County Bumping Table. — This table, although one of the earliest of the many forms of percussion tables, is still one of the most commonly used, and the original design has remained practically unchanged. The table is typical of the end-bumping class. The construction is as previously described. The cam for driving the table may be either above or below. The return is accomplished either by flat steel bar springs or by coiled springs, rigidly fixed at one end, which are forced back by the table on the back stroke, and spring forwards as soon as the cam releases them. The double table is about 4 feet wide and 7 feet long. There are two plates, one on each side of the bumping beam, set on a slope of ¾ inch per foot for the greater portion of their length. On the last 3 feet, at the head of the table, the slope is increased to 1 inch per foot.

The pulp is run into the rectangular feed box, towards the head of the table, and flows gradually on to the table. The wash-water box is set above the feed box in order that the concentrates may be washed clear of slime and gangue before passing over the discharge. The head of the table is always set away from the battery, and the concentrates, which are comparatively dry, are allowed to accumulate on the floor at the head of the table. The capacity of a table is from 10 to 20 tons per day.

4. Perfection Concentrator.—This machine is merely a modification of the Gilpin County bumping table. The bumping beam is placed underneath the table, and one single sheet of copper is substituted for the two plates in the Gilpin County bumping table. The height and slope of

FIG 1

the table can be adjusted by means of nuts on the hangers. The coiled-spring return is used. The table is shown in Fig. 1. Its capacity is about equal to that of the double Gilpin County bumping table, but it weighs only about two-thirds as much.

5. Rittinger Side-Percussion Table.—The side-percussion table is but little used in America, though it is quite common in Europe. The table is hung on a slope, which is adjustable between 3 degrees and 6 degrees. As the name indicates, the swing and shock are lateral, the beam running across the table on the under side. The table is driven by cam and springs, as in the end-bump table, the amount of the swing being also adjustable. The pulp is run in at the upper corner of the table, on the side opposite the buffer, and flows downwards in a thin stream. The wash water is also introduced at the head, between the pulp feed box and the buffer side of the table. As the pulp and water flow downwards in a thin sheet, the heavier mineral sinks to the bottom and is gradually worked over towards the buffer side by the side jars. Those particles of mineral which, in proportion to their weight, present the least surface to the down-flowing stream naturally require the longest time to traverse the length of the table, and are consequently exposed to a greater number of shocks than the lighter particles and are worked farther over towards the buffer side. In this way the discharge over the lower end of the table may be divided into any number of classes desired, ranging from very poor tailings on the feed side of the table to rich concentrates on the buffer side. Each class discharges into a separate trough or compartment. The tables are usually double and are divided by a strip down the middle, each side having a separate feed. Many materials have been tried for the surface, but cast-iron, slate, and marble slabs are found to give the best results. The table performs a very good separation, but the product of pure concentrates is comparatively small, and the formation of middlings is undesirable in the class of work for which bumping tables and vanners are used in America; moreover, the product is mixed with large quantities of water and requires settling after it comes from the table, so that the side-percussion table can hardly hope to supplant the end-bump table to any extent in America. The wear on the cams and tappets of percussion tables may be reduced by the use of roller tappets.

VANNING MACHINES

6. The principle of all vanning machines is the same as that of the gold pan and batea, namely, separating the heavier mineral from the lighter by gently shaking or vanning the pulp, the mild agitation keeping the particles of the lighter mineral in suspension, while the particles of the heavier mineral sink to the bottom. A constant automatic discharge is accomplished with vanning tables by suspending the table and giving it a horizontal jerking motion by means of cams and springs or by eccentrics. The table moves slowly out to the end of the stroke, and returns with a jerk which gradually works the heavy mineral which has settled to the bottom of the stream of pulp running over the table, along or across the table, in a direction opposite that of the jerk. The Wilfley table is the best example of this type of machine. In the belt vanners, the operation is made continuous by precipitating the mineral on a slowly moving endless rubber or canvas belt, so that the whole precipitating surface is constantly advancing, carrying with it the mineral accumulated on it. Of this type are the Frue, Embrey, and Lührig vanners and the Woodbury concentrator.

VANNING TABLES

7. Wilfley Table.—The Wilfley table, shown in Fig. 2, is a flat linoleum-covered table, 16 feet long by 7 feet wide,

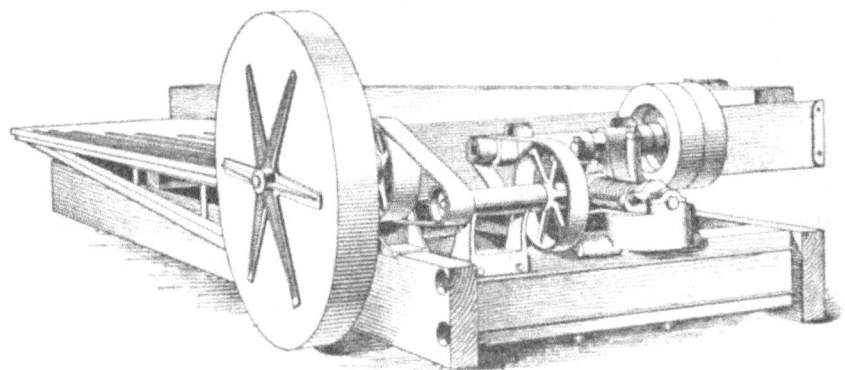

Fig. 2

set on rollers, and slightly inclined from front to back. The table is moved forwards by a togglejoint arrangement, and is brought back by springs with an endwise jerk, which gradually works the concentrates down to the discharge end.

The feed box at the back of the table extends from end to end, and is divided by a movable gate, the pulp being fed in one end and clear water along the rest of the length, keeps the headings clean, so that the operator can tell at a glance how the table is working.

A set of cleats, 2 to 7 in number, according to the character of the ore, is nailed along the table parallel to its length. These cleats taper gradually from $\frac{1}{2}$ inch thick at the upper, or tailings, end of the table to a feather edge at their lower end. The first and longest cleat is put on towards the lower edge of the table and runs up to within 2 feet of the head. The other cleats are successively shorter, the top one being about 4 feet long. The pulp is fed in as near the tailings end as possible, and the heavier mineral sinks to the bottom and clings there. The cleats prevent it from sliding straight across the table and off, and at the same time allow the particles of lighter minerals, which are held in suspension in the water by the jerk of the table, to pass over and off. The tapering of the cleats provides for a considerable range in the size of the particles of gangue and renders the extremely careful sizing, which must be done for most concentrators in order to get the best results, altogether unnecessary. The finer gangue, which is held in suspension in the stream of water, is carried over the cleats with it, but the coarser particles sink. At each jerk of the table, however, they are thrown momentarily into suspension, and as they work towards the end of the cleat, they finally come to some portion which is low enough to allow them to pass over. This operation must be repeated at each cleat. The space between the end of the lowest cleat and the head of the table allows the middlings, or unseparated ore and gangue, to pass over the edge of the table into a middlings trough, through which they flow to a wheel conveyer, which raises them into a launder returning to the feed trough, to be retreated. This insures a clean heading and at the same time prevents the loss of the mineral in the middlings. An inclined shield prevents the tailings from entering the trough. This table has shown a remarkable saving, taking the ore right from the stamps, like the bumping table, without previous classification; and it has been proved that it can compete, at least on even terms, with the best of the belt vanners on coarser products and has a much larger capacity—15 to 25 tons in 24 hours, or equal to the best bumping tables, with a much better separation.

8. Cammett Concentrator.—This machine is constructed on the Rittinger model and has recently received

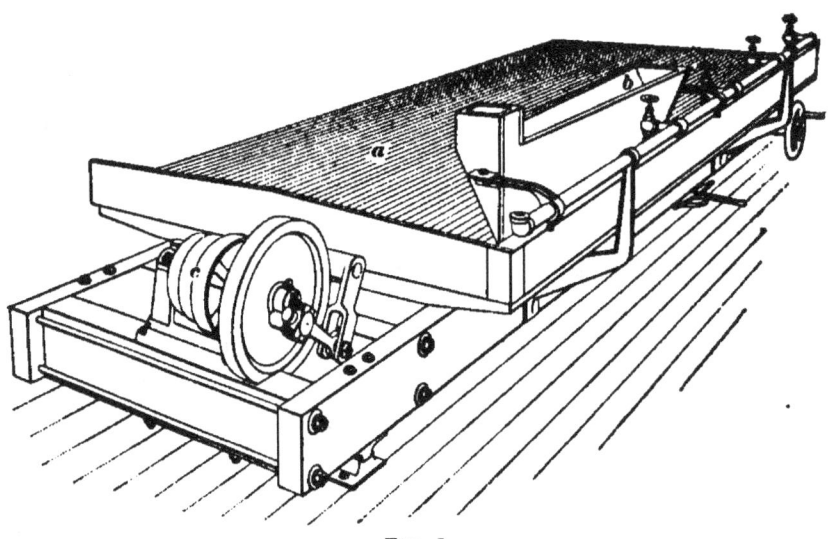

FIG. 3

much praise for accomplishing good work. It has also come into prominence owing to its being able to closely separate different metal minerals, such as zinc, iron, lead, and copper sulphides, from each other. The designer of the Wilfley concentrator imagined that the Cammett was an infringement upon his table, and so brought suit against the latter. While the suit was pending the Wilfley people bought out the Cammett people, whereupon the Court decided in favor of the former against the latter. The table is shown from the operating side in Fig. 3.

9. Construction of the Cammett Table.—The top of the table consists of redwood boards *a*, in which longitudinal riffles are cut. These riffles are said not to warp, break off, turn up at the ends, or split. It will be noticed that the riffles are continuous from end to end of the table and that they gradually become flattened until, at the discharge end, the grooves are scarcely noticeable. The pulp distributing box b is constructed on the principle of delivering the coarse ore at the head of the table and the fine ore near the tail end. The box is suspended on brackets and moves with the table top, thus insuring agitation necessary for classifying and to prevent clogging. The table is moved by the cone pulley c, which is connected by a belt 2 inches wide to a similar pulley on a countershaft, which should make about 250 revolutions per minute. The floor space occupied by this table is 6 feet 4 inches wide by 16 feet 1 inch long. The height of the table over all is 34 inches.

To further insure that the table top retain its original shape while in motion under its load of pulp and water, the proper points at which to place the bearings were determined by a system of balancing, by which parts remote from the bearings balance each other, producing equal pressure on each bearing and relieving the middle portion of the table top from strains. The satisfactory result obtained can be seen in the absolutely quiet state of the water on the table top when the table is in operation.

The pulp distributing box is one of the special features of the table and has been designed after many extensive experiments and practical working tests. This distributing box is a modification of the well-known spitzkasten, which classifies the pulp, delivering the coarse at the head end of the table and the fine near the tail end. The box is suspended on brackets and *moves with the table top.* This insures the necessary agitation requisite for a classifying action and also prevents clogging at the discharge outlets, a grave difficulty encountered in other forms of distributing boxes. With each box is provided a set of various sized outlets, made of a non-wearable material.

The wash-water pipes, three in number, are connected to the main water supply pipe by angle valves, enabling the flow of wash water to be regulated to any degree and for any part of the table. These pipes discharge the water into separate compartments of a wooden trough, which distributes the water in a uniform sheet over the table surface.

It requires one-half horsepower to drive a Cammett concentrator by an electric motor, and this power will give about 10 tons of fine slimes in 24 hours, while on coarse ore the product will be very much increased. The size of the pulp may be from 3 mesh to the finest slimes, although such a mixture cannot be treated without previous sizing, as slimes are more difficult to concentrate than coarser material. The water required for this table will vary from 5 to 20 gallons per minute. Middlings must be recrushed before placing them back on the table.

10. Belt Vanners.—Belt vanners are of two types: the side-shaking and the end-shaking. The principle, however, is the same in both. The end-shaking vanner is comparatively little used, as the side-shaking machine is the better, both in principle and construction. End-shaking machines, however, are still used in some mills where the conditions are such that they do practically as good work as the side-shaking machines; but they require a larger amount of water, a greater inclination of the belt, or a more rapid shaking motion than the side-shaking machines, in order to do the same work.

11. Frue Vanner.—The Frue vanner, shown in Fig. 4, is the original side-shaking machine, and is typical of the class. It consists essentially of a continuous rubber belt traveling slowly up a slight incline and shaking rapidly back and forth sideways. The belt is usually 4 feet wide—though 6-foot belts have been used—and has elastic raised edges. It runs over two large galvanized-iron rollers *A*, 13 inches in diameter and 12 feet apart from center to center, set at either end of a slightly inclined frame. This frame is supported from the fixed frame or table of the machine by eight flat steel rods or springs, which allow it to swing back and forth laterally, and eccentrics on a

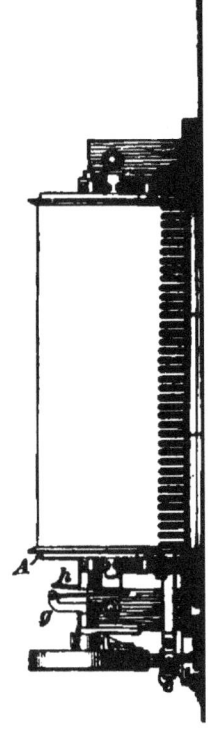

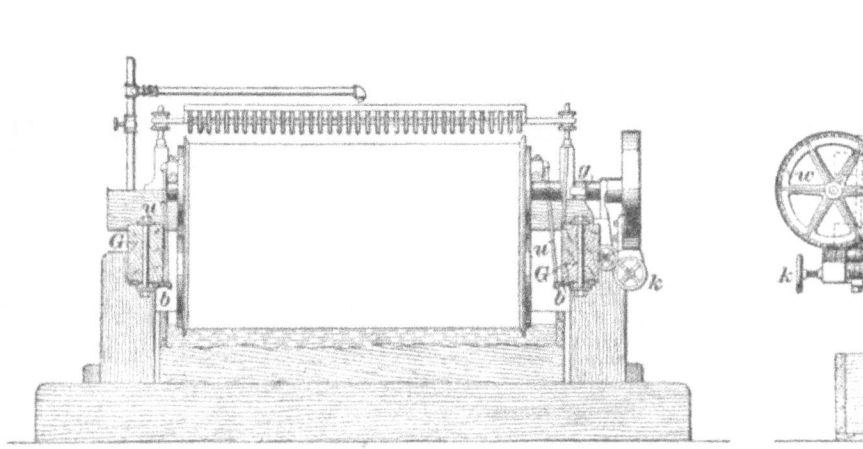

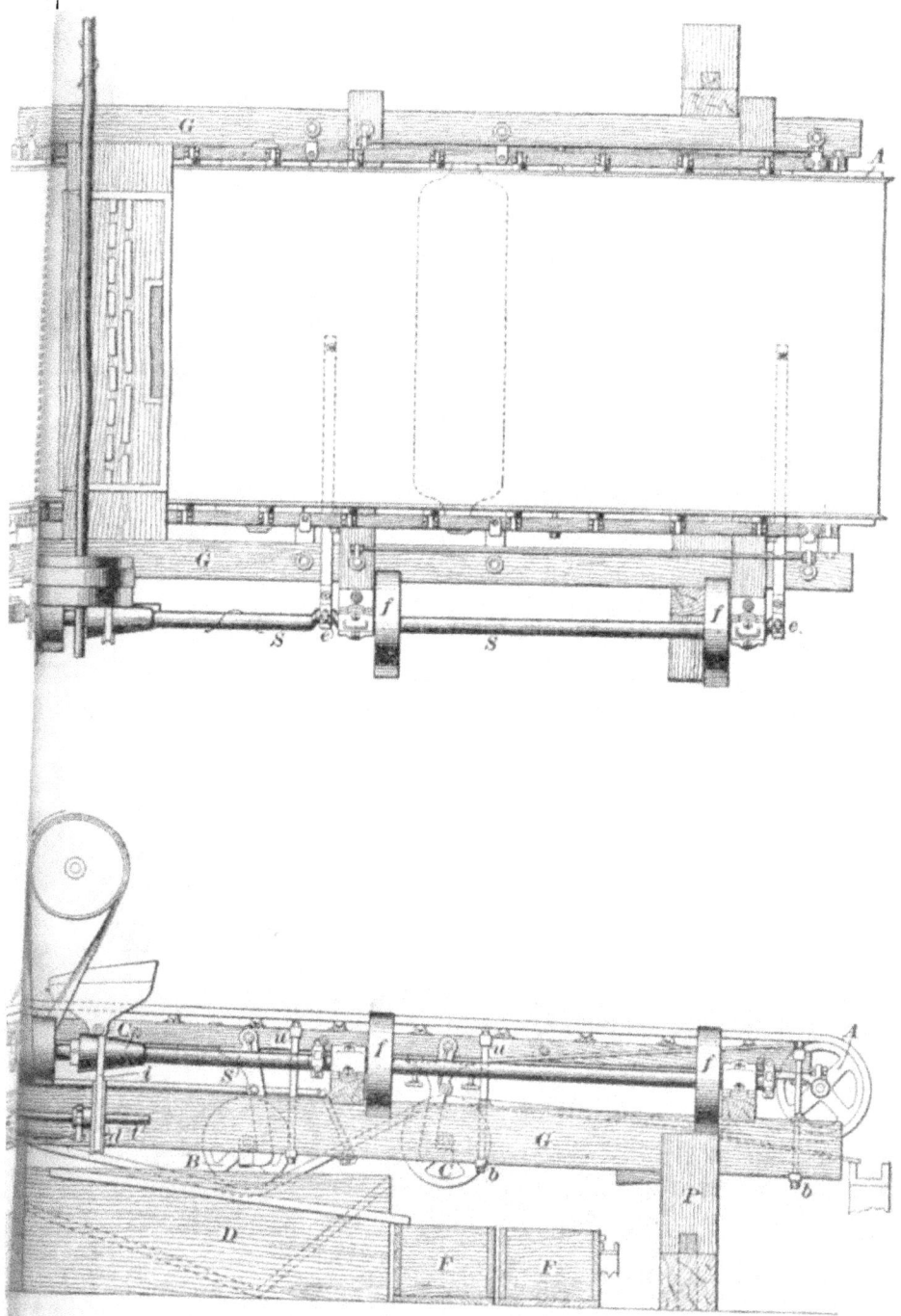

FIG. 4.

crank-shaft give it a rapid side shake—about 180 to 200 double strokes per minute being the average speed, the displacement or throw being 1 inch. A number of small rollers along the top of the shaking frame support the belt between the main rollers and keep the surface smooth and even. The pulp flows on to the belt from a distributing box about one-fourth of the way down from the head, and flows downwards in a thin sheet.

The heavy mineral settles to the belt and clings there, while the gangue mineral is carried down with the stream into the tailings launder, the separation being greatly facilitated by the rapid side shake. The belt moves slowly upwards, carrying with it the clinging particles of heavy mineral. These pass through the wash water, which is delivered in a series of jets across the belt just below the head roller, cleaning the headings so that the working of the machine may be easily watched, and such particles as withstand this are carried over the head of the machine into the concentration box D, where the concentrates are washed off and settle to the bottom, from which they are scraped every three or four hours into the box E, which is sometimes set on wheels for convenience in removing the concentrates. The guide rollers B and C carry the belt in and out of the concentrate box and also control the tension of the belt, being adjustable vertically on either side. A second series of wash-water jets is sometimes played against the belt from beneath as it leaves the box D, to clean off any mineral which might cling to it, and the overflow settling boxes F are set after the concentration box. The faces of all the rollers except C are slightly longer than the width of the belt; C, however, bears on the upper or working surface of the belt, which must run between the flanges, and it is consequently made narrower, with its corners rounded or beveled off.

12. Driving Arrangements for Frue Vanner.—The stationary frame of the vanner consists of two long timbers G, bound together by three cross timbers. The cross timbers are extended on one side to form a support for the crank-shaft S. This frame rests in shoulders cut in the four uprights P. The shoulders in the posts at the lower end of the machine are deeper than in those at the head so that the whole frame has an inclination from head to foot, and this inclination is further adjustable by means of wedges underneath the lower end of the frame at the shoulders. The eight bearings b for the rods u supporting the shaking frame are bolted underneath G, the bolt holes in the bearing being oblong, so that the bearing can be adjusted. The end bearings each have two bolt holes, the intermediate bearings one each.

The vanner is driven by a belt with a quarter twist from the countershaft pulley to the crank-shaft pulley p. On the crank-shaft are three small cranks c, each $\frac{1}{8}$ inch out of center, which connect by flat steel rods or pitmans to the middle of iron pipe girts extending across the shaking frame. The crank-shaft also carries two small flywheels f. The driving arrangement for the vanner belt is peculiar. A narrow belt i passes from a cone pulley C on the crank-shaft to a flanged pulley d on a worm-shaft t. The worm on t slowly turns a worm-wheel w driving a short shaft, which is in the same line as the axis of the head roller of the belt. On the end of the worm-wheel shaft is a crank g, which connects with the free side of a flat spring h, the other side of which is firmly fixed to the end of the roller shaft between the two shafts. This spring forms practically a flexible crank connection between the two shafts and yields to the swinging of the frame. The worm and worm-wheel are covered by a cast-iron casing, which has a limited motion about the worm-wheel shaft on an independent bearing bolted to the stationary frame. This casing also forms the bearing and support of the worm-shaft and its adjusting rod. When the machine is idle, a hand screw draws the casing around and throws on the casing the whole weight of the worm-shaft, flanged pulley, and adjusting rod, but when it is in operation, the screw is loosened and the weight allowed to fall on the belt i, keeping it tight.

13. Pulp Distributing Box.—The distributing box for the vanner is shown in Fig. 5. It is attached to the frame and shakes with it, the feedpipe being flexible. The spreading blocks are fastened to the top board of the spreader, shown upside down at (*a*). The distributor should be close to the surface of the belt,

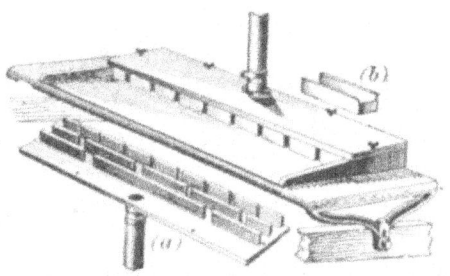

FIG. 5

in order to get a gentle feed and avoid washing away mineral. In treating the pulp from an amalgamating battery, a silvered copper plate is sometimes used, which sets in the bottom of the box and catches nearly all the amalgam and mercury coming over in the pulp. Or, again, the mercury and amalgam may be caught in a copper well shown at (*b*) which sets in the box directly under the pipe, so that all the pulp from the battery must fall into it. This well can be removed and emptied at any time.

The wash-water distributor is usually a narrow wooden trough with holes 3 inches apart, through which the water discharges on to the belt. Iron troughs are also used, with spouts of brass $1\frac{1}{2}$ inches apart; by stopping up every other hole, the effect can be made the same as that of the wooden trough. The wash water should fall upon the belt from a height of not less than $1\frac{1}{2}$ inches, in order to secure the best effect. The trough is supported on standards on the stationary frame and the height is adjustable by hand screws.

14. Vanner Belts.—There are two styles of belts in use on Frue vanners, the plain and the corrugated. The latter, called by the manufacturers the *improved belt*, has a series of low, flat corrugations or riffles across its working surface. It is claimed by the makers, with apparent justice, that this belt doubles the capacity of the machine, or that one improved vanner, which is only a little more expensive than the ordinary machine, will do the work of two of the latter. New corrugated belts are much more expensive than plain belts, but wear well; so that, when the capacity of the mill warrants it, the improved vanner should be given the preference. Practically the only difference in construction between the two forms is in the belts, but this necessitates several slight changes in the adjustments. The improved vanner allows the use of a steeper grade and requires more wash water in proportion to the increased capacity. The general grade of the old vanner, or the grade of the frame, is usually from 3 to 4 inches in the length of the frame, with 6 inches as a maximum, and in addition to this the head-roller bearing is about ⅜ inch higher, with reference to the frame, than the lower bearing, and the small guide roller next to the head is also raised a little, slightly increasing the grade at the head of the belt. With the improved belt, the average grade is 5½ to 6¼ inches, the shoulders in the lower posts being cut correspondingly deeper, and the head roller is raised ⅜ inch above the tail roller, increasing the grade at the head by that amount. The plain-belt vanner is better for saving very fine slimes than the improved or corrugated belt vanner.

15. Adjustments of the Frue Vanner.—*Plain Belt:* A plain-belt Frue vanner running on ordinary ore should have a speed of about 190 shakes a minute and a belt travel of 28 to 34 inches in the same time. The grade of the frame should be about 3 or 4 inches. The amount of wash water used varies from 1 to 1½ gallons per minute— just sufficient to keep the field between the water and pulp distributors covered, with no projecting fingers of sand—and there should be from 1½ to 3 gallons per minute in the pulp. The belt should be smooth and even and should run true on the rollers; there should be a slight corner of sand along each edge of the belt, as sloppy corners cause a loss. If the corners are sloppy, there is too much water in the pulp, and either the supply must be diminished or less water must be drawn off with the pulp from the classifiers. If they are too heavy, however, more water must be added to the pulp coming into the distributor. If one corner is heavy while the other is sloppy, the distribution of the pulp is uneven. This may be due to looseness of some of the parts, causing a jar, but if everything is working noiselessly and the feed is even across the belt, the fault lies in the adjustment of the latter, and is corrected by driving the slotted bearings *b* of the flat steel uprights *u*, supporting the shaking frame, either in or out by light blows of a hammer, until an even distribution of the pulp is secured. The same result may be accomplished by bending the end of the driving spring in the collar over towards the side of the belt having the heavy corner. The adjustment of the guide rollers also has a slight effect on the corners.

The condition of the sand corners is also affected by the grade and travel of the belt and the speed of the shake—a slow travel or shake or a slight inclination tending to give a heavy corner, and a swift travel or shake or a high inclination tending to give a sloppy corner; but the grade, travel, and speed of belts are determined by the amount and character of the mineral in the ore and the size of the particles, and the foregoing adjustments refer to the working of the machine after the grade, travel, and speed of the belt have been fixed. The speed of the side shake of the machine depends on the coarseness of the material, varying usually from about 180 strokes per minute for fine slimes to 200 or 210 for coarser sands (30 to 40 mesh). While vanners will handle ore, with good results, directly from the stamps, it is always best, if the capacity of the mill warrants it, to classify the pulp and carry each class to a separate vanner, as the best possible conditions for working the machine are thus obtained.

The upward travel of the belt should be adjusted according to the amount of mineral in the ore. The belt is supposed to carry off only the pure mineral in the concentrates. The rate of deposit of the concentrates upon the belt depends upon the grade, the pulp, the wash-water feeders, and the side shake. These being adjusted, the travel of the belt must be made just sufficient to carry the headings off the table at the same rate as they are deposited.   If the travel is too fast, barren sand is carried over with the headings; if it is too slow, the mineral will accumulate on the belt and some of it will wash over with the tailings.   As the headings are carried up through the wash water, they are cleaned of the last gangue matter and are washed up into little longitudinal piles between the jets, the piles varying in size with the proportion of mineral in the ore.   These headings can be watched as they go over the head of the vanner and should be free from gangue.   There should be a slight " head " or ridge of mineral just below where the wash water strikes the belt.   If the belt is traveling too fast and is discharging a greater weight of material in a given time than there is mineral in the pulp treated, sand or gangue will be found in this head and in the concentrates as they go over the head of the vanner.   If, on the other hand, the travel and discharge of the belt is too slow, the head below the wash jets becomes heavy and gradually extends down towards the pulp feed, and even through it, and a great deal of mineral washes over with the tailings.   The travel of the belt is regulated by the adjusting rod and hand wheel k, the thread on which passes through a tapped hole in the worm casing.   Turning the hand wheel carries the flanged pulley d and with it the belt i up or down the cone pulley.   Thus, if it is desired to increase the travel, the worm-shaft and flanged pulley are drawn back towards the head of the vanner and as the belt creeps farther up the cone pulley, the distance through which it travels with each revolution of the crank-shaft is increased; the opposite effect is obtained by moving the flanged pulley in.

The grade of the belt is adjusted by means of the wedges under the stationary frame, where it rests in the shoulders of the posts.   Increasing the grade of the belt gives a thinner and more swiftly flowing stream of pulp down the belt and a cleaner heading.   The feed, grade, and side shake should be adjusted so that the mineral does not pack on the belt below the pulp feed; but if the fingers are placed in the stream on the belt, the coarse sands can be felt rolling slowly downwards.

Any looseness of the belt is taken up by the guide rollers *B* and *C*. By lowering or raising *B*, the length of the belt in the concentrate box *D* may be increased or diminished, thus regulating the time for washing the concentrates.

The bearings of the head and tail rollers at opposite ends of the shaking frame are bolted to it through slotted holes and can be drawn in or out by adjusting screws. If the belt shows a tendency to creep over to one side, the bolts of one or both of the bearings on that side are loosened, the bearings drawn out as far as desired by screwing up the adjusting screws, and then the bolts tightened. Or, if this would make the belt too tight, the bearings of the other side may be let in a little instead.

16. Corrugated Belt.—The mechanism of the corrugated-belt vanner is the same as that of the ordinary machine, and the adjustments are performed in the same manner, but the working of the machine is quite different, on account of the shape of the belt. The grade is steeper than in the ordinary vanner and about double the quantity of wash water is necessary. The heavy mineral settles in the corrugations and remains there until it passes over the head roller into the concentration box, the light sands washing down over it. The speed of shake should be just sufficient to settle the mineral and keep the sands in suspension, not allowing them to pack on the belt. There should be slight indications of sand corners in the corners of the belt—neither too decided nor, on the other hand, too sloppy. The methods of adjusting to meet the various conditions are exactly the same as in the ordinary vanner.

17. Speed of Vanners.—The speed of the vanner is usually predetermined by the coarseness of the material, and when once set up, nearly all the regulation of the machine consists in adjusting the grade to fit the character of the pulp. The amount and cleanness of the concentrates are regulated by the travel and wash water. An experienced vannerman can tell at a glance how his machine is working, and if anything is wrong, knows just how to correct it; if there are several ways in which this can be done, his experience will indicate to him the most suitable for the case. This is a knack which can be acquired only by experience. A good rule to observe in the care of vanners is to keep all parts of the machine clean. There should be no splashing of pulp over the sides of the belt. All working parts in particular should be gone over frequently with cotton waste, to prevent any grit getting in the bearings. Good care results in a considerable saving of power. If the power of the mill, and consequently the speed of the vanner, is constant, vanners once adjusted will run right along with very little attention except that necessary to keep them clean, and one man can tend to as many as sixteen machines; if, however, the power is constantly changing, *one* machine will sometimes give a man more than he can do. To get any machine to work properly, it must work under the proper conditions.

There are many other concentrators of this type, all more or less of the same design, but as none have succeeded in displacing the vanner to any extent, and as they all are really forms or imitations of the vanner, they will not be described.

18. Lührig Vanner.—The Lührig vanner is a recent machine of a type intermediate between the side-shaking and end-shaking machines. It is practically a continuous-belt bumping table of the Rittinger type. The belt is not flanged and is horizontal in the direction of its travel, with a slight inclination sideways at right angles to the travel. End blows are delivered to the vanner at the rate of from 150 to 210 strokes per minute, according to the ore, the bumping mechanism being similar to that of the bumping table. The stroke can be varied from $\frac{1}{4}$ inch to $1\frac{1}{2}$ inches, according to the nature of the ore and the size of the particles. The belt is of rubber, 4 feet wide and 19 feet long (total), and has a travel of 18 to 20 feet per minute. The pulp is fed in through a distributing box at the upper right-hand corner—the belt traveling from right to left—and flows down the belt; wash water is distributed from a perforated pipe running diagonally across the belt. The heavy mineral sinks to the belt and clings to it, and the end jar and the travel of the belt combine to carry it along to the left-hand end; at the same time the mineral works slowly downwards, under the combined influence of the wash water, the jar, and the inclination of the belt. In this way the particles of the greatest specific gravity are carried nearly or all the way down the lower end of the machine before they discharge over the side of the belt. The tailings, which remain in suspension or cling very lightly to the belt, discharge near the right-hand end, while one or more classes of middlings discharge at different points along the belt, according to their specific gravity. The receiving trough along the lower edge of the belt is divided by movable gates or partitions, so that the amount and range of each product can be readily adjusted. This production of middlings is one of the principal features of the Lührig vanner. The capacity is about equal to that of the Frue vanner, and the machine uses about twice as much wash water. The relative merits of these two machines are still a subject of dispute. The Frue vanner maintains the supremacy in this country, very few of the others being used, while the Lührig is rapidly gaining a foothold abroad, where the production of middlings is not considered an objection.

19. Embrey Concentrator.—The Embrey concentrator or vanner is typical of the end-shaking vanners. Like the Frue vanner, it consists of an endless rubber belt, running over rollers on a shaking frame, the essential difference being in the direction of the shake. The driving shaft is placed across the lower end of the stationary frame, as shown in Fig. 6. On it are a tight and a loose pulley (to which the main driving belt extends from a parallel countershaft above), two flywheels, two eccentrics *c*, *c* for driving the shaking frame, and a cone pulley *o*, from which a narrow belt extends to a second cone pulley on a shaft below the

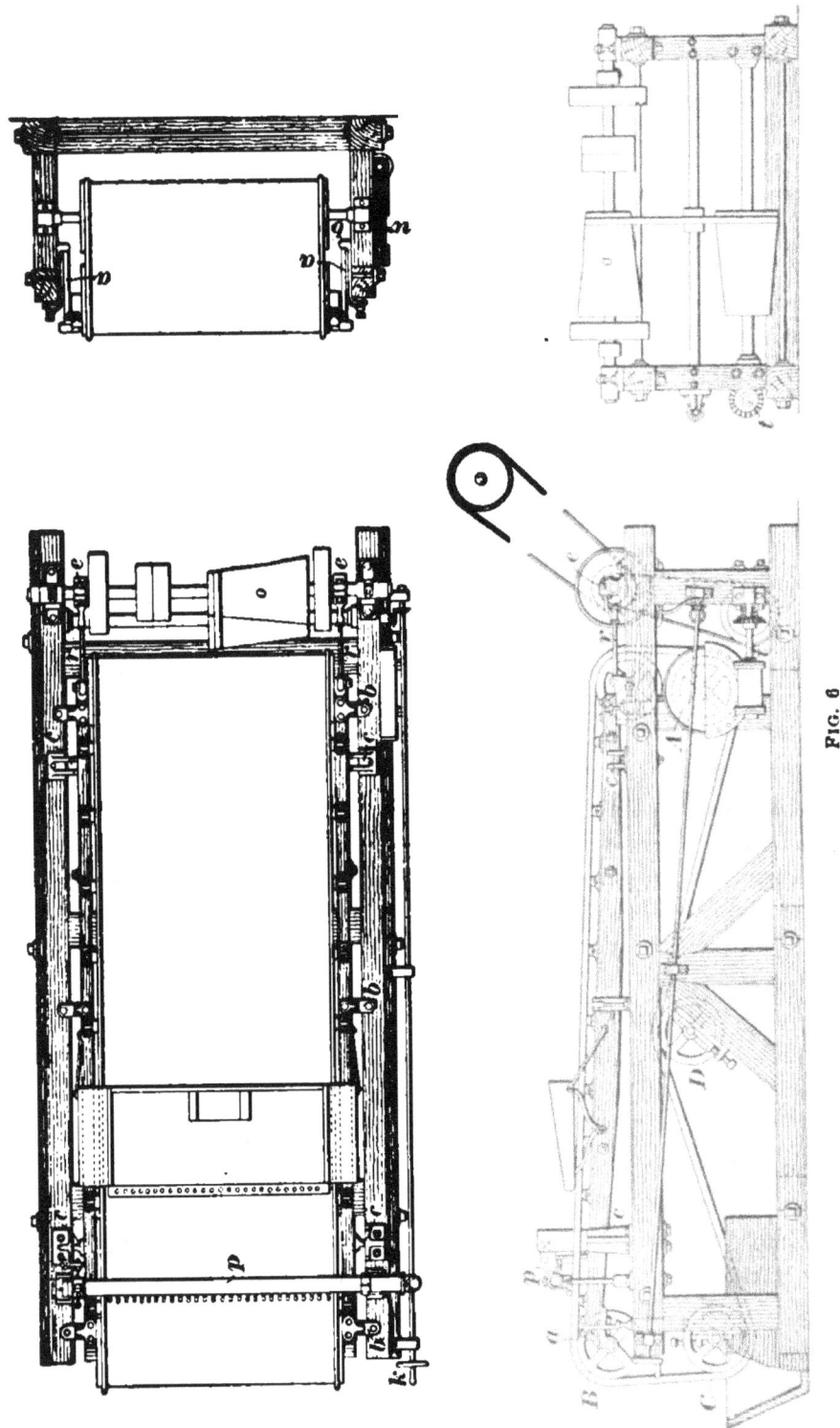

FIG. 6

driving shaft. This shaft drives the worm-shaft t through a bevel gearing, and the worm-shaft in turn drives the worm-wheel w and the driving roller A, which gives the belt its forward motion. The belt is kept tight by the adjustable roller D. The bearings of B and C are also adjustable. The shaking frame is supported by six upright legs or toggles a, three on each side, resting in stirrups b fixed to the stationary frame. The frame is driven by the two eccentrics c, which have a throw of about $\frac{3}{4}$ inch and are connected by the rods r with the tail-roller bearings. The legs supporting the head end of the shaking frame are longer than those at the tail end, giving the belt a uniform grade of about 3 inches in its length. This grade can be increased or diminished at will by means of wedges under the ends of the table.

The shaking frame is kept in line by four cast-iron standards c bolted to the stationary frame, with projections on the inside pressing against the shaking frame. One of these standards at the head of the table is lengthened and serves as a support for a bell-crank that gives a slight motion across the belt to the water pipe p, which is supported on spring legs; the other end of the bell-crank is connected by a strap to the shaking frame. The pulp distributor is practically the same as that on the Frue vanner. In another form of Embrey vanner, the crank-shaft passes under the shaking frame, upon which falls the entire weight of the belt and lower rollers, causing the machine to run more heavily.

20. The **Triumph concentrator** is an end-shaking machine, somewhat similar to the Embrey in construction and operation. The travel of the belt is regulated by a friction roller, however, instead of by cone pulleys. The distributing trough is of iron and contains quicksilver. All the pulp passes over this quicksilver, which is kept agitated by a shaft, with stirrers attached, running through the distributor.

21. The **Woodbury concentrator** is another end-shaking vanner, similar to the Triumph in construction, but having the rubber belt or apron divided into seven parts by longitudinal partitions.

The end-shaking concentrators all require a more rapid shaking motion than the side-shaking machines to accomplish the same work and are run at an average speed of 220 to 240 double strokes per minute.

22. Care of Vanners.—It is absolutely necessary for the efficient operation of concentrating machinery that it be kept clean. The operator should go over all working parts frequently with cotton waste, so that no dust or grit will have a chance to work into the bearings, and the entire frame should be wiped off at least once a day. Belt vanners are not suitable for working sizes coarser than 30 mesh, while a still smaller maximum size, say 40 mesh, is preferable. As the sizes become finer, the slope of the belt and the speed of the shake are diminished.

DRY CONCENTRATORS

23. Centrifugal Dry Concentrator. — The Clarkson & Stanfield dry concentrator separates ore from gangue through the agency of centrifugal force. The carefully sized ore is fed on to a rapidly revolving horizontal disk or shallow pan, the rim of which is perforated with a large number of small holes. The ore is thrown through these holes by centrifugal force and falls into a series of annular troughs surrounding the central disk. As centrifugal force varies directly as the mass (or as the specific gravity), the particles will be discharged from the plate with a force proportionate to their specific gravity, and the heaviest particles will therefore be thrown farthest. In addition to this sorting effect of centrifugal force, the particles of heavy mineral present less surface to the resistance of the air, in proportion to their weight and momentum, than the gangue and middlings particles; consequently, to a certain extent, the law of equal falling particles enters into the action of the concentrator; but this is much less marked and important than in the hydraulic concentrators, air being a much rarer (thinner) medium than water. This concentrator has given good results in experimental runs, but has not been adopted in practice to any extent.

24. Wood's Dry Placer Miner.—Wood's dry placer miner is a form of dry concentrator designed for treating gold-bearing placer sands without the use of water, the absence of which has heretofore rendered worthless many otherwise valuable placer fields. The machine is allied to the pneumatic jig, the sand being separated from the gold by a blast of air.

The material must be quite dry in order to obtain satisfactory results. It is passed over a grizzly, which removes the larger stones and boulders. The finer material falls through into the disintegrator trough, through which run two shafts carrying paddle blades or beaters. These blades are curved like the blades of a screw propeller, so that the dirt is fed ahead as well as broken up. From the disintegrator the material passes on to the inclined table on which the separation is performed. This table consists of a perforated metal sheet, forming the stationary cover of a bellows, and is covered with a blanket or carpet of such texture that it allows the passage of air without permitting any dirt to fall through into the bellows. Copper riffles are placed at intervals across the table and secured in position by the side boards, which are held in place by clamping screws, so that they may be readily removed for cleaning up. A sheet-steel cover slides up between the side boards, about 6 inches above the surface of the table, to prevent the escape of dust. The bellows and table are hung from flat steel-spring standards and are given a short longitudinal shake by means of eccentrics on the driving shaft, which runs in bearings across the lower end of the frame. The disintegrator and bellows are driven from the same shaft through a system of gears and countershafts. The bellows are operated by connecting-rods to slotted crank plates at either end of a countershaft. The throw of the cranks, and consequently the stroke of the bellows, may be regulated.

The blast of air from the bellows through the holes in the table blows away the dust and carries the coarse, light sand to the top, like a hydraulic jig, and the shaking of the table carries it down and discharges it. The grains of gold, however, sink to the bottom and catch on the blanket and behind the riffles, together with more or less sand and heavy minerals. These concentrates are cleaned up at least once a day, by removing the cover, side boards, and riffles, and brushing the concentrates from the blanket, and are cleaned by panning or amalgamating. The machines are provided with link-belt feed and tailings elevators, if desired. The capacity is given as from 8 to 12 tons per hour. A smaller machine is also made for prospecting, with the construction modified to adapt it to hand labor, with a capacity of from 1,500 to 2,000 pounds an hour.

These machines have given satisfactory results in test runs, and their invention may mark the birth of a new era for the hitherto worthless desert placers of Arizona, Utah, New Mexico, and other Western States and Territories.

25. Pneumatic Jig.—The Krom pneumatic jig, shown in Figs. 7 and 8, is essentially a dry concentrator, the gangue being separated from the ore by means of rapid puffs of air up through the ore bed. No water at all is used, and the ore must be perfectly dry. The ore bed *o*, Fig. 7, is only about 5 inches wide and extends the full length of the machine. The thickness of the bed is regulated by the height of the tailings-discharge dam *a*, which extends along the front the entire length of the bed, as shown in Fig. 8. The feed is regulated by a similar vertical gate *b* in front of the opening of the hopper *h*, extending, like *a*, the full length of the machine.

The sieve compartment is connected with the fan chamber by a narrow vertical slit *c*, Fig. 7, extending the full length of the bed. The sieve is made up of inverted troughs of wire gauze, open at the end next to *c*. These troughs are placed from $\frac{3}{16}$ to $\frac{3}{8}$ of an inch apart, according to the size of the ore to be jigged.

The fan f is a flat, horizontal vane, extending the full length and width of the fan box. There are several flap valves in it that open on the down stroke, preventing suction and reducing the resistance. It is keyed to a rocker-shaft g in the back, upper angle of the box, and this shaft is

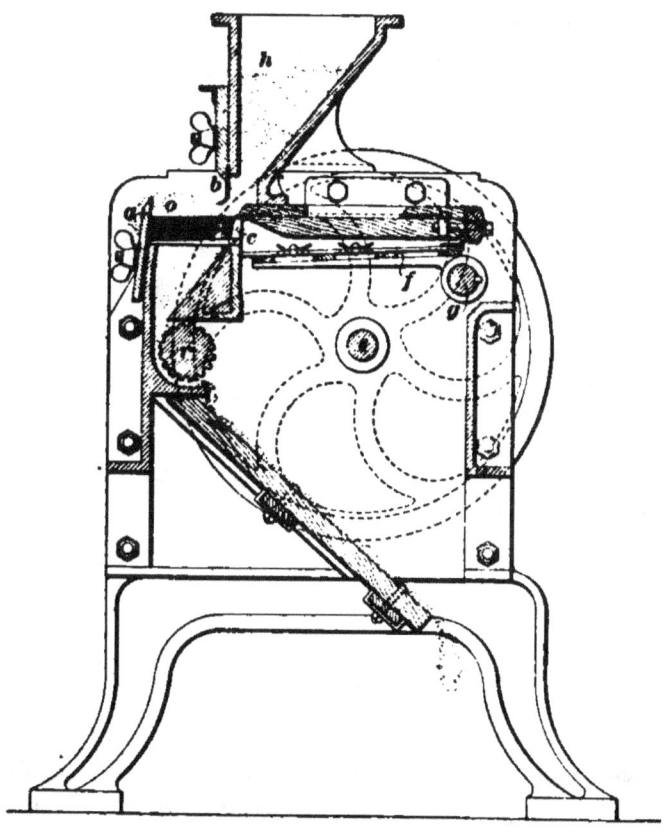

FIG. 7

operated by a lever l, Fig. 8. A projecting roller tappet on the lever is held by a spring d against a ratchet wheel t on the end of the driving shaft s. As the driving shaft revolves, the lever is forced back until the roller passes the ratchet tooth, when the spring draws the lever sharply back into position, throwing the fan upwards. A strap c checks the lever from striking the wheel on the in stroke. The trip wheel has six teeth, so that for every revolution of the driving shaft the fan makes six strokes. The shaft is driven at a speed of from 80 to 90 revolutions per minute; consequently the machine gives from 480 to 540 puffs per minute.

The rate at which concentrates discharge beneath the screen is controlled by a long, horizontal grooved roller r, Fig. 7, having a fine-tooth ratchet at the lower end. This ratchet is driven by pawl p from the trip wheel t. The bearing pin of this pawl is fixed in a radial slot in the trip

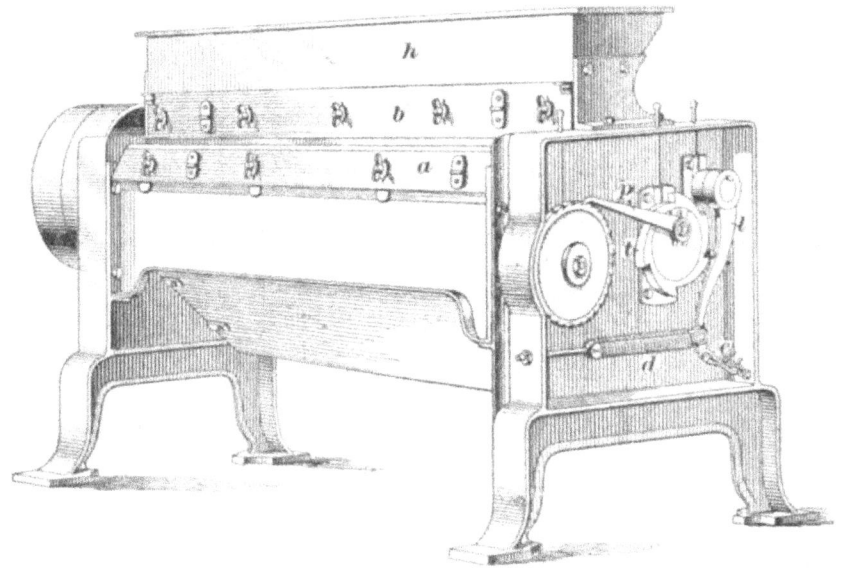

FIG. 8

wheel, so that the throw of the pawl, and consequently the rate of discharge, may be adjusted independently of the speed of the other parts.

Though this jig has received the endorsement of many good engineers and has given good experimental results, it has not succeeded in displacing the water jig to any extent. It works best on very fine sizes, much below that at which the efficiency of the hydraulic jig ceases. It may eventually find a field of usefulness in regions where water is very scarce.

26. The Hooper Pneumatic Concentrator. — This machine, like the Krom pneumatic concentrator, will not treat impalpable powders. Fig. 9 shows a perspective view of the machine set up ready for operation. It will separate at one operation as many as five minerals, some of whose variations in specific gravity may be less than one point; for instance, feldspar or quartz from mica. One separation gave a 98-per-cent. pure product between quartz and corundum. In the illustration the concentrating table and

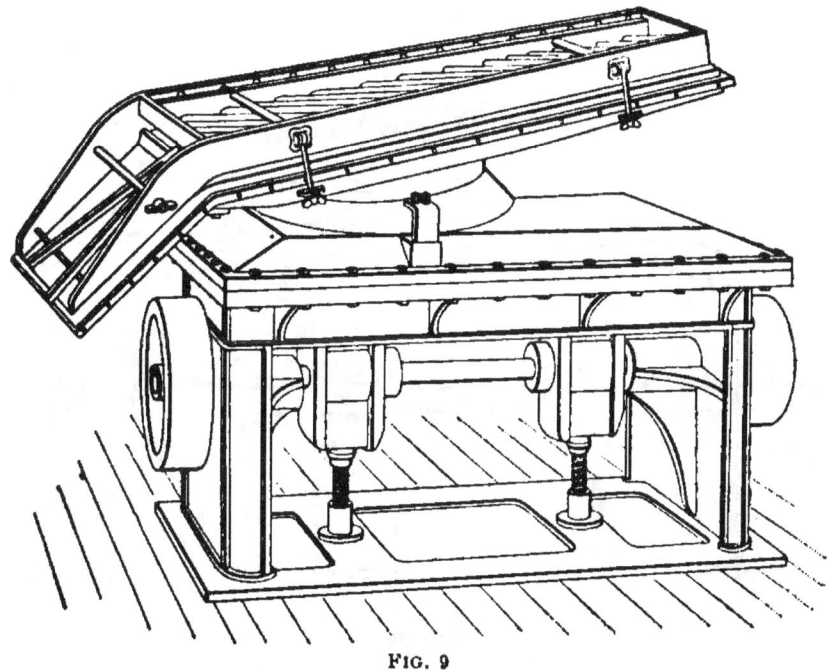

FIG. 9

skimmers are shown on top. The rods on the apron are movable and may be adjusted to guide the different products of a machine to their respective bins. As the rods are placed in the illustration, the products will be concentrates, which form at the right of the machine looking from the apron end, next the middlings, and to the extreme left the tailings. The machine is made of iron throughout, with the exception of valves, diaphragm, and dividers, but there is no reason why it should not be so constructed that no section should weigh more than 125 pounds.

N. M. III.—14

A sectional elevation is given in Fig. 10 (*a*), in which *a* is the machine shaft having a belt-driven pulley *b* at one end and a balance wheel *c* at the other. The eccentric *d* shown in Fig. 10 (*c*) and (*b*) is rigidly fastened to the shaft *a*, while an eccentric sleeve *c* may be turned and fastened to vary the stroke from 0 to 1¼ inches. The sleeve *c* to lengthen or shorten the stroke is turned to correspond with the marks on the dials, Fig. 10 (*c*), and then held in the desired position by a setscrew, shown in Fig. 10 (*a*) and (*b*). The

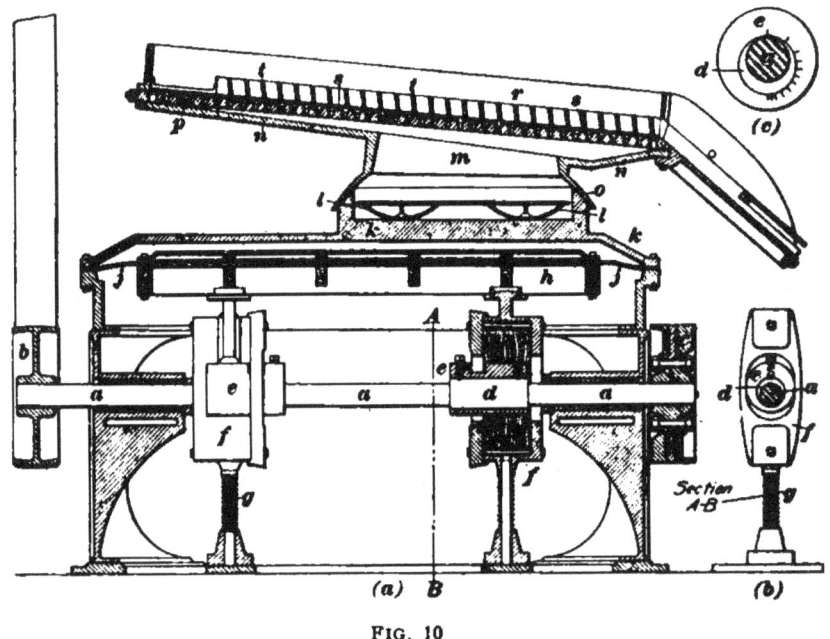

FIG. 10

eccentric collar in revolving on the shaft works the rod *f* up and down at a speed of from 350 to 450 strokes per minute. The rod *f* works on the spring *g* at the lower end and is attached at its upper end to a frame *h* having a perforated leather diaphragm *j* upon which sheet rubber is fastened transversely to form five double flat valves that open on the down stroke to admit air and close on the up stroke to compress the air between the cast-iron cover *k* and the diaphragm *j*. The cover *k* is also partly perforated and supplied with rubber valves *l*, raised by the compressed air above the diaphragm *j*, admitting that air into the chamber *m*. It is evident that the air comes into the machine from below on the down stroke and escapes above the machine on the up stroke. The concentrating table *n* forms a universal joint with the cover *k* at *o*. The table bed consists of a cast-iron grating *p*, shown in Fig. 11, with its bars and air space arranged at an angle of 45° with the sides of the table. Around this grating a fine woven broadcloth is stretched and held in place by a frame shown in Fig. 9. Upon this cloth the frame *r*, Fig. 11, is placed, which is crossed diagonally with strips of steel $\frac{3}{16}$ of an inch in height, spaced so as to rest directly upon the grating *p*

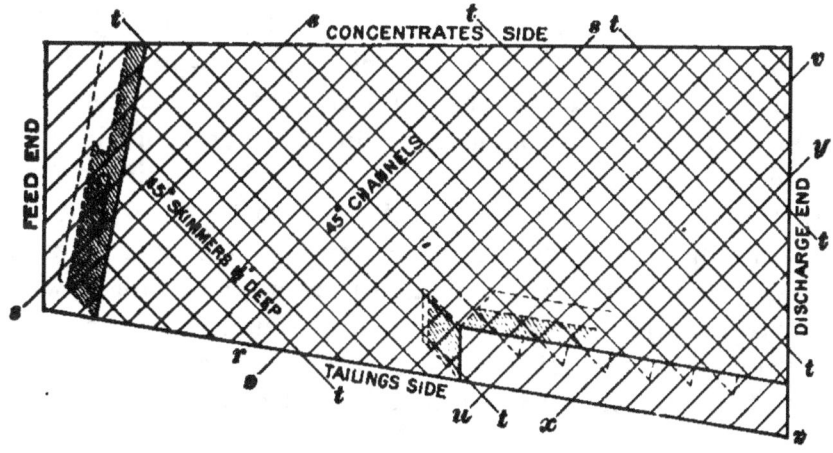

FIG. 11

beneath the cloth. The weight of the frame *r* is sufficient to form a tight joint and cause the cloth between the two bars to vibrate with each pulsation of the machine and allow the compressed air from chamber *m* to escape through the meshes of the cloth. The strips *s* form riffles in which the heavier mineral particles gather and move towards the right-hand side of the table. At right angles to the strips *s*, and above them, the skimmers *t* are fastened. These are given an angle of 45° with the sides of the frame and are made of cast iron $1\frac{1}{2}$ inches high, thus forming diagonal channels through which the lighter minerals travel to the left or tailings side of the table. The crushed ore is fed from the hopper at the upper end and passes on to the cloth, which, being agitated, throws the mineral slightly upwards. The air escaping as the mineral settles prevents the lighter particles from reaching the cloth as quickly as the heavier. The heavy particles, therefore, arrange themselves next the cloth and fill up the riffles *s*. The lighter particles accumulate on top of the heavier, and being unable to settle in the riffles, they follow the channels formed by the skimmers *t*. By the time the ore has reached the line *u v*, Fig. 11, all the minerals of the same specific gravities have separated and the heaviest will be found on the concentrates side, the next heaviest will follow approximately the line *x y*, while the tailings will be discharged between *y* and *z*, provided there are but three minerals to separate. The table is given an inclination upwards, 11° being the most allowable, to facilitate the passage of the ore from the feed to the discharge end. The side inclination has a direct effect upon the character of the concentrates, some minerals requiring more than others; this matter, therefore, is to be determined by examining the concentrates as they pass over the table. The table may be adjusted when running, and should, when in proper adjustment, give a clean product. The capacity of this concentrator varies greatly, but actual working results gave from 8 to 22 tons daily per concentrator.

MAGNETIC CONCENTRATORS

27. Magnetism of Minerals.—All substances of whatever nature are to some extent sensible to magnetism; that is, they are either attracted or repelled by the poles of a magnet. In the case of only a few, however, is this property appreciable under ordinary conditions. Substances that are attracted by the magnet are called *paramagnetic;* those that are repelled, *diamagnetic.* Iron is notably paramagnetic; nickel, cobalt, and chromium are feebly, but appreciably, attracted by a hand magnet. Manganese, titanium, cerium, platinum, palladium, uranium, and osmium, though generally considered non-magnetic, are paramagnetic, and can be sensibly attracted by very powerful electromagnets. All the other metals are more or less diamagnetic. The minerals of the different metals show the same magnetic characteristics as the metals themselves, but in a much smaller degree. Of the iron minerals, magnetite is quite strongly attracted by a hand magnet, and pyrrhotite somewhat less strongly; while the rest, with occasional exceptions, are apparently non-magnetic. Some of the nickel, cobalt, and chromium minerals are very feebly attracted by a hand magnet, but most of them are apparently non-magnetic.

The apparently non-magnetic minerals may be made quite strongly magnetic by either an oxidizing or a reducing roast. This fact has been taken advantage of in the concentration of lean hematite, limonite, siderite, and pyrite ores. The roasted ore is passed through the fields of strong electromagnets, separating the magnetized mineral from the non-magnetic gangue, and thus raising the grade of the ore to a point where it can be profitably smelted. The tendency at present, however, is to do away with the trouble and expense of roasting and to concentrate directly, by increasing the strength and modifying the design of the electromagnets. Magnets are now made which will attract any of the iron minerals except pyrite without previous roasting, and manganese minerals have been successfully concentrated by magnetic concentration.

When concentrating an ore of iron by means of magnets, not only is the percentage of iron in the resulting product raised, but in some cases deleterious elements, such as phosphorus or sulphur, may be removed. This is especially true in the case of ores which are naturally magnetic and in which the phosphorus occurs in small crystals of apatite and the sulphur in crystals of pyrite.

When the magnetic process is used for concentrating any mineral, the material should be crushed to the size of the average particles to be separated before it is fed to the machine. The results will be high-grade concentrates, practically barren tailings, and a middle product which will require further crushing and reconcentration.

28. Among the successful applications of magnetic concentration may be mentioned the following: The separation of magnetic iron ores from gangue materials; the separation of minerals containing iron from pure or nearly pure zinc minerals; the separation of the various materials in monazite sands; the separation of garnetiferous rocks where it is desired to obtain pure garnet; and the removal of garnet from corundum ores.

By a proper adjustment of the machines, it is possible to make separations between two compounds having different magnetic properties.

29. Comparison of Magnetic Concentrators.—New machines are constantly being brought out for use in connection with magnetic concentration. But all machines can be divided into two general classes: one in which the material is brought within the field of the electromagnets, and the magnetic material thus deflected far enough from the path of the main stream to be deposited in a separate receptacle; and the other in which the magnetic material is actually picked out of the stream of ore, carried off, and deposited.

The first class is used only for quite strongly magnetic material, such as magnetite or roasted ores. In a typical machine of this kind, the stream of ore falls past a series of electromagnets, set off to one side, and the attraction of these magnets draws the magnetic material somewhat out of the path of the main stream, so that it falls into separate concentrates chutes.

In the machine of the second class, the material falls or is conveyed by a belt conveyer into the field of the electromagnets, which pick up the magnetic material. The nonmagnetic material goes on and is discharged practically freed from iron. The magnetic material is carried along on the under surface of a continuous traveling belt or on the surface of a cylinder encasing the magnets, until it passes out of the field of the magnets, which remain stationary, and then it drops off into a concentrates receptacle or on to a conveyer of some sort.

30. Theory of Magnetic Separation.—The chief minerals of importance to which magnetic concentration is applied are zinc and iron. Iron is not wanted in zinc concentrates, and as it exists in the form of pyrites in connection with zinc sulphide, it becomes necessary to reduce the iron sulphides to oxides by roasting and without appreciably affecting the blende.

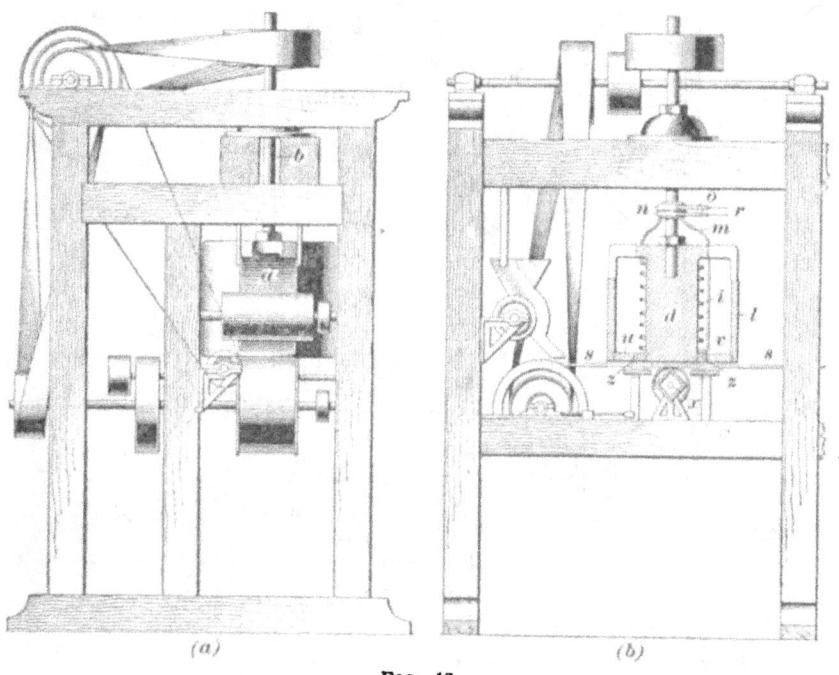

FIG. 12

The intensity of a magnet for the purpose of magnetic separation depends upon the "lines of force" radiating from a unit surface of area, rather than upon its size. To obtain this, specially constructed machines, of which the Wetherill and the Cleveland-Knowles are types, must be used and they belong to the second class of magnetic concentrators mentioned.

31. Fig. 12 shows two different views of the Cleveland-Knowles magnetic ore separator. Fig. 12 (*a*) is a side

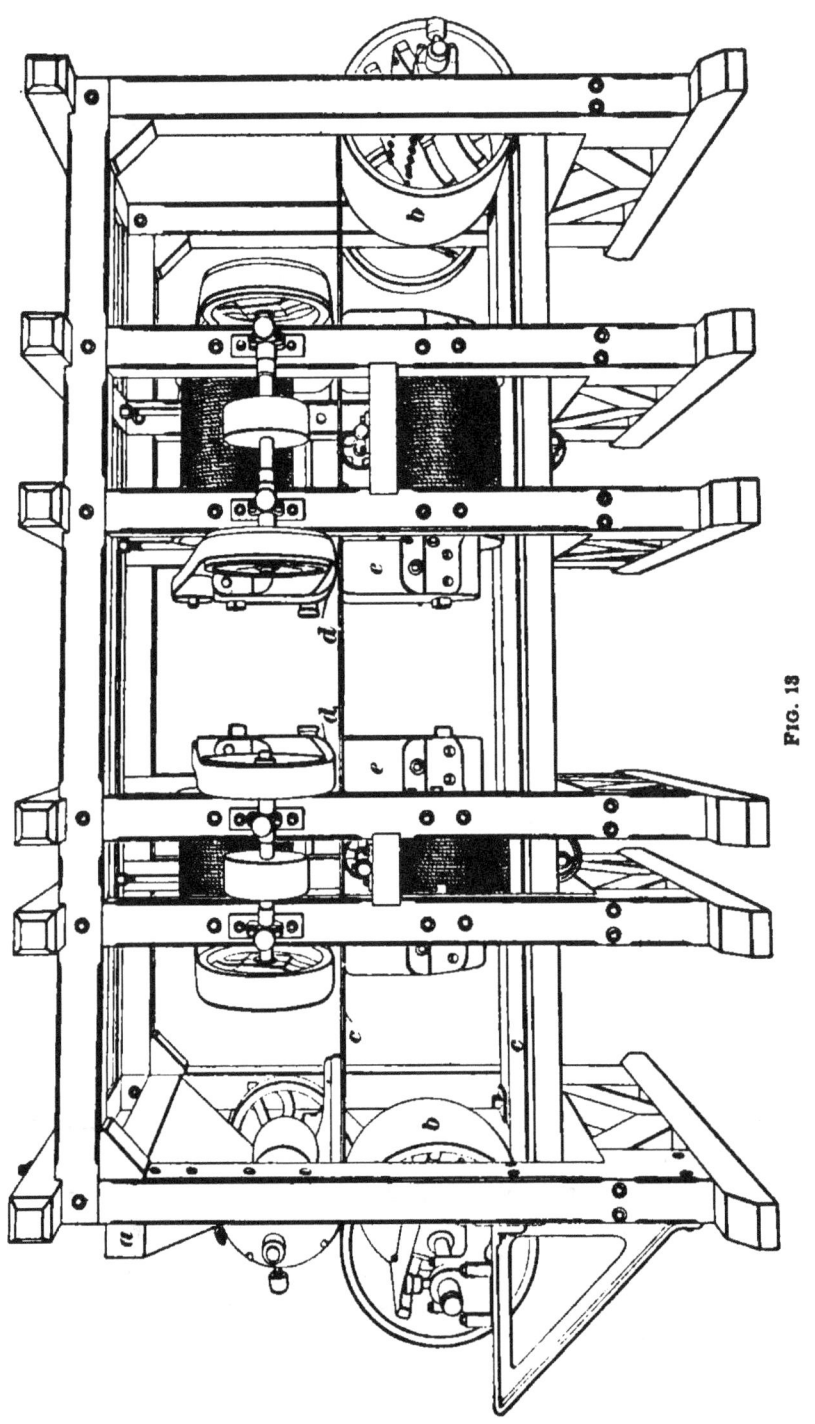

FIG. 18

elevation, in which *a* represents the magnets, which are suspended by shafts *b* and which are rotated by means of suitable belts and pulley wheels. A cross-section of one magnet is shown in Fig. 12 (*b*) and consists of a core *d* and a casing *l*. The space between the core and the casing is filled with a coil *i*; the ends *m* of the wire which forms the coil are carried up through the top of the magnet and are connected to the slip rings *n*, which make contact with the brushes *o*, to which are attached feeding wires *r*, leading to the dynamo. Beneath the magnets and in proximity to their lower faces an endless belt conveyer *s* passes so as to leave a magnetic gap *z*. The ore to be separated is fed into a hopper which distributes it evenly over the belt in a thin layer. The belt carries it under the first magnet, where, owing to the intense magnetic field formed between the two poles *u* and *v*, all the particles of sufficient permeability are attracted as the magnets rotate and are carried to one side of the plate, where they are removed by revolving brushes, not shown in the figure, but nevertheless situated just above the angular revolving drum *x*. As the belt passes over the angular drum *x*, material clinging to the belt is detached by the corners of the drum striking against the belt, and thus particles which have been covered up and so prevented from adhering to the magnet will be brought to the surface and adhere to the magnet towards which the belt moves. So far, this separator seems to have given general satisfaction in several different localities.

32. The Wetherill Magnetic Concentrator. — This machine is fairly well known and is used in Europe, Africa, and America. The concentrator gives general satisfaction in zinc separation in New Jersey. The machine shown in Fig. 13 was built for the De Beers Consolidated Mines, Ltd., Kimberley, South Africa, where it is employed to remove magnetite, ilmenite, chromite, garnets, olivine, etc. from diamonds.

The ore is delivered crushed to a proper size from a feed hopper *a* on a main conveyer belt *c*, which passes through the magnetic field of the magnets *c*. The magnetic minerals are attracted and raised to the upper traveling cross-belt *d*, and by it are removed from the magnetic field and fall into a hopper not shown in the cut.

By properly adjusting the current strength and by regulating the distance between the poles, the ore on the main belt passes from weaker to stronger magnetic fields, so that a separation of the minerals of different magnetic attractability is readily effected. A separator like that shown in the illustration treats from 2.5 to 5 tons of Broken Hill tailings per hour, removing about 25 per cent. of garnets as a magnetic product.

MISCELLANEOUS CONCENTRATORS

33. Pan and Batea.—There are, besides the concentrators already described, various concentrators of miscellaneous forms and minor importance. In placer mining and in cleaning up around amalgamating plants, the pan and batea, both in the hand and as mechanical forms, are frequently used for cleaning the amalgam from the plates.

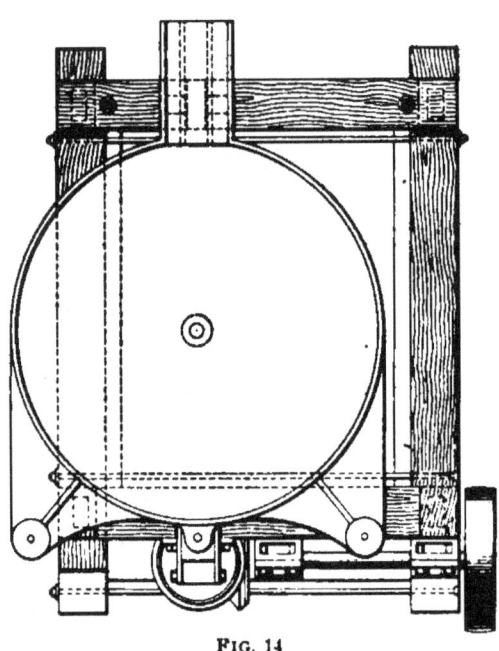

FIG. 14

The hand pan is usually made of either Russia iron or agate ware. It is a shallow pan, 10 to 12 inches in diameter on the bottom, 17 to 20 inches at the top, and 2½ to 3 inches deep, pressed out of a single sheet of metal. The amalgam is placed in it, softened with quicksilver, and washed with an excess of water. The pan is grasped by both hands, on opposite sides, and given a gentle, circular, swinging motion, under the influence of which the mercury and amalgam settle to the bottom, while the lighter material rises and may be poured off with the water.

The **batea** is a shallow wooden bowl, usually about 20 inches wide and $2\frac{1}{4}$ inches deep, used by Mexicans and South Americans. It is made out of a solid block of green wood, the cavity being dug out with an instrument resembling a shoemaker's hammer. The bowl is buried until seasoned, after which it is smoothed and sandpapered. It is shaped somewhat like a sheet-iron or agate-ware gold pan and is used in the same manner and for the same purpose.

34. Mechanical Pans. — The object of mechanical pans and bateas is to imitate as closely as possible the motion and action of the hand articles in the treatment of much larger quantities of material than can be handled in the latter. This is accomplished by various mechanical devices for giving the pan either a shaking or a gyratory motion. The device for driving the mechanical batea, described in the next paragraph, is of the latter class.

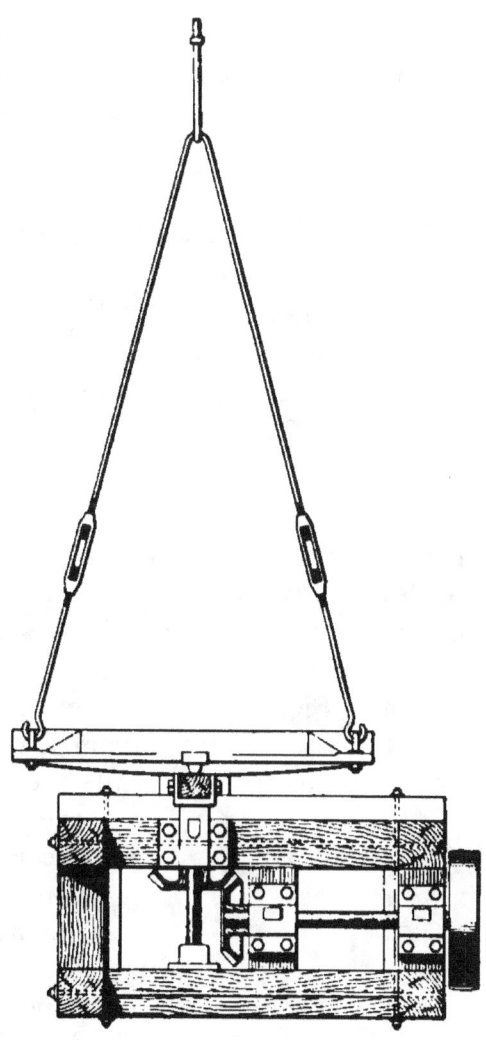

FIG. 15

The mechanical batea is shown in plan in Fig. 14 and in elevation in Fig. 15. The pan is of cast iron, about 4 feet wide, with a rounded bottom and a plug in the center. The front end or spout is set on a roller, allowing the pan to slide back and forth, while the back is supported by two light iron rods, allowing the pan to swing freely. The pan is given a rapid gyratory motion by a crank at the rear, on a short, vertical shaft driven by bevel gearing.

35. Log Washer.—When ores are encased in clay or other adhering substance, they may be separated from it by means of the log washer shown in Fig. 16. This consists of cast-iron or wrought-iron trees upon which blades are bolted in such a way as to form a screw conveyer. One end of the

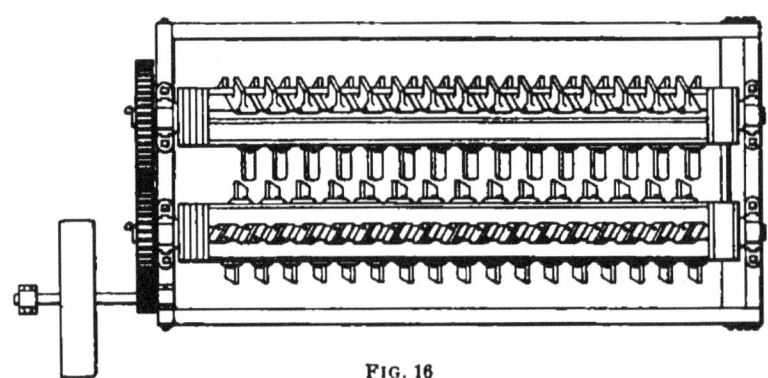

FIG. 16

log (so called because the original apparatus was a log with blades) works in a gudgeon placed below the water in the box containing the ore to be washed; the other end works in journals. The logs, which are driven by gear-wheels, as shown in Fig. 17, work the ore towards the head of the box and discharge it into a bin. Water is introduced at the upper end of the box, while the ore is fed at the lower end;

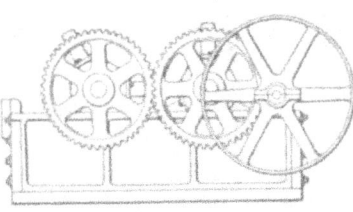

FIG. 17

the clean water thus meets the ore and discharges when it becomes dirty at the lower end, taking away at the same time the clayey sediment. There is no general standard for these washers. The box is about 4 feet deep

at one end and two feet at the other, according to the length
of the logs, which vary from 16 to 30 feet and are pitched at
an angle sufficient to give a rise of 1¼ inches to the foot. A
pair of logs usually work together and can wash from
one to two hundred tons of iron or phosphate rock per day,
with from 50 to 300 gallons of water per minute, according to
the nature of adhering material.

AMALGAMATION

PROPERTIES OF MERCURY

36. Amalgam.—Mercury (quicksilver) rapidly dissolves
gold at ordinary temperatures, forming an amalgam which
is liquid, pasty, or solid, according to the proportion of
mercury. An amalgam containing 90 per cent. of mercury
is liquid, while one containing 85 per cent. crystallizes in
yellowish-white prisms. Silver and also the base metals,
copper, lead, and zinc, are likewise soluble in mercury,
and any or all of these metals can exist in an amalgam
at the same time as gold. The excess of mercury in a
liquid amalgam may be strained through cloth or buckskin,
leaving the hard, dry amalgam behind. The mercury from
amalgam may be distilled off by heat—applying at first
a gentle heat and gradually increasing it; if the heat is
stopped at any point, the distillation ceases, but recom-
mences if the heat is again raised; at a bright-red heat all
but a mere trace of the mercury is expelled, and if the heating has been gradual, the vaporized mercury carries off but
little of the precious metals with it. A piece of gold readily
absorbs mercury and becomes brittle, this brittleness sometimes remaining after the mercury has been volatilized by
heat. The fumes of volatilized mercury are exceedingly
poisonous, producing serious salivation if breathed even in
comparatively small quantities; hence the distillation should
always be performed in strong, hermetically sealed retorts,

and the fumes condensed by passing them through a water-cooled condensing tube or coil.

The amalgamating property of mercury, together with the ease with which it can be separated from the amalgamated metals by distillation, condensed, and used over and over, makes it a very important factor in the metallurgy of gold and silver, particularly the former.   Amalgamation is one of the principal processes of recovering free gold and silver from their ores.   The ore is crushed up fine enough to free the metals from their gangue, and the pulp, mixed with water, is passed over copper plates which have been coated by quicksilver, or over a bath of liquid mercury; or the gold and silver may be amalgamated by grinding the ore or pulp together with mercury, in machines like the arrastra, amalgamating pan, and Huntington mill, or the mortar of a stamp mill.   Mercury is also used in placer mining, being placed in the sluice riffles to catch fine gold, which might otherwise be carried on down the sluice by the swift current and be lost.   Chloride of silver is also soluble in mercury, and in the case of silver ores in which the silver is present as a non-amalgamable compound, the ores are crushed, roasted with common salt to bring the silver into the form ·of a chloride, and the chloridized ore, mixed with warm water, is introduced into cast-iron pans, around which revolving arms carry mullers or mixers; a little clean mercury is put in the pan and amalgamates the silver chloride, the mullers insuring the contact of all the ore with the mercury.

In America, mercury is sold in "flasks" containing 76½ pounds; the Australian "bottle" of mercury contains 75 pounds of the metal.

37. Losses of Mercury.—The mercury used for amalgamation should be as free as possible from base metals. A little gold or silver in the mercury is a decided advantage in amalgamation, as mercury containing a trace of gold or silver amalgamates much more rapidly than perfectly pure mercury; this advantage is so marked that, when a new lot of mercury is received at a mill, it is either mixed with old mercury that has been strained off of amalgam and always retains a little gold and silver, or, if there is no old mercury obtainable, as in starting a new mill, a little silver is dissolved in the new mercury. Copper plates are silver-plated for practically the same reason.

38. Sickening.—Mercury is said to be "sick" when it contains some substance which coats it with a film and prevents it from amalgamating the gold and silver. When the mercury becomes separated into fine drops, this film prevents their reuniting, and they are broken up finer and finer, and finally washed away with the tailings, the film preventing their catching on the apron plates. This is one of the principal sources of loss in gold and silver amalgamation. Sickening is caused by certain base metals and their compounds in the mercury. Lead, copper, tin, and zinc in mercury oxidize rapidly and cover the mercury with a film of their oxides. Lead and copper are the worst of these, because the most common. They also make the amalgam pasty, necessitating the use of inordinate quantities of mercury, and consequently increasing the loss. If an ore contains soluble salts of any of these metals, it is very apt to be unfit for amalgamation, at least by the ordinary methods, such as amalgamation in the stamp battery or pans or on copper plates, as the iron and copper precipitate the metals from solution by galvanic action, and the precipitated metal is at once dissolved by the mercury, soon causing it to sicken. For this reason, the old-fashioned stone arrastra is better suited to working some ores than the improved modern machinery.

Arsenic, antimony, and bismuth also cause a great deal of trouble. Whether they occur in the metallic form or as compounds, they are dissolved by the mercury, and then a black, crystalline coat of the metallic element forms on the surface of the mercury. Easily decomposable sulphides and sulphates reduce and form sulphide of mercury. This is the case with sulphides of arsenic, antimony, and bismuth, and more or less with the sulphides of lead, copper, and silver. Clean iron pyrites is not affected by mercury.

Sickening, when due to the formation of metallic oxides, may be remedied by the addition of a little sodium amalgam. This is prepared by adding metallic sodium, in pieces about the size of a pea, to a bath of mercury heated to about 300° F. Each piece of sodium causes a slight explosion and a bright flash of flame. The reaction becomes less violent after about 3 per cent. of sodium has been added, and the amalgam is then poured into a shallow pan and allowed to cool, becoming solid when cold. It is then broken up and kept under naphtha in closely stoppered bottles, to prevent oxidation. A little of this amalgam, added to a lot of sick quicksilver, will reduce the coating of oxide, the oxygen combining with the sodium and forming a soluble salt, while the metal is absorbed into the mass of the mercury, leaving the surface clean and lively. This will cure nearly all sickness of the quicksilver except that due to sulphides of antimony and bismuth, but is applicable only to pans, mercury wells, and riffles, and not to plate or battery amalgamation, where the mercury is in a thin film.

39. Flouring.—When mercury is broken up by the machinery into very fine globules, while a film of air around each globule prevents their reuniting and permits them to float away in the waste water, flouring is said to occur. If this film of air be broken in any way, the particles instantly unite; but a considerable proportion of the total mercury unaccounted for is lost in this way, particularly in stamp milling. Flouring can be prevented to a considerable extent by passing the pulp through mercury wells or troughs, where the air films around the minute globules are broken by agitation and the globules unite with the mass of the mercury.

LOSSES OF GOLD

40. Float Gold.—Clean gold is readily amalgamated by mercury. Very fine gold particles, however, may become surrounded by a film of air (like floured mercury), which prevents their being amalgamated, and floats them away with the tailings. Finely divided pyrites are floated off in the same way, but cannot be strictly considered as "float gold," although frequently no distinction is made. This floating away of gold and mineral is one of the chief sources of loss in gold milling.

41. Non-Amalgamable Gold.—Some gold that is not directly amalgamable, being coated with or surrounded by some substance which prevents it coming in contact with the mercury, is also a source of loss. This includes gold in pyrites, "rusty" gold (gold surrounded by a film of some oxide or sulphide which prevents its amalgamation), greasy gold (the particles of gold being prevented from amalgamating by a film of greasy material), and gold in chemical combination, as tellurides.

42. Gold in pyrites is mostly, if not wholly, in the form of native gold in minute crystals in the cleavage planes of the pyrites. Some authorities claim that the non-amalgamable gold in pyrites is in the form of a sulphide, but the majority of the evidence is in favor of its being native gold. Examined under a powerful microscope, the gold can always be seen in the cleavage planes, gilding the edges of the little streaks or striæ on the sides of the crystals marking the edges of the cleavage planes. By leaching with potassium cyanide, this gold can be dissolved, leaving microscopic crevices between the cleavage faces and pitting the faces themselves. The crystals of gold are so minute that a large proportion of the gold remains with the pyrites, even when the ore is crushed very fine. The very finest slimes of pyrites, even when ground in pans with mercury, will seldom yield more than 40 per cent. of their gold by amalgamation. If the pyrite is oxidized, either by weathering or roasting, the proportion of amalgamable gold is considerably increased; but, on the other hand, a great deal of the gold is apt to become coated with a film of iron oxide, particularly when the pyrite is oxidized by weathering and is rendered unfit for amalgamation. The plan usually followed with such ores is to save what gold is possible by amalgamation in the stamp battery and on the plates, and then concentrate the pyrites and extract the gold from them, either by smelting or by chlorination or bromination. In the latter two processes, the pyrites are first roasted, to drive off the sulphur, and then moistened and treated in closed vats or barrels with chlorine gas or a solution of bromine. The action is very similar in both cases. The gold is brought into the form of a soluble chloride or bromide; this is leached out of the pulp by water and the gold then precipitated as a brown powder, usually by ferrous sulphate (copperas), sulphureted hydrogen, or charcoal, and the precipitated gold is melted and molded into bars.

Even when concentration is employed, more or less pyrites is inevitably carried over with the tailings; but by careful work this loss can be reduced to a minimum.

43. Rusty gold is native gold the scales of which are coated with a thin film of some mineral substance which prevents them from amalgamating. The film is usually oxide of iron, silica, or some sulphide or arsenide. The films are frequently perfectly transparent, so that the gold appears to be perfectly clean and pure, and the existence of the film is indicated only by the failure of the gold to amalgamate. If the scale of gold be broken or cut so that the least surface of clean gold is exposed to the mercury, the scale will amalgamate at once, but otherwise it will be carried over with the tailings.

Roasting or calcination is a process usually beneficial in the case of rusty gold. Calcination is merely roasting to drive off water or decompose carbonates, and is so called to distinguish it from the oxidizing roast to drive off sulphur and arsenic. If the coating of the gold scales is a sulphide or an arsenide, it is readily decomposed by an oxidizing roast. A calcining roast is frequently effective in treating limonite (hydrated iron oxide) gold ores, in which the particles of gold are coated with a film of the iron oxide. The heating drives off the water and leaves the ore open for amalgamation. Instances are known where ores in the raw form would yield only from 30 to 40 per cent. of their gold by the most careful and elaborate amalgamation, but after calcining would yield from 80 to 90 per cent. With ores of this type, however, it is usually advisable to employ some other process, such as chlorination or bromination.

44. Clayey Gold Ores.—When the gangue of an ore is talcose—i. e., like talc (soapstone)—the gold scales are apt to be coated with a slime which prevents their amalgamation and floats them off. A somewhat similar effect results when grease or oil gets into the ore. The scales of gold become coated with a thin film of grease—a condition known as *greasy gold*—and refuse to amalgamate. A very small quantity of grease will do the mischief—a little tallow from the miners' candles or a few drops of oil in the mortar or pan being sufficient to cause considerable loss and trouble. The effects of the grease may be counteracted by the use of a little potassium cyanide, caustic soda, or potash in the water, to cut the grease.

45. Telluride Gold Ores.—Tellurides of gold are of a greasy, talcose nature, and are exceedingly difficult to amalgamate. They frequently contain free gold, but even this is apt to be coated and non-amalgamable. They may be rendered amalgamable by roasting, which volatilizes the tellurium; but the volatilized tellurium carries off with it a great deal of gold—altogether too much for practical working—so that amalgamation is practically out of the question for telluride ores. Chlorination and bromination are open to the same objection, as the ore has to be roasted preparatory to leaching, and even smelting is unsatisfactory on account of the large gold loss by volatilization. In some cases the cyanide process, in which the gold is dissolved out of the ore by leaching with a very dilute solution of potassium cyanide, has given the best results on tellurium ores. There are so many factors entering into the case that it is impossible to make any general statement as to which is the best method for telluride ores.

46. Loss of Amalgam.—More or less gold is lost in the shape of finely divided particles of amalgam. This can be remedied by careful working, not allowing the amalgam to get too hard or to accumulate in too large quantities, and by placing amalgam traps below the apparatus, such as mercury wells, shaking copper plates, or the amalgam savers on vanners—vanners themselves would catch most of the fine amalgam, but as it would go in with the concentrates, the mercury in it would be lost; the amalgam saver is not expensive, in view of the saving it will accomplish.

AMALGAMATING APPARATUS

PRIMITIVE AMALGAMATION

PRIMITIVE APPARATUS AND METHODS

47. Arrastra.—The arrastra is similar in construction to the Chilian mill, except that, instead of stone rollers, flat

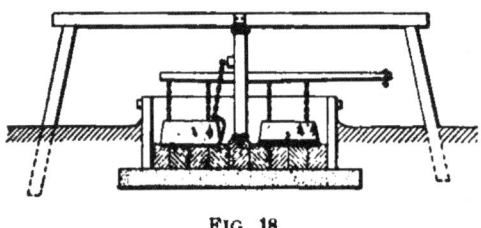

FIG. 18

stones, usually four in number, are dragged around the pan by the radial arms, to which they are fastened by ropes or chains. The front ends of these stones or mullers are raised a few inches from the floor by the suspending ropes or chains, insuring their riding on the ore and grinding it instead of merely plowing through it. The mullers, which weigh from 600 to 1,500 pounds each when new, are used till they wear down to about 400 or 500 pounds; they are then renewed one at a time, so that there are always old mullers in the pan. Power for the arrastra is obtained from a mule at the end of a pole or from water or steam. Fig. 18 shows a mule-power arrastra. Small portable arrastras with iron pans are made for prospecting purposes.

In many ways the arrastra is an ideal amalgamating machine, in spite of its crude construction, and on certain classes of ore it will save a larger proportion of the values than the more modern machinery, notwithstanding the loss of mercury and amalgam through the stone bottom. If the gold is rusty, for instance, the grinding action is almost certain to break the film surrounding the scales and give the mercury a chance; and, again, in the case of ore containing easily reducible salts of lead, copper, or other base metals, as there is no metal about the machine with which the pulp comes in contact, there is no reducing action and the metals remain in solution instead of reducing and sickening the mercury. As the ore is ground to an impalpable pulp, there is considerable flouring of mercury and amalgam, but subsequent settling and washing save a great deal of this.

The arrastra is too slow in its working to be applicable to any but very rich ores. The ordinary arrastra will grind from 800 to 1,500 pounds of ore in twenty-four hours, using about twice that amount of water in the pulp. About one ton is usually charged at a time in an ordinary arrastra. Enough mercury is added to have the amalgam contain not more than 20 per cent. of gold and silver. The mercury is usually alloyed with silver, copper, or zinc, both to keep it from breaking into globules and running into the crevices and to make it amalgamate more rapidly. In starting up an arrastra, about 5 or 10 pounds of mercury are added at once, and then about half a pound is added every other day. The amalgam is cleaned up at intervals, varying from twice a month in the rudest arrastras to twice or four times a year in those of the best construction. It is washed, with the addition of fresh mercury, then strained and retorted.

48. The Patio Process.—In the patio process for the extraction of silver the ore is first crushed in Chilian mills and then ground in the arrastra, where any free silver and gold or other directly amalgamable forms of the metals, such as horn silver (silver chloride), bromides and iodides of silver, etc., are extracted by amalgamation. The slimes from the arrastra are then carried to settling pits, where the pulp is freed from surplus water, and from here are conveyed in the form of a liquid mud to the *patio*.

The **patio** is simply a court or enclosure, usually from $\frac{1}{4}$ acre to $1\frac{1}{2}$ acres in extent, which has been carefully graded and paved with stone, cement, asphalt, or even match-boards, made as nearly as possible impervious to mercury. The court has a slight inclination, so that the water readily drains off from the pulp beds. The pulp is brought on to the floor from the arrastra in a semi-fluid state, made up in piles containing from 30 to 130 tons, and allowed to drain and dry for several days, dams of sand, wood, or stone being built around the pile to prevent it spreading all over the floor; in large works, permanent circular walls or dams are sometimes built for this purpose. The piles are about 1 foot thick and 20 to 50 feet in diameter. When it becomes stiff enough to work, it is spaded over thoroughly, and then from 2 to 5 per cent. of salt is scattered over its surface and worked in thoroughly by spading and by mules or horses driven around in it. This operation of spading and treading the pile or *torta* is known as *repaso*.

After the salt is thoroughly worked in, the pile is once more spaded over and the *magistral* is added. The essential constituent of the magistral is copper sulphate. Magistral was formerly made by roasting copper pyrites, converting the sulphide to sulphate of copper, or by roasting together iron or aluminum sulphate and insoluble copper salts, converting the copper into sulphate and spreading this over the torta. Of late years, however, it has become customary to leach the copper sulphate out of the roasted ore, crystallize it, and use it pure as magistral. The office of the magistral is to convert the non-amalgamable silver salts into an amalgamable form. The reaction is as follows: The copper sulphate reacts with the salt (sodium chloride), forming copper chloride and sodium sulphate. The copper chloride then reacts on the silver salts, forming chloride of silver, which is readily amalgamable.   The copper sulphate of the magistral also reacts upon any soluble lead and zinc minerals in the torta, converting them into insoluble forms, and thus preventing them from being reduced and sickening the mercury.   The amount of magistral added is carefully proportioned to the character of the ore and its silver contents; any excess of magistral causes a loss of mercury.   The proper proportion may be determined by the way the torta works.   The magistral is spread over the surface of the torta and another repaso made; and this is repeated every second or third day, about eight hours at a time, until the operation is finished.   The chemical action of the magistral generates heat, and the proportion of magistral necessary is indicated by the temperature of the pile.   If the pile gets too hot and steams, there is an excess of magistral and a loss of mercury; while if there is not enough magistral, the pile is too cold and the action slow.   An excess of magistral may be corrected by the addition of a little finely ground ore containing oxide of copper, or by adding lime or wood ashes to decompose the excess of copper chloride.

About 6 or 8 ounces of mercury are added to the torta for every ounce of silver it contains.   Usually one-half to three-quarters of the total mercury is added at first, along with the magistral, and the remainder added in small quantities from time to time, always sprinkling through a strainer in order to get the mercury thoroughly distributed in as small globules as possible.   The mercury is spaded in, a hot solution of copper sulphate added, and the pile trodden for 8 or 9 hours the first day, and again the next day, and then every second or third day until the pile is finished. The average time is about 20 or 25 days.

When the tests of the pulp and amalgam show that the amalgamation is completed, a considerable excess of mercury is thrown in to thin the amalgam and catch floured mercury, and the treading is continued for a while longer. The pulp and amalgam are then washed, usually in very primitive box settlers built of stone or equally primitive tubs or pans. The pulp is kept in motion in the settlers by men dancing and wading in the water. The tailings from the settlers are further concentrated on inclined planes of masonry. The amalgam from the settlers is collected, strained, and retorted.

The losses in the patio process are very high. There is always a loss of mercury of at least 1 ounce for every ounce of silver in the pulp, and even the best work seldom saves more than 60 or 65 per cent. of the silver. The system is only applicable to fairly rich ores, in remote regions, where labor is very cheap and machinery very expensive and difficult to obtain.

49. The **cazo** is a round pan or tub, about 1 meter (39.37 inches) in diameter, made entirely of copper, or of wood or stone with a copper bottom. This is set over a rude fireplace, the thin pulp with 5 to 15 per cent. of salt added, and the mixture brought to a boil; the mercury is added as soon as the salt is dissolved, and a man stirs the pulp with a piece of wood, rubbing the bottom to keep it clean of amalgam and assist in the amalgamation of the rich silver minerals.

50. The **fondon** is a form of the cazo, but much larger, being about 7 feet in diameter, and the mulling is done by two copper mullers revolving about a spindle in the center of the pan. If the mercury is added carefully and the mullers kept moving at the proper speed, the amalgam will not cling to the copper bottom and mullers of the fondon.

51. The **tina** is even more like the modern amalgamating pan; in fact, it is practically identical with it, except that the bottom and mullers are of copper instead of cast iron and wood, and are operated by a bevel gearing above instead of from beneath. The operation is the same as that of the fondon and cazo. These machines are suited only to very rich, easily amalgamable silver ores, such as chlorides, bromides, and iodides of silver.

MODERN AMALGAMATING MACHINERY

52. Amalgamating Pan.—The amalgamating pan is an adaptation of the principle of the arrastra to the requirements of modern metallurgy. The prospecting arrastra previously mentioned is really a crude amalgamating pan, a connecting link between the arrastra and the modern pan. The amalgamating pan is essentially a cast-iron pan in which the pulp is ground and amalgamated by means of mullers of cast iron traveling about a vertical shaft or spindle, passing up through a hollow cone in the middle of the pan and driven by bevel gears underneath. The grinding faces are replaceable. The bottom, or muller path, is made up of a series of dies, which fit together to form a closed ring, the joints between them being filled in with hardwood strips. They are usually held in place by dovetail lugs on the lower side, which set into corresponding grooves in the bottom of the pan and tighten automatically, the lugs and grooves being narrower at the end towards which the dies are drawn by the motion of the muller, or they may be bolted down. The muller shoes are usually fastened into the muller ring in the same manner, but with the narrow ends of the slots in the opposite direction, so that the forward motion of the muller tightens both the shoes and dies in their places, while a reversal of the motion will loosen them when for any reason it is desired to remove them. The muller ring is a wide, horizontal flange on the bottom of the hollow cone or hub on the driving spindle, through which the motion of the spindle is transmitted to the muller.

The muller must be attached to the spindle in such a manner that it can be readily raised from the dies without throwing the machine out of gear while charging the pan or while amalgamating, when a stirring action only is desired.

A number of wings of iron or of amalgamated copper, usually something in the shape of inverted plowshares, are set around the sides of the pan, above the level of the mullers, and deflect the pulp inwards and downwards as it rises towards the rim of the pan by centrifugal force, carrying it in through the openings in the muller cone and under the muller again and again, till it is all thoroughly reduced and amalgamated. In the Stevenson pan, curved mold boards, which guide the pulp upwards and inwards, on the principle of a screw, are used instead of wings.

The pans are made either entirely of iron or with a cast-iron bottom and wooden-stave sides, bound together by iron hoops; the latter form is used more particularly when working strongly acid ores which would corrode iron pans. The pans are usually about 5 feet in diameter and $2\frac{1}{2}$ to $3\frac{1}{2}$ feet deep. The shoes used in the different pans vary considerably in shape, size, and number. The minimum number is three; the maximum, twelve. They are usually from 2 to 3 inches thick, and are so designed as to draw the pulp under them.

53. Heating Pulp in Pans.—The pulp in the pans is sometimes heated nearly to boiling point (to about 200° F.), the heat being supposed to assist greatly in amalgamation. The heating is done either by passing live steam directly into the pulp or by exhaust steam in a space between the bottom of the pan and a false or steam bottom. The sides, too, of iron pans are sometimes jacketed. The use of steam passed directly into the pulp is gaining popularity in the later models, as it accomplishes the heating so much more rapidly. It is, however, open to some objections, which are made the most of by the advocates of the other system. In the first place, it requires the use of live steam right from the boilers, as exhaust steam always contains more or less oil, and grease of any sort in the pulp will cause a great deal of trouble. The second objection, which is not a very serious one, is that the water from the condensation of the steam thins the pulp and makes it too thin to work properly. The use of jackets allows the utilization of waste steam for the heating, but it is rather slow and unsatisfactory in its operation. A compromise between the two methods is a good idea, the pulp being brought up to the proper temperature by direct steam and maintained there by the heat from the exhaust steam in the false bottom and steam jackets, if the latter are used. When live steam is used, the pans should be sealed with a cast-iron or wood cover, the steam being introduced through a hole in the top. Covers are useful in any case, as they retain the heat and confine the splash.

54. Forms of Pans.—There are a number of different forms of pans in use, though the idea is the same in all. The pan is essentially an amalgamating machine, not a crusher, and the grinding action should be of secondary importance. By far the greater part of the wear on the machine arises from the grinding, and for this reason it is advisable to crush the ore as fine as practicable before it enters the pan. Moreover, the longer the pulp has to be ground, the greater is the loss of mercury from flouring. In some mills the ore is ground or crushed to the desired fineness before entering the amalgamating pans, either by the ordinary crushing apparatus or, as in the Boss process, in special grinding pans; in such a case, the iron shoes in the amalgamating pans may be replaced by wooden ones.

Pans were formerly made with conical bottoms, with the idea that they required less power, but this construction is now practically abandoned, as it has been proved that for the work done the conical-bottomed pan requires as much power as the flat-bottomed pan, and that flat bottoms wear more evenly, and consequently longer, than conical bottoms, and are more easily repaired. When the sides of the pan are of wood, the bottom is usually cast with a rim inside of the staves, against which they are drawn very tightly to prevent leakage of mercury and amalgam, and the pan is sometimes also covered on the outside with a tight casing of sheet iron. The pans discharge at the bottom. A common and convenient scheme is to have the discharge through a flexible hose, which can be tied up against the side of the pan when not in use. In some mills the mercury and amalgam are withdrawn by themselves into an amalgam kettle, and the pulp is then discharged into the settlers; in others, the entire contents of the pan are conveyed at once into the settlers, and all the amalgam and mercury collected there. With most flat-bottomed pans there are always 50 or 60 pounds of mercury left in the bottom of the pan; but this is really an advantage rather than a disadvantage, as the mercury is not lost, but is utilized in the next charge, and the silver and gold it may contain make it amalgamate more readily and alloy any new mercury which may be put in the pan.

55. The **Wheeler** pan, though one of the oldest forms, is still quite extensively used. The general details of the construction are the same as those previously given in the general description; the method of suspension, however, differs somewhat. The pan is shown in section in Fig. 19. The upper portion of the spindle *g* is threaded, and a corresponding thread is cut in the muller nut *n*, so that it can be screwed up or down on the spindle when cleaning up, putting in new shoes, etc. A keyway cut in the muller nut receives a key which locks the muller at any desired height on the spindle. The position of the muller on the spindle is only altered when a large movement is required, as when the muller must be raised above the top of the pan, in cleaning up or in order to change shoes or dies. All small alterations in the height

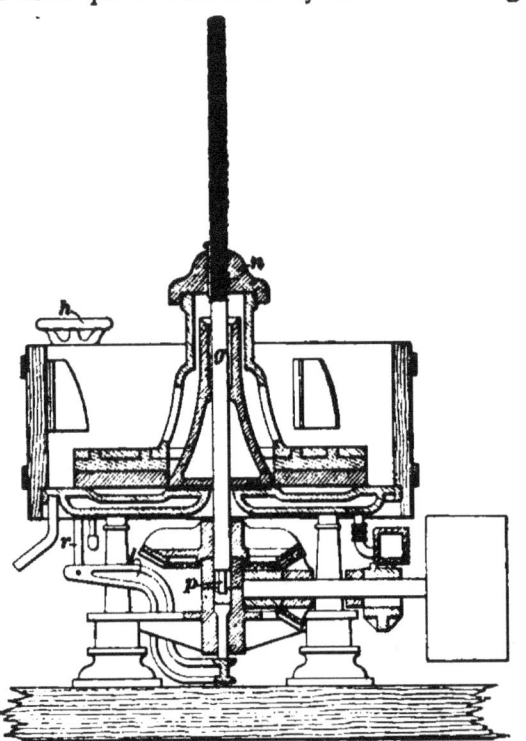

FIG. 19

of the muller, such as raising it from the dies when charging
or amalgamating, are made by means of a hand wheel h on a
vertical rod r at one side of the pan; the rod r connects with
one end of a bent lever l, underneath the pan, the other end
of the lever being suspended from the frame of the pan.   The
lever passes under the center of the pan and forms the support for a pin p, which in turn supports the lower bearing of
the spindle g.   The bevel gear-wheel on the spindle is attached
in such a manner as to allow the spindle a slight vertical
displacement without affecting the gears, and the shaft and
muller can be raised or lowered as desired by screwing the
hand wheel to the right or left, respectively.

The sides of the pan are variously made of cast or wrought
iron or wood; in the latter case, the bottom is let slightly
into the sides.   The pan has a double, or steam, bottom for
heating the pulp.   The shoes are from 6 to 12 in number,
and are wedged into the muller plate with wooden wedges.
The dies, 4 to 12 in number, are in the shape of sectors of a
circle, and are fastened in the bottom of the pan by dovetail
wedge joints; the radial spaces between the dies are usually
filled in with strips of hard wood.   The shoes and dies usually
last from 3 to 6 weeks.   Very hard dies may last considerably
longer, particularly with good care, but many millmen prefer
rather softer shoes and dies, made of a mixture of equal parts
of white and soft iron.   The space left between the muller
and the sides of the pan is considerably larger than in most
of the other pans.

56. The **Varney** pan is somewhat similar in construction to the Wheeler.   The shoes, however, are much larger
and of a peculiar spiral outline, and the shoes and dies are
bolted to the muller and the bottom of the pan, respectively,
making them rather inconvenient to insert and remove.
The suspension of the muller also differs somewhat from
that employed in the Wheeler pan.   The muller is merely
keyed firmly on the spindle and not threaded on, and when
for any reason it is desired to raise it any considerable distance from the dies, the key must be loosened and the muller

raised by overhead tackle. Any small vertical movement of the muller, however, is accomplished by the hand wheel and lever underneath the pan, as in the Wheeler pan. The pulp is heated by direct steam, there being no steam bottom or jacketing. The guide wings are not attached to the sides of the pan, but are suspended from rods passing through the wooden cover of the pan and through cast-iron sleeves bolted to it, and are raised and lowered by means of hand wheels threaded on the upper portion of the rods. The cover is bolted to a flange on the rim of the pan when the machine is in operation.

57. The **Horn, Greeley, Patton,** and **McCone** pans are all double-bottom pans, and differ essentially from the Wheeler pan only in the method of suspension of the muller, which is the same in all. The muller ring and the suspending cone or driver are cast separately and either bolted or wedged together in the different pans. The shoes and dies are held in place by self-tightening wedge joints. The apparatus for raising or lowering the muller is all above the pan, as shown in the illustration of the McCone pan,

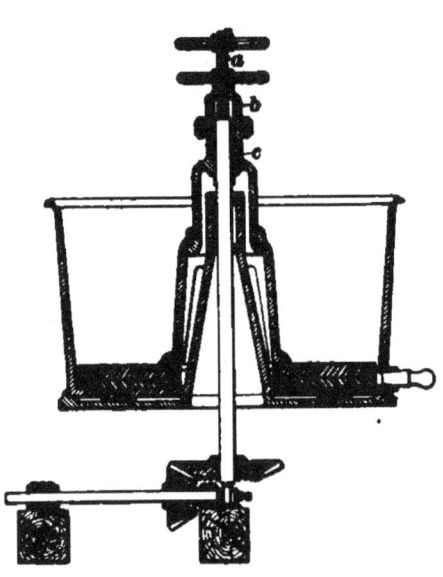

FIG. 20

Fig. 20. It consists of a vertical screw a, turned by a hand wheel at the top. The lower end of this screw rests upon the top of the driving spindle, and a thread is cut on the inside of the muller nut b, corresponding to that on the screw, and the muller is raised and lowered by turning the wheel. The muller is free to slide up and down on the spindle, but is caused to rotate with it by a vertical key fixed in the spindle and working freely in a slot in the muller hub *c*. The muller in the Horn and Patton pans is not fastened tightly to the driving cone or hub, but is caught in grooves, so that it tightens when the driver is turned forwards and is loosened when the motion is reversed, and can be readily detached from the driver. All these pans differ more or less from one another in unimportant details, such as the shape of the false bottom and muller, the shape, weight, and number of the shoes, etc., the difference being in many instances barely sufficient to characterize the pan. The Fountain and Stevenson pans dispense with the ordinary guide wings, having instead flaring lips on the muller which guide the pulp downwards through radial slots in the plate beneath the lips. A similar device is sometimes used on the McCone pan. A method of suspending the muller sometimes used is similar to that on the Wheeler pan, except that, instead of keying the muller at the desired height on the spindle, it is locked in position by a hand wheel threaded on to the upper portion of the spindle and screwed firmly down on the muller nut when the latter is in the desired position. The muller nut is made as a hand wheel, for convenience in raising and lowering the muller.

58. The Boss pan, used in the Boss continuous process, presents some characteristic features. One of the most important of these is the extension of the steam bottom up into the central cone of the pan, as shown in Fig. 21, thus greatly increasing the heating surface. The cone of the false bottom is extended up beyond the top of the main-bottom cone, and forms the sleeve and upper bearing of the spindle. A rust joint is made between the two cones. The steam enters on one side of the bottom and exhausts on the other, and is regulated by a horizontal valve operated by a hand wheel on a rod extending out to the side of the pan. The pans are set in series, the pulp flowing from one to the other, and the driving shafts of the entire series are coupled together and driven as a whole; and any pan in the series can be thrown in or out of gear by a friction clutch

without disturbing the rest.   The friction ring which the
clutch engages is cast separately from the driving bevel
gear and bolted to it, so that it can be replaced independently
when worn out or broken.   The step bearing of the spindle

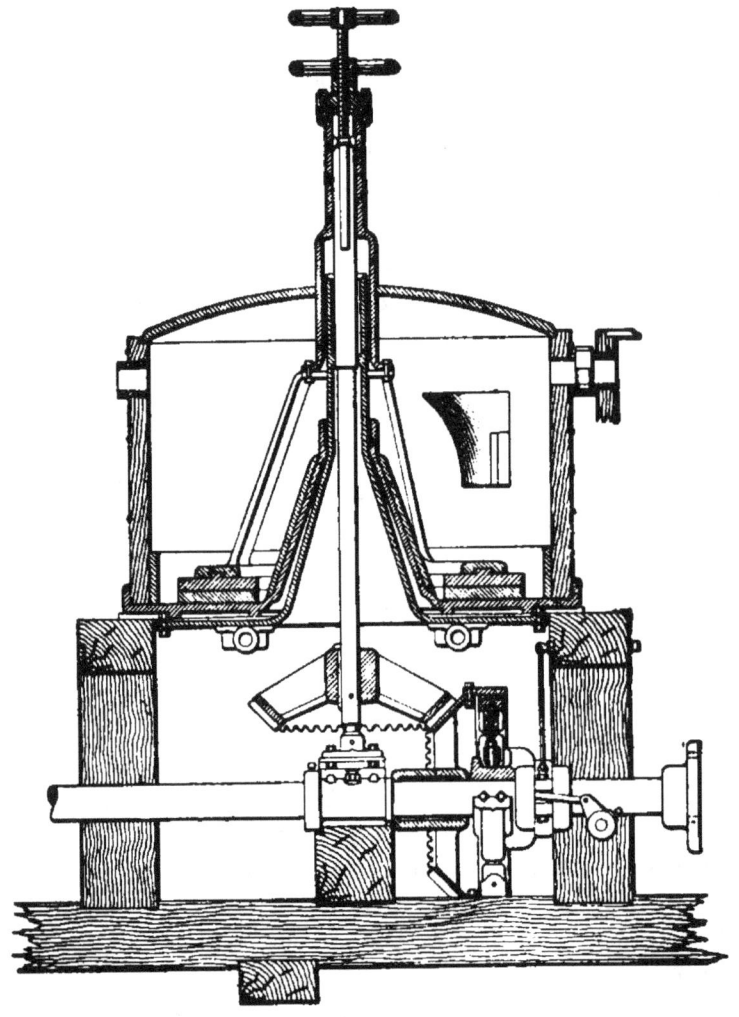

FIG. 21

is carried on a bracket cast on the driving-shaft box in such
a way as to allow the removal of the shaft without disturbing
the spindle.   The mercury bowl in front of the pan (see
Fig. 22) has a siphon arrangement, by means of which the
pan may be readily drained of amalgam and pulp, if desired.

59. Chemicals.—In the early days of pan amalgamation, all sorts of mixtures were employed in pans, with the idea of assisting the operation. These nostrums were mostly harmless, but some of them were highly ridiculous and nearly all of them absolutely useless. They were hit upon with about the same authority as the "charms" of a negro "voodoo doctor"; and once an old-time millman conceived the idea that tobacco juice, sage tea, or some other equally nonsenical substance was an aid to amalgamation; it was as difficult to shake his faith in it as it is to undermine that of a superstitious negro in his "charm." However, new blood and modern education have now almost completely displaced this class of "rule-of-thumb" metallurgists, and out of all the motley collection of nostrums formerly used, only those remain whose chemical reactions are known and which are used to obtain definite effects. Thus, salt and "bluestone" (sulphate of copper) are always used, for exactly the same reasons as they are employed in the patio process. Lime is used to counteract the sickening effect on the mercury of too much acid in the water; dilute sulphuric acid is used when the ores are too strongly basic; lye and potassium cyanide and niter are used to cut grease, and cyanide and niter to clean rusty scales of gold and silver. Sodium amalgam is used, as in all processes of amalgamation, to clean and enliven sickened mercury.

60. Charging the Ore.—While charging the pan, it is customary to raise the muller about ¼ inch from the dies and use it at first merely as a stirrer or mixer. If the charge is of dry-crushed or roasted ore, water is first run into the pan until the muller is covered. The muller is then started up, running at a speed varying, in different mills and with different pans, from 60 to 90 revolutions per minute, and the ore is then dumped or shoveled in. *The muller must be in motion while charging,* or the ore will pack on it and hold it down, and it will either be impossible to start it or the force required will be so great that there will be considerable risk of breaking the machine. The charge varies from 800 or 1,000 pounds, in the old-type Wheeler and Varney pans, to 4,500 pounds or even more in large pans treating slimes. Modern practice favors the use of pans of large capacity, as they do nearly, if not fully, as good work as the smaller pans, require but little more time to grind and amalgamate the charge, and the additional power necessary to drive them is small in proportion to the increased capacity.

The pulp in the pan, when the charging is finished, is about the consistency of batter and fills the pan about half full. The motion of the muller causes the pulp to rise nearly to the top of the rim of the pan, sloping inwards towards the center, and the guide wings or mold boards, as the case may be, throw it back again to the center. As soon as the charge is all in, the muller is lowered until the shoes and dies almost touch and the ore is ground to a fine pulp. The grinding usually requires from 1 hour to $1\frac{1}{2}$ hours, the end of the operation being recognized by the feeling of the pulp when rubbed between the thumb and finger. Rebellious ores sometimes require as long as 4 hours. Towards the end of the grinding, steam is turned into the pan or into the steam bottom, as the case may be, and the charge brought up to the proper temperature, about 200° F., and this temperature is maintained throughout the amalgamating operation.

To obtain the best results in grinding, the pulp should be moderately thin, while the best amalgamation is obtained with the pulp somewhat thicker—about the consistency of honey. The pulp, for good amalgamation, will thicken up sufficiently during the grinding if not made too thin at the start.

61. Charging the Mercury.—There seems to be no general rules for the addition of the mercury, either as to the amount or as to the time of addition. The charge of mercury is usually a fixed weight per pan, the amount varying in different mills from 100 to 350 pounds per ton of ore in the charge; as a rule, the greater the capacity of the pan, the smaller the proportion of quicksilver required. In some mills, indeed, the amount of quicksilver added is proportioned to the amount of silver in the ore, which is determined by assay; but even these rules are empirical and applicable only in the practice of these particular mills. The addition of mercury is generally made after the grinding is completed, and the muller is raised from the dies during the amalgamation. Occasionally, a mill will be found where the mercury is charged and ground with the ore, but this is extremely bad practice under ordinary circumstances, as it does not appreciably increase the amalgamation, and it causes excessive flouring and loss of mercury and amalgam. The mercury may be all added at once, or a part of it reserved till the amalgamation is completed and then added to thin and collect the amalgam.

The mercury is scattered over the top of the pan charge through a cloth strainer placed over the top of the flask, or poured from the flask between the closed fingers, or in some mills it is squeezed through canvas, the object in all cases being to break it up into very fine globules. The motion of the pulp and the muller further breaks up the mercury, the thick pulp holding the minute globules in suspension, and they gradually circulate throughout the entire charge, amalgamating all the gold and silver they meet. The amalgamation usually requires from 4 to 5 hours; the additional saving from lengthening this period seldom compensates for the loss of time. The entire time of the ore in the pan is usually from 4 to 6 hours; in some cases it runs as high as 8 hours.

62. Charging the Chemical Reagents.—The essential chemical reagents—salt and bluestone—are charged at different periods in the operations in different mills, and in quantities varying with the character of the ore. The amount of each reagent necessary must be determined by experiment. The charge of bluestone seldom runs higher than 4 pounds per ton of ore; the salt is about twice or three times as much. They may be added with the mercury or at any time previous. Some millmen charge the salt and bluestone with the ore; others charge the salt with the ore, and the bluestone at some time during the grinding; there is no fixed rule, and the time of the addition seems to make very little difference with the amalgamation. The longer the chemicals are in the pan, the greater will be the corrosion of the ironwork—a strong argument for postponing their addition till towards the end of the grinding operation, when there will still be plenty of time for their chemical action on the ore.

In combination mills, working ores containing both gold and silver in amalgamable form, and amalgamating both metals in the pans, the mercury is allowed sufficient time to amalgamate the gold before the addition of the bluestone, as the latter hinders rather than aids the amalgamation of gold, though it is practically indispensable in the amalgamation of silver minerals.

The *auxiliary* reagents—lime, cyanide, sodium amalgam, etc.—are added as required. When the character of an ore requires the constant use of any of these reagents in definite quantities, they may be made up into a stock mixture or solution with the salt and bluestone, to save time and trouble. The quantities used are, relatively to the size of the charge, usually very small—a few pounds to the ton of ore being sufficient in most cases to produce the desired result.

63. When the amalgamation in the pans is completed, the speed of the mullers is usually reduced to about 40 revolutions per minute, the steam shut off from the pans, if it is on, and the pulp thinned with water to cool it and allow the suspended globules of mercury and amalgam to settle. They are run for 15 or 20 minutes in this way, and then the mullers are stopped and the pans drained into the settlers. The amalgam in the bottom of the pans may either be withdrawn separately or the entire contents of the pans may be run into the settlers, where the mercury and amalgam remaining in suspension are settled out.

64. Settlers.—The settlers used for this purpose are merely large tubs of wood or iron—or with a cast-iron bottom and wooden or sheet-iron sides—usually 7 or 8 or even 10 feet in diameter, in which the thinned pulp is slowly agitated by stirrers resembling the mullers of amalgamating pans. The whole construction of settlers, indeed, is quite similar to that of amalgamating pans, except that, as no grinding is required, the shoes on the stirrer are usually made of wood, and no dies are used on the bottom of the pan. The general construction of the settler is also considerably lighter than that of the amalgamating pan, as the speed of the stirrers and the resistance of the pulp are both very much decreased.

The stirrer usually has four radial arms, on which the shoes are arranged as in Fig. 22; the path of each shoe on the short arms thus lies between the paths of two other shoes on the long arms, and the width of the shoes being equal to the width of the spaces, the entire bottom of the pan comes under the action of the shoes, up to the base of the central cone. The stirrer is hung like the muller of an amalgamating pan and is raised and lowered by a hand wheel. The shoes never quite touch the bottom, but are worked to within ¾ inch or less of it, and the currents set up by them keep the pulp from packing on the bottom.

Settlers vary little in their essential details; in fact, most of the ordinary forms are practically identical in construction, the only radical departure from the construction here described being in the case of the Boss settler.

65. Discharging Settlers.—The pulp is discharged from ordinary settlers through a number of orifices in the side of the pan at different levels, as shown in Fig. 22, the lowest being about 8 inches above the bottom of the pan; these are closed by wooden plugs, and are opened one at a time, beginning with the top one. The pulp being discharged by layers, the lower portion is undisturbed by drawing off the upper portion, so that settling and discharge go on together, thus saving considerable time in the operation. The discharge holes are placed diagonally down the side of the pan, so that the lower plugs will not become covered

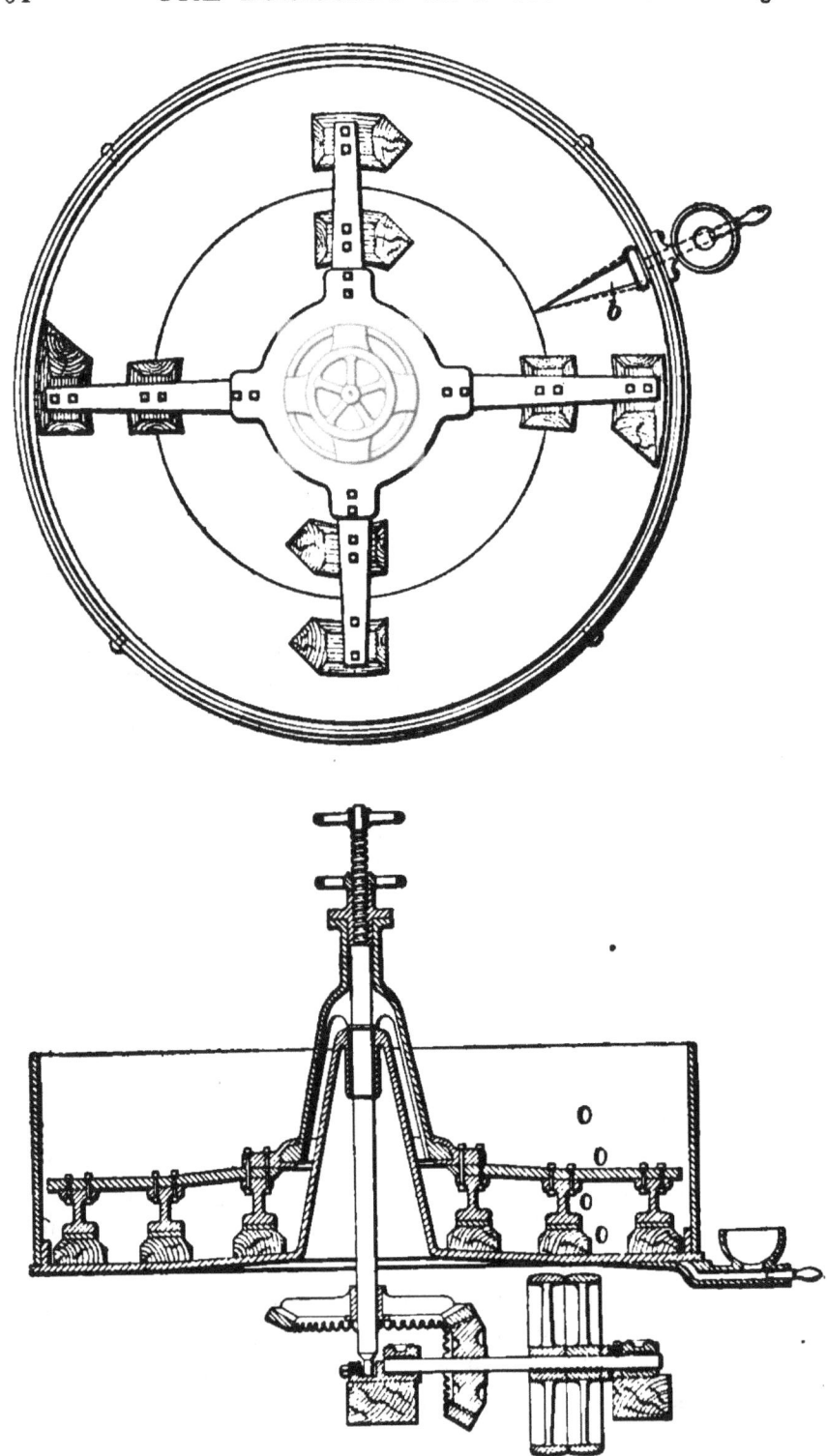

FIG. 22

with the slime from those above. The bottom of the settler usually slopes from the center. A groove *b* in the bottom of the pan collects the amalgam and mercury, from which it can be tapped as desired, into the mercury bowl in front of the pan. In some of the later designs a small bowl or hollow in the bottom of the pan, near the side, replaces the outside mercury bowl, and the amalgam discharges automatically, through an inverted siphon, as fast as it forms. The excess of mercury in the amalgam is strained off and returned to the reservoir for further use.

66. The pulp is ordinarily discharged from the settler through the side holes alone, leaving about 8 inches, more or less, of the heavier sand and pulp packed in the bottom of the pan; and the next charge from the amalgamation pans is turned in on top of this. In starting the stirrer on a new charge, it is raised so that the shoes are at about the level of the lowest side hole, just clearing the top of the packed sand from the previous charge, and is run in this position for about half an hour before adding any water, in order to get the heavy sand into suspension; or water may be added slowly, in a series of fine jets, about 1 inch apart, from the under side of a pipe extending radially from side to center of the pan. The stirrer is then gradually lowered, reaching its lowest position about 2 hours after starting. The addition of water through the jets is continued meanwhile, and is kept up until the contents of the pan are within about 6 inches of the top, when it is shut off. The stirrer is run in its lowest position from 1 to $3\frac{1}{2}$ hours with the pan full; the top plug is then removed and clean water allowed to flow through the hole for about half an hour; then the next plug below is removed, and so on, the stirring continuing all the while. The entire time consumed in the settling is gauged to correspond with that required for the amalgamation, so that neither the pans nor the settler will be obliged to stand idle, waiting for one another. Each settler usually handles the pulp from two pans, though sometimes a single small settler is used for each pan.

67. Clean-Up.—Once a week the settlers are completely drained through a hole in the bottom of the mercury well, and all the amalgam carefully collected in an iron vessel and washed. In this time from 300 to 400 pounds of mercury and amalgam will have collected, unless a continuous siphon discharge is used.

The amount of water used in the settlers should be carefully regulated, for if the pulp is too thick the mercury will not settle completely, and if it is too thin, the coarser sand will settle along with the mercury and prevent the globules from uniting. Beyond a certain point all further addition of water is useless, as at this point the mercury separates as readily from the pulp as it would if the pulp were thinner, and any further addition of water only thins the pulp unnecessarily. With ordinary care, the settler is not apt to clog; if, however, the sand shows a decided tendency to pack on the bottom, the settler should immediately be cleared, to avoid any risk of breaking the apparatus.

68. Agitators.—The pulp from the settlers is sometimes run into still other settlers, called *agitators* or *dolly tubs*. These are simply large wooden tubs, from 8 to 20 feet in diameter and from $2\frac{1}{2}$ to 4 feet deep, with four radial arms, hung and driven like the arms of a settler, revolving about in the tub at the rate of from 10 to 20 revolutions a minute. Each of the arms carries from six to eight vertical wooden staves, reaching nearly to the bottom of the tub, and these keep the pulp in a state of gentle agitation, allowing the fine shots of mercury and the coarse sand to settle. A constant stream of water is kept running through the agitator. Every three or four days the accumulated material is shoveled out and worked over in pans. One agitator will handle the pulp for five or six settlers.

These are now seldom used, having given way to modern concentrators or to various modifications of the ordinary settlers.

69. A few years back it was customary to build long blanket sluices below the agitator, through which the tailings from the agitator were run. These sluices resembled the undercurrents used in hydraulic mining. They usually consisted of a number of shallow troughs, 18 to 20 inches wide and 2 or 3 inches deep, placed side by side, on a grade of from 6 to 10 inches in 12 feet, and varying in length from 75 to 1,700 or 1,800 feet. The bottoms of the sluices were covered with blankets, tarred on the under side to prevent their rotting; or, in the short sluices, riffle bars were sometimes used. The sluices were cleaned at intervals, either by sweeping or by removing the blankets and washing them. Both blanket sluices and agitators have of late years given way almost entirely to improved modern slime washers, such as buddles and vanners.

ORE DRESSING AND MILLING
(PART 4)

AMALGAMATION

SILVER AMALGAMATION

BOSS-PROCESS MILL

1. Boss Continuous Process.—The Boss continuous process for the amalgamation of silver ores is a comparatively new process, which will in time probably supersede all the older processes. The adoption of the Boss system eliminates a considerable portion of the hand labor ordinarily required about a silver mill. The operation is entirely continuous, and, except where the ore requires roasting, it need not be handled from the time it is fed into the stamps until the mineral and amalgam are cleaned out of the settlers. When the ore is roasted, the pulp from the stamps (crushed dry) is conveyed to a roasting furnace, and from there to the cooling floor, where it is spread out and allowed to cool before charging into the grinding pans, in which the ore is mixed with water and ground to the proper fineness before entering the amalgamating pans. A Boss-process mill section is shown in Fig. 7.

2. Grinding Pans.—The grinding for this process is done in small iron pans 4 feet in diameter and 1 foot 4 inches

§ 28

For notice of copyright, see page immediately following the title page.

deep, constructed somewhat similarly to the grinding and amalgamating pans previously described. The driving arrangement is the same as that used on the Boss amalgamating pans, with this exception: there is a compressed spring in a sleeve around the muller nut, which keeps the adjusting screw pressed down firmly on the spindles.

The muller ring and driving cone are cast in one piece. The ring is a flat, vertical rim, connected to the cone by horizontal spokes, and the shoe is bolted to lugs projecting outwards from the sides of the rim, opposite the ends of the spokes. The upper edge of the rim is turned inwards at right angles, forming a flat, horizontal flange, or lip, about 5 inches wide. The shoes and dies are both solid, flat rings, but have oblique slots on their inner edges, extending a short distance into the rings, in order to get the same effect of suction that is obtained when the shoes are in segments, with oblique slots between them. The pulp is fed into the muller ring, and is obliged to pass under the muller in order to get to the outside of the pan, since the joint between the muller ring and the shoe is made water-tight by a rubber gasket, and the flange at the top of the ring prevents the pulp from splashing over the sides and turns it back to the middle of the pan. Grinders are usually placed in pairs, two for each ten stamps, one being set slightly lower than the other. The pulp from the entire ten stamps passes into the first pan, where it is ground; thence it passes on into the second pan, where it is still further ground; from here it discharges into the first amalgamators—or in some mills into a *chemical mixer*, which is interposed between the grinders and the amalgamators, and in which the chemicals are mixed with the ore before it enters the pans. The grinding pans are on a higher level than the amalgamators, and behind them, and are driven by friction clutches on a separate shaft, the clutches being thrown in and out of gear by levers operated from the pan floor.

3. The **amalgamating pans** used in the Boss process are described in Art. **58,** *Ore Dressing and Milling*, Part 3.

The number of pans in the series depends both on the capacity of the mill and on the character of the ore. Thus, the greater the capacity of the mill, the larger must be the number of pans in the series, since the ore must be in the pans a certain length of time in order to obtain a good percentage of amalgamation. As the operation of the mill is perfectly continuous and all the pulp has to pass through every one of the pans in the series, the greater the quantity of pulp, the less time it will be in each pan; consequently, if the amount of ore handled by the mill is increased, the number of pans in the series must be increased proportionately, in order to have the material exposed to the action of the mercury for the proper length of time. Or, again, if one ore amalgamates more readily than another, it will require less time in the amalgamating pans, and allowing both ores the same time in each pan, the first would require fewer pans in the series to treat the same amount of pulp than the second. The same is true of the settlers. The pans, settlers, and—when one is used—the chemical mixer are all in the same line and on the same level and are driven from the same shaft, each machine being thrown in or out of gear independently of the rest by means of the friction clutches shown in the illustration of the pan in Fig. 21, *Ore Dressing and Milling*, Part 3.

To avoid the necessity of stopping the whole mill when pans are cut out for cleaning or repairs, steam siphons are used, which carry the pulp past the idle pans and into the next pans beyond them, and the operation of the mill proceeds without interruption. The shaft is made in sections— one section to each pan—coupled together. Every other coupling is a clutch coupling and the rest are ordinary flange couplings. Between the faces of each flange coupling a ring or washer is inserted, the thickness of which is a trifle greater than the distance it would be necessary to draw the two portions of the clutch coupling apart in order to disengage them. When it is necessary to remove any section of shafting, the cover of the shaft box is removed, the flange coupling at one end is unbolted, the washer

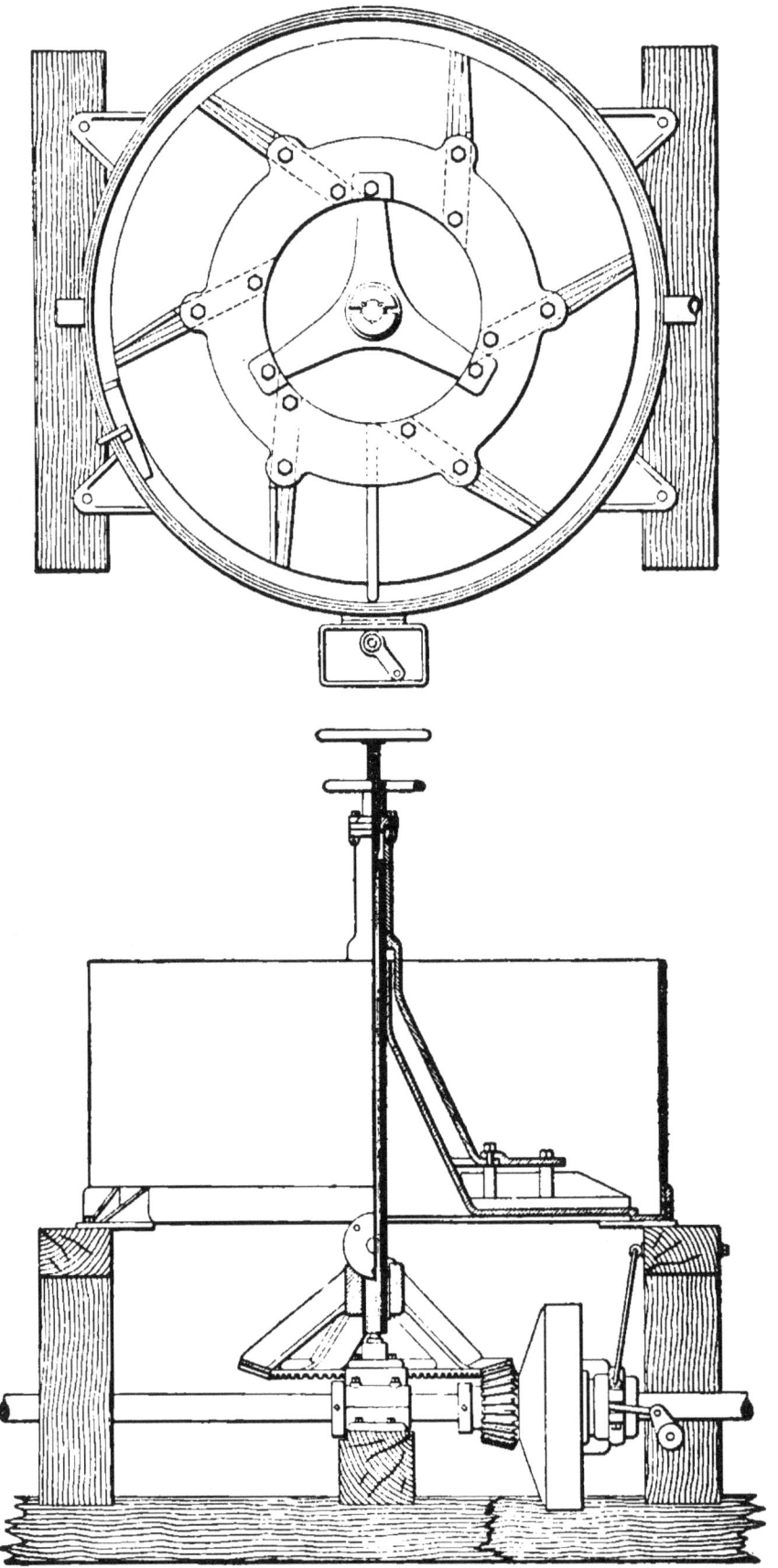

FIG. 1

removed, and the section drawn back till the clutch coupling disengages, when the shaft can be lifted out of the box.

4. The **Boss settler,** which is used in the Boss continuous process, is shown in Fig. 1. This settler differs from those previously described mainly in the shape and arrangement of the stirrer and shoes. The pan is 8 feet in diameter, with a cast-iron bottom and wrought-iron sides. The bottom of the pan is flat, and the central cone has a much wider base than is usual in the ordinary forms. There is a slight trough in the bottom, next to the sides of the pan. The shoes are of iron and are much larger than those used in the ordinary settler, each shoe extending across the bottom from the base of the cone to the edge of the outside groove, at a slight inclination from the radial line instead of radially. They are bolted to a muller ring or plate, which is in turn bolted to the three legs of the driver. The shoes do not touch the bottom of the pan, but work quite close to it. The angle at which the shoes are set induces a strong current on the bottom and keeps the pulp from packing. The driving and following gears are proportioned so that the stirrer makes about 20 revolutions a minute. (The speed of the amalgamator mullers is about 60 revolutions.) As a rule, no additional settling apparatus is used in Boss-process mills.

5. The **chemical mixer,** which is frequently used in the treatment of somewhat refractory ores, is of the same size as the settler and nearly identical in construction. It has, however, wooden sides, as they withstand the reaction of the chemicals much better than wrought iron; and a steam cone is run up inside the main-bottom cone, the heat from it greatly assisting the action of the chemicals. The chemicals are relatively light and the motion of the stirrer slow, so that the solution is strongest in the top of the charge, where the pulp meets it on entering the mixer; and the discharge is practically from the bottom, a vane or wing just in advance of the discharge pipe deflecting the

pulp into it as it rises from the bottom at the outside of the pan. In this way, the chemicals are retained longer in the pan and the necessary chemical action on the ore is completed sooner. The number of mixers used in the series depends on the character of the ore and the capacity of the mill. The more refractory the ore, the longer must it be exposed to the action of chemicals; consequently, very refractory ores sometimes have to be passed through several mixers in order to give the chemicals sufficient time to act on them. The chemicals are fed into the mixers by an automatic chemical feeder. When mixers are not required, the chemicals (salt and bluestone) necessary for amalgamation are fed into the first pan by one of these feeders, which is placed between the first two pans of the series, and if any additional chemicals are necessary, they are fed in by another feeder farther down the line, usually between the last two pans.

6. The **mercury system** is entirely mechanical. The mercury is stored in an iron reservoir and is run into the pans through pipes having inverted siphon traps, where they connect with the pans to exclude the pulp. The amount being charged into each pan is regulated by a cock in the pipe near the pan. The total amount of mercury used is shown on the dial over the reservoir. The shaft which carries the hand on the dial has a small sprocket on it, over which passes a link-belt chain, one end of which is fastened to a cast-iron float on the mercury in the reservoir, while the other end carries a counterweight which keeps the chain taut. As the mercury falls in the reservoir, the float sinks with it and draws the hand around on the dial. The mercury bowls of the pans and the first settler are connected by pipes with a receiving tank, into which the amalgam may be run by merely withdrawing the plugs in the bottoms of the bowls. The amalgam can be drawn at will from the receiving tank into canvas straining bags, from which the excess mercury drains off. The strained mercury runs through pipes into the boot of a small link-belt elevator, by which it is raised and dumped back into the storage reservoir, to be used again.

7. In the Boss process, when the ore is crushed wet, the pulp runs directly from the stamps into the grinding pans. The proportion of water ordinarily used in wet crushing in stamp mills would render the pulp altogether too thin for amalgamation; so the size of the screen meshes is increased considerably and the water cut down to the proper proportion for pan pulp. In this way the same crushing capacity and the average size of the product is practically maintained, while the pulp can pass to the grinding pans without intermediate settling.

When the ore is very refractory and has to be crushed dry and roasted, the roasted ore is mixed with water and fed continuously into the grinding pans by a screw feeder. Ordinarily the roasted ore is spread out on cooling floors to cool before going to the pans, but in the more modern continuous-process mills there is no handling of the ore from the beginning to the end of the process. The ore passes from the bins into revolving drying furnaces; from the driers it goes directly into the automatic feeders of the stamp battery; the pulp from the battery is elevated by a continuous-belt elevator into the hopper of the roasting furnaces—which are generally of the revolving type (Howell-White continuous or Brückner cylinder)—where the salt is added to chloridize the ore; after passing through the furnace it goes to the mixing pan, from which the screw conveyer carries it to the grinders; and from here on the operation is as previously described. In old-style mills the roasted ore is spread out on a cooling floor and cooled, and is then carried to the mixer by hand.

8. Clean-Up Pans.—Clean-up pans are small amalgamating pans, used in gold and silver amalgamating mills for cleaning dirty or impure amalgam before retorting, and for working up small quantities of heavy blanket concentrates, battery sands, etc. They are made in various sizes, from

15 inches to 5 feet in diameter, 4-foot pans being the most commonly used in large mills. The construction and mechanism are the same as those of ordinary amalgamating pans For cleaning amalgam, wooden muller shoes are generally used, as they are required more for stirring than for grinding. The dirty amalgam is charged into the pan with enough additional mercury to make it perfectly fluid and it is then thoroughly stirred. The foreign matter rises to the top of the liquid bath of amalgam and is washed away by water running through the pan.

For treating concentrates and battery sands, iron shoes are used, as it is necessary to grind the pulp, and iron shoes not only wear better but clean the surface of the gold, so that it amalgamates more readily. When the pans are used for both purposes, the deep wooden shoes are shod with cast iron. The wooden blocks extend above the surface of the amalgam, so that no amalgam will be deposited on top of the shoes when the pan is drained.

In many small mills, amalgam is still cleaned by hand, grinding with more mercury in iron pots or hand mortars, and concentrates and battery sands are panned with an ordinary miner's gold pan, to recover free gold and amalgam. All modern mills, however, are fitted with some form of mechanical clean-up apparatus—usually pans. A clean-up pan known as the **Berdan pan** is considerably used in Australia and New Zealand. It is merely a revolving pan, slightly inclined from the horizontal, and containing an iron ball, which naturally stays at the lower edge as the pan revolves, grinding the ore as it is carried around by the pan.

9. Clean-Up Barrel.—The clean-up barrel is made and operated in the same way as the amalgamating barrel shown in Fig. 2, but is unlined, and not so large—being usually 3 feet in diameter and 4 feet long. The amalgam and mercury are charged into the barrel; the barrel is then nearly filled with water, closed, and revolved for several hours at about 20 revolutions a minute. Scrap iron is sometimes added, but this flours the mercury and does not materially assist the operation, and its use is now being generally abandoned.   At the end of the agitating period the barrel is opened and washed out with water—the

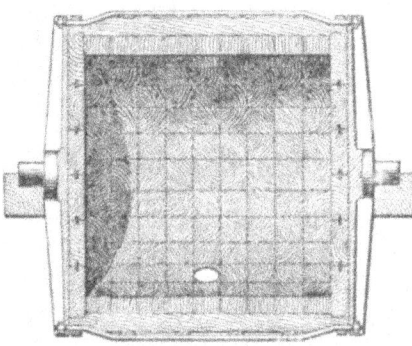 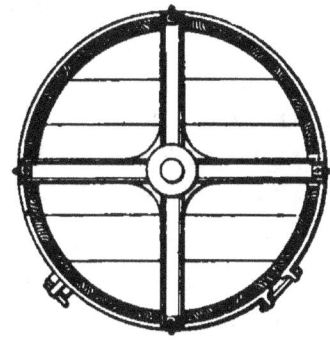

Fig. 2

tailings being run over amalgamated plates or through some other form of amalgam saver—while the amalgam in the barrel is removed, strained, and retorted.

10.   Barrel Amalgamation.—The Freiburg, or barrel, process of amalgamation is not much used in America. A few attempts at barrel amalgamation have been made, but in most cases they have given way to the pan process, which is somewhat quicker, and by most metallurgists considered superior to the barrel process.   However, a description of the barrel and a brief outline of the process will be given.

11.   The Barrel.—The amalgamating barrel is cylindrical in shape, usually about 4 or 5 feet long inside and the same in diameter.   Fig. 2 shows one form of barrel.   Some barrels have a replaceable lining of wooden blocks, about 5 inches square and 3 or 4 inches thick, set on end, as shown in the illustration; this lining can be replaced when it wears out, and the barrel will last indefinitely.   The barrel is made of soft pine staves, 2 or 3 inches thick, bound together with iron bands; and the joints between the planks forming the heads are grooved and fitted with tongues of hard wood.   When the barrel is not lined, the staves are from 4 to 6 inches thick when new, and are replaced after they have worn down to about 2 inches. The ends of the barrel are strengthened and braced by a cast-iron spider, through the ends of whose arms are passed tie-rods which draw the barrel heads firmly up against their seats. These spiders are cast in one piece with the trunnions on which the barrel revolves. In some mills steam is used in the barrel to heat the pulp, in which case one of the trunnions is made hollow and the steam pipe is passed through it with a gland to prevent leakage.

12. The general plan of operation in the barrel process is practically the same as in pan amalgamation—drying, crushing, roasting, and amalgamation—but the details of the process are entirely different and necessitate much expensive machinery, while the process extends over a longer period than would be required to treat the same ore by the pan process.

13. Drying.—The ore from the mines is run through a jaw crusher and then conveyed directly to the drying floor or kilns, where it is spread out to dry as rapidly as possible. The drying floor is usually a floor of cast-iron plates, forming the covering of a series of flues about 8 inches deep, through which the smoke and waste gases of the roasting furnace pass; or the floor may be heated by a special fireplace. The ore is constantly raked and turned on the drying floor until perfectly dry.

14. Roasting.—From the drying floor the ore is carried to some form of dry-crushing apparatus, usually a stamp mill, where it is reduced to a maximum size of 40 mesh, or, in some cases, to 60 or even 70 mesh. The crushing and screening to such fine sizes naturally consume considerable time. From the crushers the ore goes to the roasting furnaces, which are usually either reverberatory or cylinder furnaces. The ore is roasted with salt until the silver is all in the form of chloride, and any sulphur and arsenic are volatilized. This usually requires 6 or 8 hours; in the case of very rebellious ores, it may require as much as 16 hours or even longer. The ore is roasted for several hours before adding the salt, in order to volatilize the sulphur, etc., as the chloride of silver is slightly volatile, and a small amount of it is apt to be carried off in the fumes, particularly in the first part of the roasting operation, when copious fumes of sulphur, arsenic, and antimony are passing off.

15. Screening.—When the roasting is complete, the roasted charge is withdrawn and spread out on the cooling floor to cool as rapidly as possible, in order not to form lumps. As soon as it is cool enough, it is run through a 40-mesh screen, or finer, and all lumps which refuse to pass this screen are repulverized. The reason for this screening is that the barrel pulp is mixed with only enough water to make a rather thick mud of it, but not sufficient to soak into and disintegrate clots, so the mixing must be extremely careful. It is this fine crushing and repeated screening that consume so much time in the barrel process and make it so much slower than the pan process.

16. Charging.—The roasted ore is thoroughly mixed with sufficient water to make a fairly thick mud, and this is charged into the barrel, together with from 60 to 300 pounds of scrap iron and the barrel revolved for 2 or 3 hours, the object being to allow the iron to reduce the chlorides of silver, lead, and copper to metals, which the mercury would otherwise have to do. From 250 to 500 pounds of mercury is then added and the barrel again revolved for from 12 to 16 hours at the rate of 15 revolutions per minute. A small quantity of bluestone or magistral is usually added with the mercury for the same reason as in pan amalgamation.

17. Discharging.—When the amalgamation is complete, a plug in a hole in the barrel is removed when the hole is in its lowest position, and the mercury and amalgam are withdrawn into a receiver beneath. As soon as the pulp commences to follow, the plug is replaced. The amalgam is then removed and the pulp from the barrel is run into a large agitator, where the fine globules of amalgam are settled out. The rest of the process is the same as in pan amalgamation.

STAMP MILL AMALGAMATION

18. Amalgamating Stamp Battery.—The amalgamating stamp battery is specially designed for the recovery of *metallic* gold and silver from their ores. It is the simplest apparatus known for this purpose, and with proper care one of the most efficient. The pan and barrel systems just described are designed especially for the treatment of *silver* ores, in which the metal is in the combined form, either as minerals not susceptible to direct amalgamation—as the sulphides, arsenides, etc.—or as minerals, like the chloride, bromide, and iodide of silver, which although amalgamable, are, from their comparative lightness and other physical characteristics, not readily brought into direct contact with the amalgamated surfaces of the battery and apron plates. The slow and long-continued working of such ores in pans or barrels gradually brings every particle of ore in contact with the mercury—a thing practically impossible in the comparatively swift moving and turbulent water in the battery and on the apron plates.

19. Amalgamating Mortar.—The mortar of the amalgamating stamp battery differs very little from mortars used simply as crushers, except that amalgamated copper plates are placed *inside* the mortar. The amalgamating mortar must be made somewhat wider than is necessary if the mortar is to be used for crushing alone, in order to get the plates farther away from the stamps and dies, and thus decrease the force of the splash. The same thing may be accomplished by deepening the mortar, using a higher discharge, and setting the plates higher above the dies; but this either decreases the capacity of the battery or necessitates the use of a coarser screen or more water, and is less satisfactory generally. If the plates are too close to the stamps, the splash will scour off the amalgam and greatly injure the efficiency of the battery as an amalgamating machine.

20. Wet crushing is absolutely essential for successful direct amalgamation in the battery, as the pulp must be very thin and open to insure the gold coming in contact with the plates. Even when the pulp is very thin, the battery screen of quite fine mesh, and the gold comparatively coarse, there is always more or less gold which escapes from the mortar without amalgamating, and with it considerable finely divided amalgam; in order to catch these particles, copper *apron plates* are placed in front of the battery. Apron plates are usually thin sheets of copper, from 8 to 12 feet long and the full width of the mortar, set on an incline of about 1 in 12. The pulp falls from the battery screens directly upon the plates and flows over them in a thin, rippling stream; the gold and amalgam sink to the bottom and are caught by the amalgamated surface of the plate. If the gold is very fine or is rusty, some of it will not be caught on the plates, but will be floated on with the tailings. Formerly this loss in the tailings was overlooked; but close competition and low-grade ores have forced it upon the notice of the millman and metallurgist, so that at present there are a number of devices in use designed to save float gold.

21. Battery Plates.—Figs. 32 and 33, *Ore Dressing and Milling*, Part 1, show amalgamating gold mortars. Inside the mortar there are usually two copper battery plates—one at the back, bolted to the mortar, as in Fig. 33, or held in place by taper keys, as in Fig. 32, and one in front screwed on to the chuck block and inserted and removed with it. Occasionally end plates are used, but they are not common. Some mills, working base ores by combined amalgamation and concentration, use only the front plate. The plates are generally from $\frac{3}{16}$ to $\frac{3}{8}$ inch thick and are usually not silvered.

In most modern amalgamating mortars, the back plates are protected from the scouring of the ore, as it is fed into the mortar, by projecting shelves or lips, as shown in Figs. 32 and 33, *Ore Dressing and Milling*, Part 1, but there are still many mortars in use in which the back plates are set directly in the feed opening. When the plates are protected by a lip, the mortars are sometimes made with a second opening at the back, extending the full length of the mortar, through which access may be had to the plate in the place beneath the lip. This opening is kept closely covered while the battery is in operation by a tightly fitting splash board. On an average, about 75 per cent. of the total gold saved as amalgam is caught on the inside plates of the battery. The proportion in a few instances runs up to a little over 90 per cent. Most of the gold that escapes the battery plates is caught on the apron plates, and a little on the blankets, bumping tables, or other concentrating apparatus.

22. Apron Plates. — Apron plates are set immediately in front of the mortar, fitting snugly up under the cast-iron lip or apron. They are usually of $\frac{1}{16}$- or $\frac{1}{8}$-inch copper, and in some mills are electroplated on the upper surface with silver—from 1 to 3 ounces of silver to the square foot of surface—as the silver amalgamates much more readily than pure copper and works well from the very beginning of the run, while it requires several days' run and elaborate preliminary cleaning to get the surface of a plain copper plate into fit condition for amalgamating the gold to good advantage. Plates of Muntz metal—an alloy of 60 per cent. copper and 40 per cent. zinc, used for ship sheathing—have been tried in some localities, and on certain qualities of low-grade ores give better results than plain copper plates and do not foul so readily, probably on account of a weak galvanic current formed by the copper and zinc with the acid water. The plates amalgamate more readily than plain copper and work well from the very start, but the coat of amalgam is very thin and superficial and easily removed; hence, they are less advantageous for rich ores, as they have to be cleaned up too frequently. Further experiments with this metal are desirable.

23. The plates are fastened to carefully planed tables by copper screws or nails or are held down by cleats, and are beaten to a perfectly smooth, even surface by means of hardwood blocks, which are placed on the plate and struck with heavy mallets. It is highly important that the surface of the plate be perfectly smooth and set horizontally parallel to its width, so that the pulp will be evenly distributed over the entire surface. The table may be made of narrow widths of heavy planking, spiked firmly to the inclined framework; but a better design is made up of narrow strips of wood, set edgewise, lengthwise of the table, and bound together every two or three feet by light iron tie-rods. A table built in this manner will not warp and the even surface of the plate is more easily preserved. Tables made of sections of cast iron are sometimes used, but wood is cheaper and lighter and answers the purpose just as well.

24. California Milling Practice.—In the California milling practice, a very peculiar and unpractical arrangement of the outside plates has been in vogue for years, and, though rapidly dying out, is still retained to a considerable extent. The length of the apron plate proper, which is ordinarily from 10 to 16 feet long, is cut down to from 10 to 48 inches, and from it the pulp is run into sluices from 14 inches to $2\frac{1}{2}$ feet wide and 10 to 16 feet long, paved with copper sluice plates. Besides being much narrower than the apron plates, the sluice plates are given a slightly greater inclination, usually $\frac{1}{4}$ inch more to the foot. In spite of the increased speed and scouring action of the pulp stream after leaving the apron, as a result of the contracted channel and increased grade, the California millman of the past imagined that he was saving in the sluices gold and fine amalgam which could not be caught on the apron plates. Notwithstanding the fact that on its very face the whole idea is contrary to all reason, California millmen, being governed by precedent rather than by reason, have clung desperately to it for years, refusing to see what one would suppose impossible to overlook.

25. Preparing the Plates.—As has been stated before, quicksilver, which already contains some gold or silver, catches gold or silver more readily than mercury alone. If the surface of the plates be amalgamated with pure mercury, they will allow considerable gold to escape at first, but will steadily improve as the amalgam grows richer, until they are working normally.   But if the surface be previously coated with a gold or silver amalgam well worked in, the plates will do full duty from the start.   For this reason, new copper plates are coated with gold or silver amalgam before being used, or else plates electroplated with silver are used, the silver plating amalgamating with the mercury used in dressing the plates and giving the same effect as preparing with amalgam, with much less time and trouble.   Gold amalgam works somewhat better on the plates than silver amalgam, but is seldom used, on account of the great expense.

To prepare plates for amalgamation, they are first scoured with sand or emery paper until they are clean and bright. They are then washed with a strong soda or lye solution to remove all traces of grease.   Dilute nitric acid—a 10-per-cent. solution—or a comparatively strong solution (about $2\frac{1}{2}$ per cent.) of potassium cyanide may be used instead of lye or soda.   After washing with the chemicals, the plate is well washed with water and then a mixture of equal parts of sand and sal ammoniac and a little mercury is rubbed on with a scrubbing brush.   The sand and sal ammoniac are used to keep the plate clean while the mercury is being rubbed in.   Enough water is used to make a thick mud. More mercury is sprinkled upon the plate from a flask with a piece of cloth tied over the top, and the rubbing is kept up until the surface of the plate has absorbed all the mercury it will hold.   The plate is allowed to stand for about an hour and is then washed clean with water or with cyanide of potassium and water, and more mercury is added if the plate will hold it.   If the plates are plain copper, they are next given a coating of gold or silver amalgam and are then ready for use.   The amalgam is rubbed in with a piece of rubber belting or cloth, the plate meanwhile being kept wet with sal ammoniac.   When rubber belting is employed, a piece is fastened between two blocks of wood, so as to leave about 1 inch of the belting projecting from the wood.   The entire block, belting and all, is sometimes called the rubber. If old mercury is used in preparing the plates, the use of special amalgam is unnecessary, as mercury strained, or even distilled, from amalgam retains enough gold or silver to start amalgamation at once.

26. Cleaning and Dressing Plates.—The plates are cleaned up at intervals varying with the richness of the ore being treated, the idea being to let as much amalgam accumulate as possible without loss from scouring. The amalgam does not accumulate evenly all over the plate, but in ridges, which grow steadily, and if left too long commence to scour off. If the mill is running on low-grade ores, the outside plates are usually wiped twice a day—morning and evening—or only once a day if the ore is very poor.

As the richness of the ore increases, the interval between clean-ups shortens, and for very rich ores it is sometimes necessary to clean the apron plates every hour, or even oftener. The battery plates are not cleaned until the amalgam stands up in thick ridges; with very rich ores it may be necessary to clean them once or twice a week, or even every 48 hours, but they are usually cleaned every two weeks, when a general clean-up of the whole mill is made.

27. The plates are cleaned by rubbing the amalgam loose with a wiper made by fastening a piece of rubber belting between blocks of wood, with about half an inch of the rubber projecting, or with a whisk-broom, cut down to make it stiff. The apron plates are wiped from bottom to top, the men using bits of plank to kneel on while working. If necessary, fresh mercury is sprinkled on the plates after wiping and is rubbed in with a piece of rubber. In a general clean-up the stamps are hung up, two batteries at a time, and the screens, battery plates, and dies are removed and carefully cleaned of amalgam. The battery sands are removed and are either fed into one of the other batteries or are panned, bits of iron removed by a magnet, and the concentrates transferred to the clean-up pan ; or, in some mills, they are saved and returned to the mortar on starting

up again.   All amalgam sticking to the stamps and to the
inside of the mortar is carefully collected, and, with the rest
of the amalgam from the clean-up, is first strained to
remove the excess of mercury and then transferred to the
clean-up pan, or—in small mills—to a hand mortar, where
it is ground with fresh mercury and cleaned.   The amalgam from the pan is again strained, and the balls of dry
amalgam are then ready for retorting.

28.   Scraping Plates.—Every three to six months,
according to the richness of the ore, the plates are scraped
with a piece of steel or a spatula, leaving only a very thin film
of amalgam on the plates.   In some mills the plates are
"sweated," in order to get more gold out; but the plates
frequently have to be resilvered or reamalgamated after
sweating, and the operation involves much more labor than
simply scraping; the thin film of amalgam left after scraping
is, moreover, very advantageous in starting up again.
Sweating is merely heating to loosen the amalgam.   The
plates are removed from the tables and heated over a wood
fire, expelling most of the quicksilver, and the gold scale
remaining behind is then scraped off.   Practically the same
result is accomplished without removing the plates from the
tables by washing them with boiling water or playing a jet
of steam on them to soften and loosen the amalgam, which
is then scraped off.   In some mills, chemicals are used in
the sweating, the plate being first heated to expel the quicksilver and then rubbed with a solution of niter and sal ammoniac and again heated, when the gold rises in scales and
blisters.   The plate is sometimes plunged into a tank of
boiling water on being removed from the fire, when the gold
scales off.   "Skinning" plates in this manner is not usually
advisable during the life of the mill, as the plates are very
apt to get buckled and be irreparably ruined, and there is
then nothing left but to melt them up and get new plates.
The gold recovered from the old plates is usually more than
sufficient to pay for a new set, but it takes some time to get
new plates to work properly, and they will soon absorb

practically as much gold as would have remained in the old
plates after scraping; so that there is really little or nothing
gained financially, even if we do not consider the time lost
in changing the plates, and the time and amalgam lost in
breaking in the new plates.

29. Dressing Tarnished Plates.—New plates rapidly become tarnished from the action of the acids in the
pulp, forming a thin film of copper salts, or "verdigris,"
over the surface of the plates, which prevents them from
catching the gold and amalgam flowing over them. The
tarnish appears in spots—yellow, brown, or greenish—and
spreads rapidly if not removed at once. To remove the
stains, the battery is stopped and the spots scrubbed with a
solution of sal ammoniac; this is left on for a few moments
to dissolve the coating, then it is washed off and the plate
scrubbed with a potassium cyanide solution to brighten it;
this is finally washed off, more mercury added if necessary, and the battery again started up. After the plates
have once acquired a thick coating of amalgam they do not
tarnish very readily; for this reason silver plates give less
trouble from this source than plain copper. Plates dressed
with nitric acid tarnish more rapidly than those dressed with
soda and cyanide. In some mills, a little cyanide of potassium is fed into the mortar from time to time to prevent the
plates from tarnishing, and soda or lye is frequently used
in the mortar to counteract trouble from grease.

30. Mercury Feed and Loss.—The amount of mercury used depends upon the richness of the ore. The quicksilver does not wholly dissolve the gold scales, but only forms
a coating of amalgam on the surface; consequently, weight
for weight, the finer the gold the more mercury is necessary
to amalgamate it, as the smaller particles present more surface in proportion to their weight than the larger ones.
Generally speaking, with a clean gold ore, about 1 ounce of
mercury should be charged into the mortar for each ounce
of gold in the ore treated. Impurities alter the proportion
greatly, however, and it is sometimes necessary to charge

two or three times as much mercury as there is gold in the ore, in order to preserve an amalgam of the proper consistency. In American practice no mercury is charged on to the outside plates, except a mere sprinkling after cleanups, all the mercury for the amalgamation being charged into the mortar. It is charged a little at a time, at intervals of from half an hour to two hours. Automatic mercury feeders are made which dip up and feed small quantities of mercury into the mortar at proper intervals, but most millmen prefer hand feeding, as the feed is more easily regulated, and the work amounts to practically nothing; in fact, the regulation of an automatic feeder to suit the varying conditions in many mills would consume more time than would be required to feed the batteries by hand.

The rate of feeding mercury to the battery is regulated according to the appearance of the amalgam on the apron plates. About half of the quicksilver fed into an amalgamating mortar escapes through the screen on to the apron plate, and if the battery has no inside plates, practically all the quicksilver sooner or later finds its way on to the apron plate. This mercury catches on the amalgamated surface of the plate and then amalgamates with any free gold or amalgam coming in contact with it. The proportion of mercury fed should be kept such that the amalgam thus formed on the apron plate is pasty. If too much is fed, the amalgam becomes liquid and is apt to gather into globules and run off the plate or scour off easily; if the feed is too slow, the amalgam becomes too hard and does not catch gold well.

31. On an average, from 20 to 30 per cent. of the mercury used in stamp-battery amalgamation is lost—mostly through flouring and sickening. Of course, the loss depends largely upon the character of the ore and the experience and judgment of the amalgamator; in many instances it runs much above 30 per cent., either unavoidably or through carelessness. By careful amalgamation and the use of amalgam savers below the apron plates, it can be reduced to a minimum. The loss of mercury varies from $\frac{1}{2}$ of an ounce to 5 ounces per ton of ore crushed, and in exceptional cases it is more.

DEVICES FOR SAVING FLOAT GOLD AND AMALGAM

32. Step Plates.—One of the chief sources of loss is from float gold and floured amalgam being carried away in the pulp stream without coming in contact with the plates at all. In stamp milling, various devices are adopted to overcome, or at least to mitigate, the losses from this source. The *step plate* is one of the most common of these. The apron plate, instead of being one continuous sheet, is divided laterally into two or more segments, and between the foot of each segment and the head of the next one below there is a step or drop of 1 or 2 inches—just enough to cause a slight splash and free floating particles of gold and amalgam from their buoyant film of air, submerging them and bringing them into contact with the plates, but not enough to scour the plates.

Sometimes the plates are placed directly under one another and zigzag, the alternate plates slanting with the same gradient, but in opposite directions. The pulp stream falls from one plate to the next one below, reversing its direction each time. This arrangement allows the use of a great length of apron plates on a limited floor space, and is advantageous when the gold is very fine, as the drop from plate to plate tends to submerge float gold and amalgam, while the increased length of plate surface gives the suspended particles of gold and amalgam more time to settle. It is rather inconvenient for cleaning up, however, and has not been generally adopted.

33. Shaking Plates.—For saving fine amalgam that may escape the apron plates, shaking copper plates, placed below the apron plates, are much superior to the blanket sluices sometimes used. Sheets of copper—preferably silvered—usually 4 feet square, are set on light frames, with a grade of about $\frac{3}{4}$ inch per foot; the frame is either suspended, or, better, mounted on rocking legs, and is given a rapid side shake by connecting-rods operated by eccentrics on a belt-driven shaft at the side of the frame. The shaft has a speed of from 180 to 200 revolutions per minute. The throw of the table is about 1 inch. At the upper end of the plate, which is about a foot shorter than the table, there is generally a cleat or riffle about ½ inch high, extending from side to side. In case the amalgam on the battery plates gets too hard, any lumps of amalgam escaping from the battery will be caught behind this riffle, the shaking motion of the table rolling them up into little balls, which gradually pick up more amalgam and increase in size, like a snowball rolled in the snow. Two shaking plates will handle the pulp from five stamps, after it has passed over the apron plate.

Shaking plates need not necessarily be used after stationary apron plates, but may take the pulp directly from the battery. The *Gauthier* shaking table is designed especially for this purpose. It has an end shake instead of a side shake, and is practically only a shaking apron plate, mounted on rocking legs and driven by an eccentric on a rapidly revolving shaft extending from side to side, underneath the table. If desired, this table can be used below the ordinary apron plates.

34. Corrugated Plates.—An effort has been made to save flour gold and amalgam by the use of corrugated apron plates. The corrugations extend horizontally from side to side, and form a series of parallel troughs or traps, in which mercury settles and catches the gold and amalgam passing over it in the pulp stream. It is doubtful if corrugated plates have any advantages over the ordinary flat plates. If much gold is escaping as float gold, a second corrugated plate, amalgamated on its lower surface and placed immediately above the first, with just space enough between them to allow the pulp stream to pass through without being backed up at all, will accomplish a considerable saving by amalgamating a good deal of the float gold and amalgam on its own surface, and forcing the rest to become submerged in order to escape it; and once wet, the particles will sink quickly and amalgamate on the lower plate.

35. Mercury Wells.—Mercury wells, or traps, are merely horizontal troughs parallel to the discharge of the mortar; a bath of liquid amalgam is placed in these troughs and the pulp stream either passes over or through this bath. When it passes through the bath, a vertical iron partition is run along the middle of the trough and dips beneath the surface of the bath, forcing the pulp stream to pass under it and up through the mercury on the other side, in order to get past the trap. The mercury-trap system is open to criticism at many points. A comparatively large quantity of amalgam is used in the traps, and the amount of capital locked up in this form is considerable. Again, this disposition of the mercury is not nearly so advantageous for amalgamation as the use of amalgamated plates, since it is much more difficult to secure the proper contact of the pulp with the mercury in the baths than on the plates. This is true even of those wells in which the pulp is forced to pass through the mercury bath, as it goes through in lumps or bubbles, and only a comparatively small portion comes in actual contact with the amalgam bath. The use of mercury wells is being gradually dropped from American stamp-milling practice. Single troughs are sometimes used below the apron plates, where they save some amalgam; but in all cases they could be advantageously replaced by shaking plates. Occasionally a mill is found in which mercury wells are used above the apron plates, but this is bad practice, as it makes it much more difficult to detect overfeeding or under-feeding of mercury in the mortar, conditions which, with the ordinary arrangement of the apron plates, become apparent at once in the condition of the amalgam on the aprons.

In Australia mercury wells are still retained to a great extent. The battery pulp passes through a series of wells and then usually over blanket tables—amalgamated plates, in many cases, not being used at all. In some mills, working rich free-milling ores in which the gold is very coarse, no mercury is used at all except in the clean-up, the heavier gold settling in the mortar and the finer gold being caught on the blankets. The whole practice is rather primitive in many of its details, and not up to the American standard.

36. Swinging Plates.—A form of amalgam saver used in many California gold mills is the swinging plate. Curved plates of amalgamated copper, about 3 inches deep, are hung on wires across the sluice box, with their lower edges dipping beneath the surface of the pulp stream, and the concave, amalgamated side facing up stream. The current keeps the plates swinging gently back and forth, and float gold and amalgam are forced to pass under them to get down the stream. A great deal of the gold and amalgam is thus caught by the plates themselves, while the rest, once wet, will sink to the bottom and catch on the sluice plates or blankets. A ridge of amalgam accumulates on the bottom of the sluice immediately under each plate. The plates are placed a few feet apart along the sluice. They are comparatively inexpensive and go a long way towards correcting the faulty design of the sluice. Straight plates would answer the purpose, but a slight curvature makes them more effective, drawing the pulp into the center and creating an eddy that aids materially in submerging the floating particles.

37. Miscellaneous Appliances.—In the early days of gold milling in California, before the blanket table had given way to the amalgamated copper apron plate, **Atwood's amalgamator** and the **Eureka rubber** were quite generally used below the blankets for the purpose of saving amalgam. The former has now given way almost entirely, with the blankets, to amalgamated copper plates; and the work of the rubber is much better performed by the grinding pan. The blanket concentrates were treated in the Atwood amalgamator. This was merely an inclined table with two horizontal mercury wells or troughs, over which the pulp was run. Two revolving paddle wheels, one over each well, with their blades barely clearing the bath of mercury, forced the concentrates to pass through the mercury, where the gold, on account of its weight, sunk and was amalgamated.   After passing the amalgamator, the material ran over a simple riffle sluice, and any escaping amalgam was caught in the riffles.   The skimmings from the wells of the amalgamator and the tailings from the blankets were passed through a rubber, where they were ground between iron surfaces, cleaning the gold and freeing it from gangue; as soon as it was cleaned it was amalgamated on copper plates in the rubber.   The rubber was merely a flat box with a false bottom of alternate strips of wood and cast iron, extending across the box from side to side; above this was a muller, shod with plates of cast iron similar to the bottom plates, bolted to level blocks; narrow, amalgamated copper strips were fastened on the sides of the blocks.   The muller was hung from four swinging rods, so that the shoes barely cleared the bottom, and was given a short backward and forward motion by a connecting-rod and an eccentric on a shaft at the lower end of the box.   The stroke of the muller was about 4 inches and the faces of the shoes and dies were 4 inches wide.

38.   Besides the foregoing machines, there are numerous patented amalgamators used occasionally here and there, none of which, however, gives any promise of driving amalgamated plates and pans out of the market, although some are founded on theoretically correct principles.   A great many of these make the galvanic or electric current an essential feature, the idea being that galvanic or electrolytic actions keep the mercury clean and lively—which is perfectly true; but mechanical drawbacks prevent the general adoption of these machines.   To this type belong the **Molloy hydrogen amalgamator** and the **Bazin centrifugal amalgamator.**

HUNTINGTON MILL

39.   It is not probable that the Huntington mill will ever generally displace the stamp battery, though it may to a considerable extent.

When it is desired to amalgamate as much gold as possible inside the Huntington mill, the water-supply should be kept down low, but not so as to clog the screens. The pulp then remains in the mill longer before discharging and the gold is given more time to amalgamate. As in the stamp mill, a decrease in the water-supply is attended with a corresponding decrease in the capacity of the mill. On the aprons, a moderately thin pulp, flowing readily, is best; too thick a pulp will clog the plates with sand.

ACCESSORY APPARATUS

40. Amalgam Strainers and Safes.—In silver pan amalgamation mills, where large quantities of amalgam are

handled, the liquid amalgam is poured into an amalgam safe similar to the one illustrated in Fig. 3. The top and bottom of the safe are made of cast iron and the body of wrought iron. The top is concave, with a hole in the middle through which the amalgam drops into the conical canvas bag or strainer beneath. The hole is protected by a raised cap, cast on the cover or bolted to it, to prevent the theft of amalgam. The excess of mercury in the amalgam is strained through the canvas by its own weight and falls into the bottom of the safe, leaving the lumps of nearly dry amalgam in the bags. The strained mercury is drawn off into flasks or reservoirs, or, in the continuous-system mills, is raised by a quicksilver

FIG. 3

pump or elevator to the receiving reservoir. The cover is hinged, and can be lifted to obtain access to the strainer.

The strainer is fastened to a ring, and can be taken out and cleaned.   A door in the side gives access to the bottom of the safe without removing the strainer.   Both this door and the cover are kept locked.

41. Retorts.—Mercury is separated from the gold in the amalgam by distillation.   In small silver or gold mills,

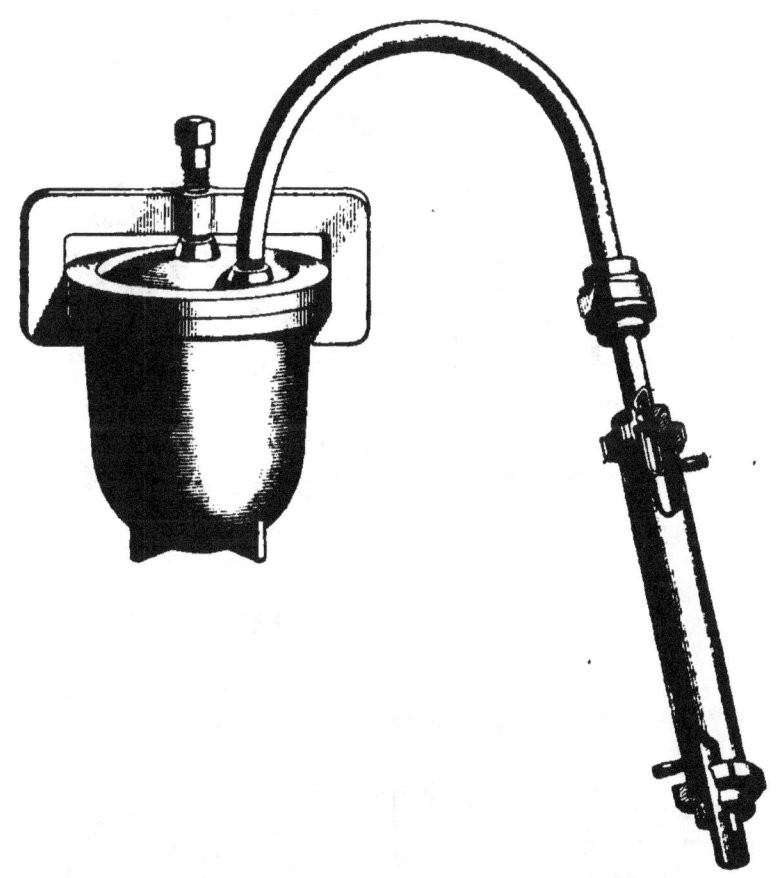

FIG. 4

where the amount of amalgam handled is comparatively small, cast-iron retorts of the type shown in Fig. 4 are generally used; but in larger silver mills, where a large quantity of amalgam is produced, retorts of the type shown in Fig. 5 are necessary.   The small pot retort, Fig. 4, does not require a special fireplace, although one is generally provided, but may be heated in a crucible furnace or blacksmith's forge. A special furnace is, however, always

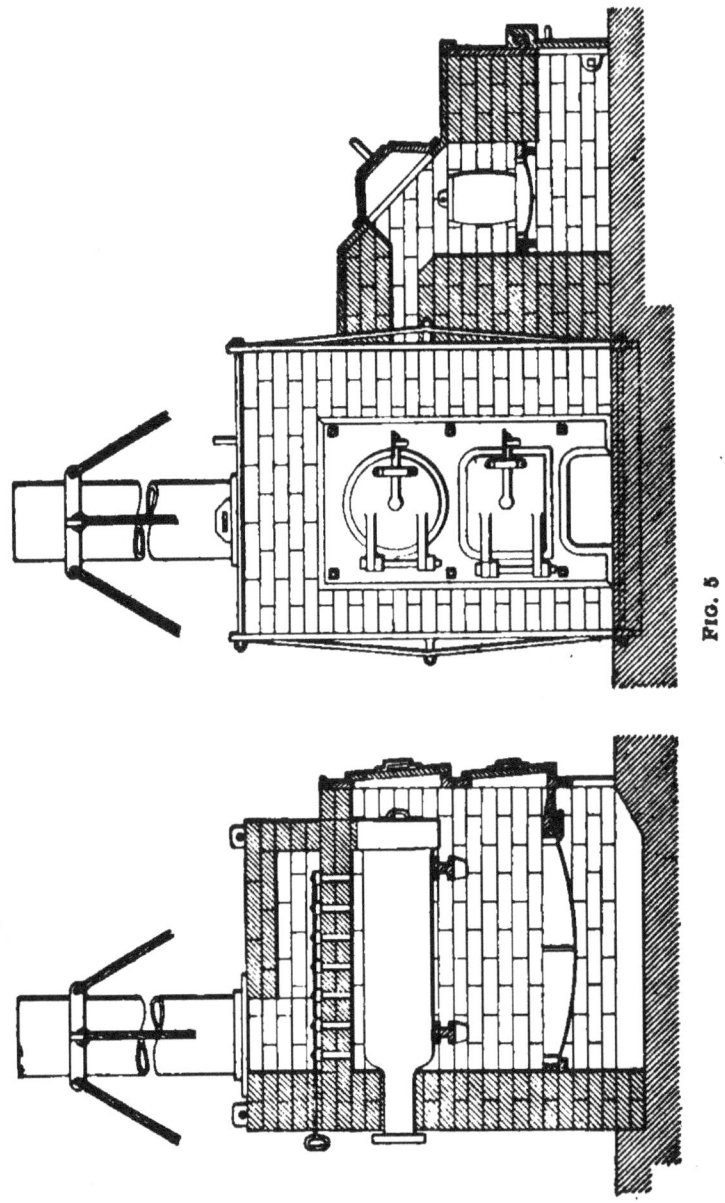

FIG. 5

provided for large retorts. A melting furnace, in which the bullion is melted to be cast into bricks or bars, is frequently built in connection with the retorting furnace, as shown in Fig. 5. The retort, 12″ in diameter, is usually placed immediately above the grate, but where large quantities of amalgam are retorted, if the furnace is left unattended for any time, a retort which is set immediately above the fire is apt to become overheated, and the weight of the metal inside then causes it to sag, ruining it completely. To prevent this, the retort is sometimes arranged with the fire at one side and a fire-bridge between, the retort being supported at several points.

In most modern retorting furnaces, a number of small rectangular openings, connecting the fire-chamber with the flue at intervals along the top of the arch, causes the heating to be distributed evenly along the length of the retort; and the draft can be very delicately regulated and the heat localized, if desired, by the use of individual dampers over these holes. Many furnaces are still built, however, with only a single flue connection at the front end.

42. Charging the Retort.—Before charging the retort with amalgam, the inside surface is chalked or coated with a thin wash of clay or is lined with a few thicknesses of paper, the ashes of which effectually prevent the gold from adhering to the sides of the retort. In large stationary retorts, the amalgam is placed in iron trays which slip into the retort and save much trouble in charging and handling. The lumps of amalgam from the strainers are broken up, placed in the retort or in trays, and pressed down firmly. In many mills the amalgam is packed with the head of a bolt, but most millmen disapprove of this practice, as packed amalgam requires longer to retort and is apt to hold some unvolatilized mercury in the center of the lumps. The condenser pipe should be carefully cleared of all obstructions, and if the amalgam is put directly into the retort, it should be spread evenly and in such manner that by no mischance can this pipe become clogged, as an explosion would be apt to result, filling the retorting room with poisonous mercury fumes and greatly endangering the health and lives of the

men. In retorting impure amalgam containing solid substances which volatilize and recondense in the condenser
tube, clogging is very apt to occur, and the condenser should
be so arranged that a rod may be slipped through the tube
from time to time to keep it open. The heating should
also be very slow at first, as a further precaution against
explosions.

After the retort is charged the cover is put on. The
cover and its seat are carefully faced, and in addition to this
a luting of clay or an asbestos gasket is placed between
the cover and the retort to prevent the escape of mercury
fumes. The cover is held firmly on its seat by clamps,
tightened either by wedges or by clamp screws. The pipe
to the condenser connects with the neck at the back of the
cylindrical retort or screws into the cover of the pot
retort. The condenser is merely a water-jacketed pipe;
a constantly changing supply of cold water keeps the pipe
cool and the volatilized mercury is recondensed and runs
into a basin of cold water at the lower end of the condenser
pipe. Thus there is very little chance of any mercury vapor
escaping condensation. Care should be taken that no water
is drawn back into the retort by sudden cooling, as the steam
generated might cause an explosion. Some millmen use a
rubber or canvas sack over the end of the condenser tube
beneath the water, to avoid risk from this source, the condensed mercury running into this sack.

43. Heating the Retort.—The heat is gradually raised
under the retort until the boiling point of mercury is reached
and active distillation commences. It is kept at this
point for one or two hours, according to the amount of
amalgam, and is then again gradually raised to a bright red
heat and held there for some time, to expel the last of the
mercury. The fire is then drawn and the retort allowed to
cool. After it is thoroughly cooled, the cover is removed
and the metal withdrawn. The trays used in large retorts
are divided into small compartments by partitions, so that
the "retorts," as the masses of retorted metal are called,

will be of a convenient size and form for introducing into the melting crucible without breaking up. The retorted metal is porous and spongy, and usually contains a considerable proportion of impurities. It always retains a small amount of mercury, which is only expelled in the final melting.

44. Melting the Bullion.—The melting is done in clay or graphite crucibles, with borax and bicarbonate of soda; and if the "retort" contains much sulphur or base metals, a little niter is also used to oxidize these impurities. The fluxes aid in the fusion and slag off the impurities. The fluxes are added a little at a time; as soon as their action has ceased and the slag becomes quiet, it is skimmed off and more flux added. This is continued till the surface of the melt remains perfectly clear and shiny, when the crucible is withdrawn and the bullion quickly poured into an iron ingot mold, previously warmed and greased on the inside with heavy mineral oil or beeswax. Most large mills do their own melting and refining, but many small mills sell their "retorts" to private refineries or directly to the mint.

45. There should be a "hood" above the melting furnace to carry away the fumes that arise when the crucible is uncovered for skimming and prevent their spreading through the room. This precaution is very frequently neglected, but the many cases of salivation among the melters are proof of the necessity of observing it.

GENERAL MILL ARRANGEMENT

MILL SITE

46. Gravity Assistance.—One of the first considerations in the erection of a gold, silver, or concentrating mill, next to the certainty of an ore supply to keep it running, is the selection of an advantageous location.

In order to avoid mechanical handling of the ore and to keep the expense of milling down as low as possible, the mill

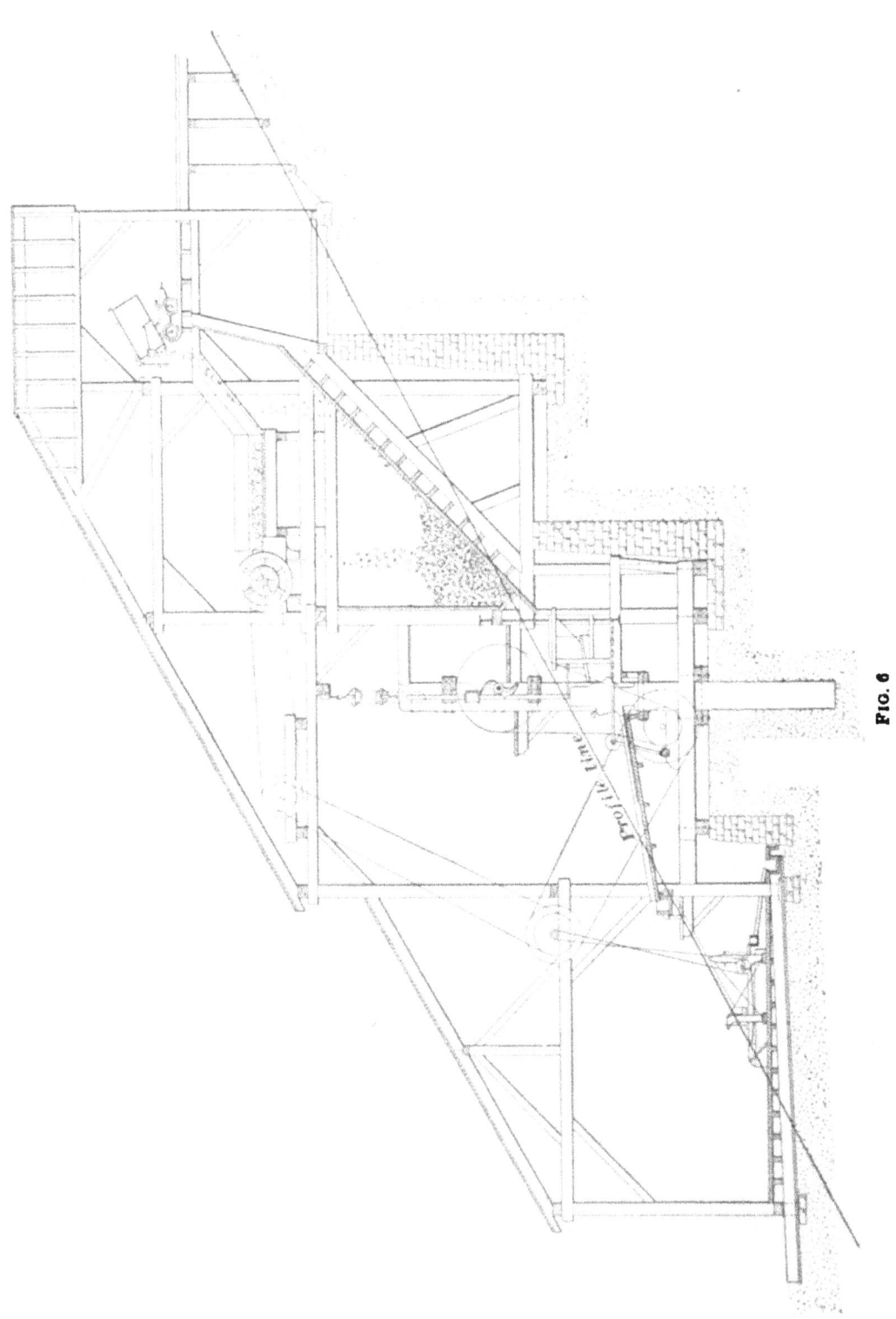

FIG. 6

designer takes advantage of the force of gravity and places the successive machines at successively lower levels, so that the material runs directly from each machine to the next one in order. To secure the necessary difference in elevation between the crusher and the final apparatus for this arrangement, without building the back of the mill very high, it is always desirable to place the mill on sloping ground. The slope should be chosen to correspond as closely as possible with the calculated slope of a line from the gates of the ore bins to the tailings discharge of the mill, in order to avoid all unnecessary building or excavation. When ground has to be cut away, strong retaining walls should be built at the back of the excavation and between the benches, as shown in Fig. 6, to prevent caving.

47. General Arrangement of Buildings and Apparatus.—Mills are usually arranged so that the mine cars or skips, or the ore wagons, if the ore has to be hauled to the mill, can run into or alongside the mill on an elevated track or staging (see Fig. 6). In most modern mills the ore is dumped on to grizzlies, and only the coarser lumps go to the rock breakers, the smaller stuff falling through the grizzlies into the ore bins below; this greatly lightens the duty of the crusher. Many mills are still found, however, where all the ore entering the mill is put through the rock breaker, regardless of its size. The coarse ore from the grizzlies passes on to the crushing floor, or, in most large modern mills, to a coarse storage bin, the gate of which opens upon the crushing floor. By keeping a supply of coarse ore in this way, the crusher may be kept steadily at work, and the power used by the mill kept more nearly constant. This is particularly desirable in concentrating mills where vanners are used, as these machines are very sensitive to change of power; a variable power makes their regulation much more difficult and renders constant attention necessary; and even if every possible precaution is observed, they will not do nearly so good work as when running under uniform power. In small mills the power consumed by the crusher is often about one-fourth of the total power of the mill, so that throwing it in and out makes a decided difference in the speed of the other machinery. With large mills this is less important than in small mills, but, nevertheless, it is a notable factor in the working of a mill.

48. Rock Breakers.—The mouth of the rock breaker is set level with the feed floor, so that the ore can be shoved into it and need not be raised, thus saving the feeder much work. Gyratory crushers are gradually displacing jaw crushers for large mills, both on account of their great capacity and the comparatively small jar and vibration. In modern milling practice, the rock breaker is frequently placed at the mine, and the ore comes to the mill bins already crushed. This relieves the mill of the strain and jar of the crusher and makes the consumption of power, and consequently the running of the mill, more uniform. The proper place for removing rock from ore is at the mine, and this can be better accomplished there when the ore is passed through a crusher.

49. Ore Bins.—The sills of the framework of the ore bins should all be on the same bench or terrace and should not be set on different levels along the slope. The bottom timbers of the bin proper are usually set sloping at an angle of about 45 degrees towards the gate, so that the ore will run down to the gate by its own weight. Bins are sometimes built flat-bottomed, but this necessitates shoveling the ore to empty the bin and thus offsets the increased capacity. The bins are double-boarded with heavy planks, usually with a layer of building paper in between to prevent the loss of fines. The inside bottom planks should be laid lengthwise down the slope, as they wear better this way and the ore slides more readily. Oak, beech, and birch make good ore-bin floors, the ore sliding over them making them smooth. The bin linings should be renewed as fast as they wear out. When large amounts of ore are handled through the bins, they are frequently lined with plates of iron. Owing to the fact that ore slides better on iron than on wood, it is possible to give the bottoms of the bins a 35° rather than a 45° slope, and hence a somewhat larger storage capacity may be had when iron linings are employed.

50. Water Tanks.—In most mills the water supply runs into wooden or iron tanks—usually circular and from 8 to 20 feet in diameter—and is drawn from them as desired. By this means a practically constant head or pressure is obtained, and there is always a reserve supply of several thousand gallons—enough to run the mill for several hours if necessary. Some mills use two tanks, one of which is filling while the other is in use. These tanks are usually set outside of the mill, on the ground. In cold countries, however, this is not always practicable, as the tanks would freeze up during the cold weather. In such cases, the tanks should be put in a separate room with its floor sills independent of the rest of the framework of the mill; or if set in the main building, they should at least be set on independent timbers; for if the tanks are set on the mill timbers, the jar of the crusher and other machinery is communicated to the water in the tanks, sets it in rhythmic motion, and the vibration of this immense weight of water is transmitted to the mill timbers, and will, sooner or later, if continued, rack the building to pieces.

51. Amalgamating Mills.—The general arrangement of the rock breakers and ore bins is practically the same for all classes of gold and silver mills. Below the bins, however, the machinery and arrangement vary with the amount and nature of the work required of the mill. Thus, the machinery of a concentrating mill differs in kind and in arrangement from that of an amalgamating mill, as will be seen by comparing Figs. 6 and 8. Of course, all mills should be designed to make the operation as nearly as possible continuous and automatic. For instance, in an amalgamating mill the ore bins discharge directly on the feed floors of the stamp battery, Huntington mill, or whatever fine-crushing machine is used; or, if automatic feeders are employed, into the hoppers of the feeders. The battery or mill discharges on to the apron plates and the pulp flows from them directly on to any subsequent gold- or amalgam-saving apparatus that may be used, and which is on a level 3 or 4 feet lower than the battery floor. If the mill is a combined amalgamation and concentration mill, as in Fig. 6, the concentrating apparatus—vanners, bumping tables, or similar machines—is put on the floor below the battery floor. If hydraulic classifiers are used, they can be suspended from the roof timbers or set on frames, usually parallel to the battery

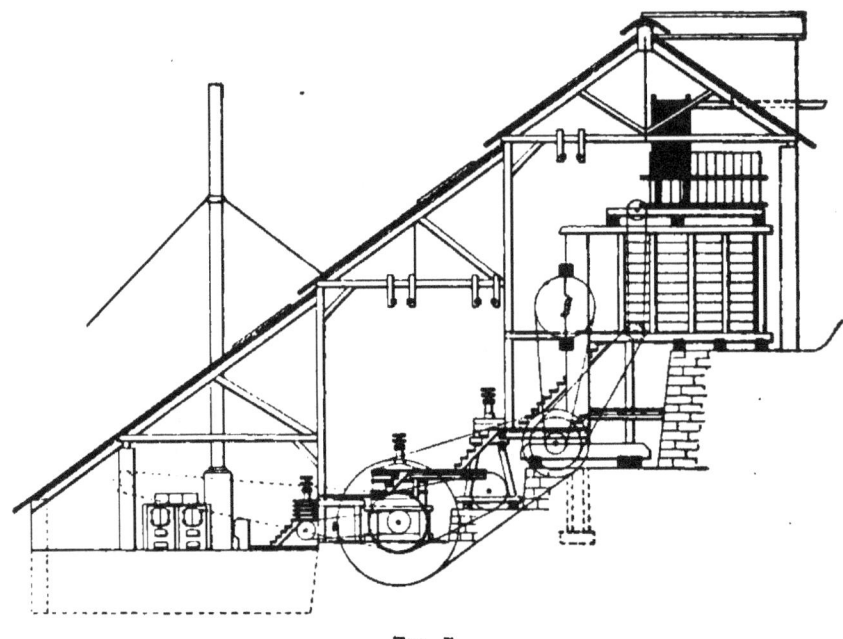

FIG. 7

discharge, and receive the pulp directly from the plates, discharging the sized ore through pipes into the distributing boxes of the concentrators. Slime-saving apparatus below the vanners or other concentrators is seldom used, but the tailings water may be run into large settling vats and the slimes settled out. This is particularly applicable in dry countries, where the water supply is limited, as the water from the settled tailings may be pumped back to the tanks and used over again, with a loss of perhaps 20 or

25 per cent. The tailings, if they contain much value, may be treated by the chlorination or by the cyanide process. In the very dry regions of Australia, the water is sometimes removed from the tailings by means of filter presses. The design of silver amalgamating mills is still different from that of gold amalgamating and concentrating mills. Fig. 7 shows a Boss continuous-process mill in section.

52. Concentrating Mills.—The object of concentrating works is merely to get the values in an ore into smaller bulk, in order to diminish the trouble and expense of shipping and further treatment, and not for the immediate actual extraction of the metals in the ore.

The operation is purely mechanical, the ore and gangue being separated by crushing, and the gangue, owing to its lower specific gravity, being washed away. This being the case, crushing and careful sizing become highly important.

53. No definite scheme can be laid down for the arrangement of concentrating mills. This depends largely upon the nature of the ore and still more upon the ideas of the designer. Several methods, each requiring different apparatus and arrangement, may be equally well adapted to the concentration of an ore, and the selection of any one method lies with the designer, who is supposed to take into consideration local conditions as far as possible. Thus, local factories, if there are such, are usually given the preference, if their machines can compete on anything like equal terms with those of outside manufacturers. The personal preferences and prejudices of designers are frequently important factors in the designing of mills.

The concentrating gold mill is for several reasons usually much simpler in design than concentrators for copper, lead, and zinc ores. In the first place, the out-of-the-way location of the average gold mill makes the freight on apparatus an important consideration in the first cost and running expenses; and, again, such mills are usually only temporary structures, doomed to abandonment as soon as

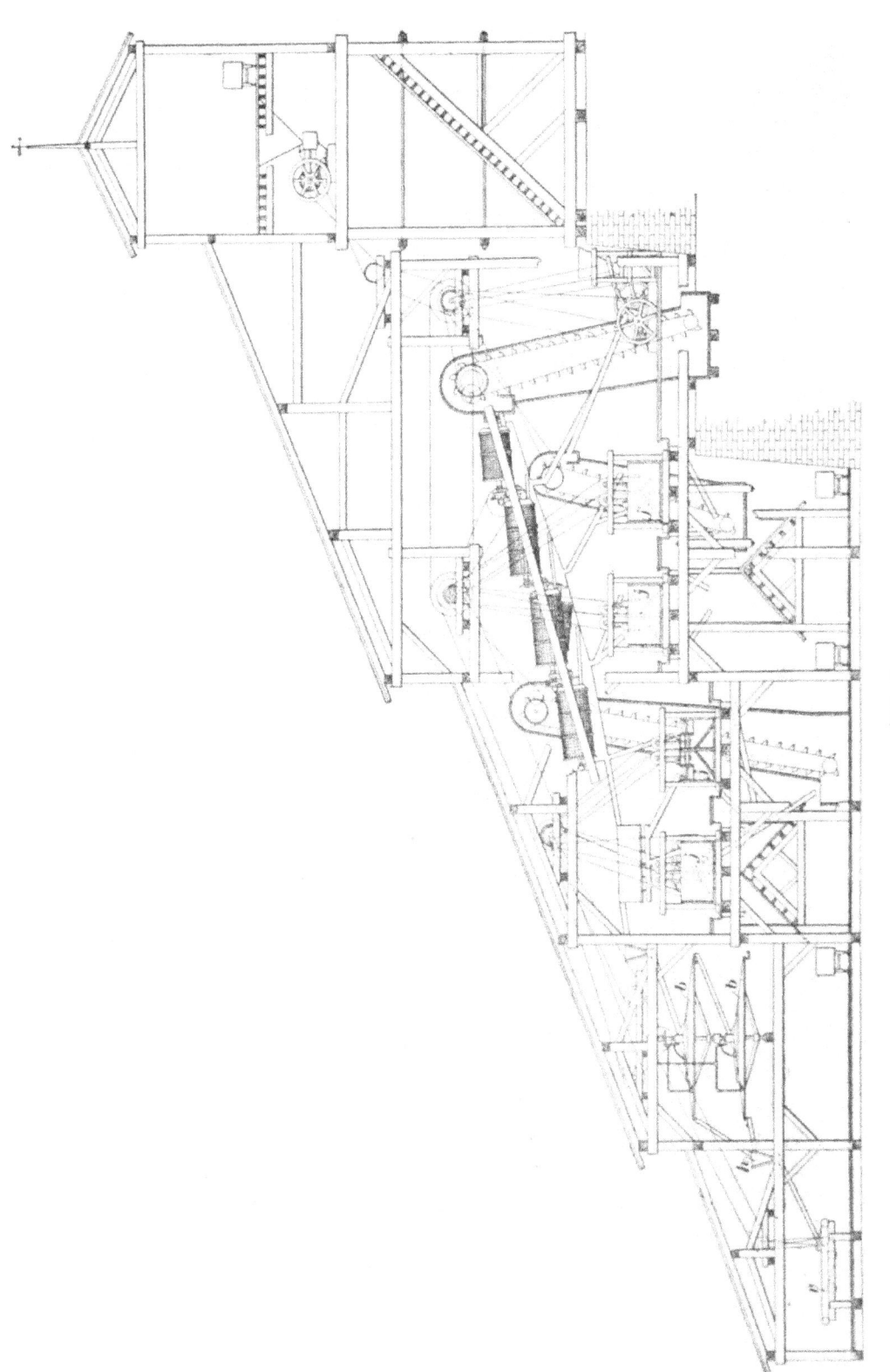

FIG. 8

the ore body is exhausted. As a rule, gold mines are exhausted after being worked continuously for a few years, and the ore body may play out unexpectedly at any time, so that it is desirable to put as little extra expense into the mill as possible. It is seldom worth while to dismantle an old mill. Nevertheless, the mistake of putting too little apparatus in a mill is much more common than that of putting in too much. Additional machines, if of good design and within reasonable limits, will usually pay for themselves.

With immense low-grade deposits, like those of Dakota, Idaho, and Douglas Island, Alaska, it will usually pay to put in more elaborate concentrating plants, as a very small increase in the saving per ton counts up rapidly where several hundred tons of low-grade material are being treated daily and the ore bodies are practically inexhaustible. Fig. 8 is a cross-section of an Idaho concentrating mill, showing jigs *j*, hydraulic classifiers *h*, buddles *b*, and vanners *v*. In the Butte (Montana) copper region, concentration has reached its highest development in America. The ores of this region, though containing small quantities of gold and silver, are essentially copper ores, the gold and silver being obtained merely as by-products.

54. Roll Crushing Concentrating Mills.—Roll crushing is almost invariably adopted in concentrators, though steam stamps have replaced rolls to a considerable extent in the Montana and Lake Superior copper regions. The usual arrangement of concentrating mills is somewhat as follows: The grizzlies, rock breakers, and bins are practically the same as for amalgamating mills. From the bins the ore goes to the coarse or *roughing* rolls or to a second rock breaker, set closer than the first, which fills the place of the roughing rolls. In many mills a second rock breaker is placed between the main rock breaker and the rolls to lighten the duty of the latter. In such a case, the product of the first rock breaker usually goes to a trommel whose meshes correspond to the maximum size of the product of the second breaker, and the material which is already fine enough to pass the second crusher is taken out, only the oversize product of the screen (the portion which will not pass through the screen) going to the second breaker.

The product of the second crusher is elevated by a belt or chain elevator back to the trommel, which is the first of a series of three or more. The undersize product of trommel No. 1 includes the undersize product from the first crusher and practically the entire product of the second crusher; this falls into the hopper beneath the trommel and passes through a chute into trommel No. 2, whose meshes correspond to the maximum size of the product of the roughing rolls. The oversize from this trommel goes to the roughing rolls, while the undersize goes to trommel No. 3, whose meshes correspond to the maximum size of the product of the fine or *finishing* rolls. The product of the roughing rolls is elevated back to trommel No. 2 and rescreened. The oversize from trommel No. 3 goes to the finishing rolls.

The product of the finishing rolls may be elevated directly back to trommel No. 3 or may be taken off by a chute— or, if the rolls are on the same level, by a horizontal traveling belt—and combined with the product of the roughing rolls and elevated with it to trommel No. 2. The latter arrangement saves one elevator, but it throws more work on trommel No. 2, without appreciably lightening the work on trommel No. 3, and necessitates a horizontal traveling belt or chute between the roughing and finishing rolls.

The undersize from trommel No. 3 goes out to the next machine. In most gold-concentrating mills this is the coarse jig, but in many large mills the undersize from trommel No. 3 is carried to a fourth trommel, only the oversize from which goes to the coarse jigs; the undersize from trommel No. 4 may be further sized by going through more trommels, each additional trommel giving another jig size, or it may be carried directly to the intermediate jigs and the work of sizing thrown upon them.

55. When ore is crushed fine and sized through screens, the undersize from the last screen usually goes to hydraulic classifiers, which remove the slimes. The spigot discharge of the classifiers is carried to the finishing jigs, and the overflow, with the slimes, goes to settlers, where the superfluous water is removed, and then to the slime concentrators—vanners, buddles, etc. In the Butte concentrators, trommels are replaced to a large extent by classifiers of the type shown in Fig. 11, *Ore Dressing and Milling*, Part 2, with two or more spigots, the discharge from the spigots going to the respective jigs or to vanners, and the overflow going either to the settling tanks, and thence to the slime concentrators, or to waste. This arrangement is common among those mills using steam stamps instead of rolls for the comminution of the ore.

56. Treatment of Jig Products.—The treatment of the jig products depends upon the character and grade of the ore. When the mineral occurs in bunches, rather sparsely scattered through a clean gangue from which it separates readily, the ore is usually crushed rather coarse, and after screening out the fines, goes to the coarse jigs. The clear mineral headings from these jigs are a finished product containing very little gangue, and go to the drying floor, and thence to further treatment for the extraction of the values. The tailings from these jigs are usually quite clean and go to the tailings dump. The middlings are mixed gangue and mineral and are recrushed in fine rolls, stamp batteries, or some patent mill, like the Heberle, Sturtevant, or Huntington, to further liberate the enclosed mineral. In the recrushing of coarse-jig middlings such as we have been considering, rolls would be preferable, and the recrushed material would be sized by trommels, the oversized going to the next jig below in the series, and the undersized going on to the finer jigs and subsequent apparatus.

57. When the mineral is distributed quite uniformly through the gangue, and particularly when the gangue is tough and intimately associated with the mineral, the crushing must be much finer to begin with than in the previous case, in order to secure a clean separation, and we will get a small headings class, a large middlings class, and a more or less rich tailings class. Frequently, in working medium-grade ores, it is found advisable to recrush the jig tailings from the coarse jig and sometimes from the fine jigs as well, in some fine-crushing machine, and then classify and concentrate them on vanners, buddles, etc. As the grade of the ore improves, other conditions remaining the same, the loss in the tailings, of course, increases.

MISCELLANEOUS APPARATUS

58. Elevators.—Belt elevators and link-belt (chain-and-sprocket) elevators are largely used in mills for automatically raising the material from the crushing machinery to the screens, or to samplers, etc., or, in general, for delivering material to higher levels. These elevators are merely continuous belts, to which sheet-iron or steel buckets are fastened at intervals. The belts in gold and silver mills are uniformly run at a speed of about 200 feet per minute, the capacity of the elevator being regulated by the size and spacing of the buckets.

The belts commonly used are 5- or 6-ply rubber belts or link belts. Leather belts are used to some extent in dry-crushing mills, but would soon stretch out of shape if used for elevating wet material. When elevating hot material, such as ore from driers and roasters, chain-and-sprocket elevators are used exclusively. The buckets are fastened to rubber and leather belts by countersunk rivets. The belts run on pulleys (or sprockets) on countershafting. The entire apparatus—unless it is to be used for handling hot material—is enclosed in a tight wooden casing or housing, to prevent splashing or dust, as the case may be, with their attendant inconveniences and loss of material. The material falls into the boot of the elevator—which is usually

made, like the rest of the casing, of wood, but sometimes of heavy sheet
iron—and the buckets
scoop it up, elevate it,
and discharge it into a
chute or spout leading
to the next piece of
apparatus, as shown in
Fig. 9.

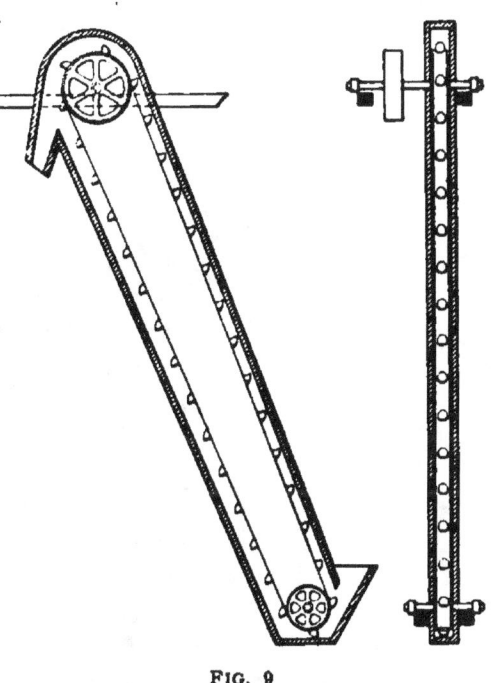

FIG. 9

Chain-a n d-sprocket
elevators are of two
general types—single-
chain and double-
chain. In the first
type, the buckets are
bolted at their backs
to the links of a sin-
gle chain, while in the
second type the buck-
ets are hung between
two parallel chains. Link-belt machinery has been generally
adopted for conveyers of all kinds.

59. Sand Wheels.—Frequently the tailings-discharge
opening of a mill or concentrating works in time becomes
blocked by the tailings backing up from the dam or tailing
dump, so that it becomes necessary to lift the tailings as
they leave the mill. This may be accomplished by a bucket
elevator similar to that shown in Fig. 9, but when an elevator is employed, it requires a belt, and the expense for the
belts in the long run is considerable. On this account sand
wheels are frequently employed. They are really nothing
but overshot waterwheels in construction, which are given a
reverse motion by machinery, so that in place of the
descending water operating the wheel, the rotating wheel
lifts the water and tailings in its buckets. The buckets are
placed on the inside of the rim of the wheel, as shown in
Fig. 10, and are filled from launders or sluices *a* at the

bottom and discharged into launders c at the top, which carry the tailings to the dump tailing dams, or lixiviating vats. The wheel in the illustration is one used in South

FIG. 10

Africa. The spitzkasten d shown in the discharge launder is for the sulphurets or heavy particles of ore to settle in, while the lighter particles go to their proper receptacles. The sand wheels are driven by belts or gearing.

60. Sand Pumps.—Sometimes tailings are discharged by means of centrifugal sand pumps, which force the material mixed with water up and into the launders on a higher level, so that it can flow away.

In cases where plenty of water is available at the mill, the tailings may be removed by means of an hydraulic elevator, which is really a water ejector, and which takes advantage of the force of a comparatively small stream of water under a great velocity and makes it move a large stream of water and sand at a comparatively slow velocity and raise it to a moderate height. This device is frequently employed for handling the tailings in placer-mining work, and will be found equally efficient for handling the tailings at mills.

61. Discharging Tailings Without Water. — In regions where water is scarce, some arrangement must be made for removing the tailings after the water has been drained or filtered out of them. This may be accomplished by means of a railroad track and cars, or by conveyers, either of the chain-and-bucket or endless-rope pattern. One great difficulty in regard to all forms of conveyers has been that as the tailings pile increased, it became necessary to lift either the track or conveyer so as to keep it from being buried under the

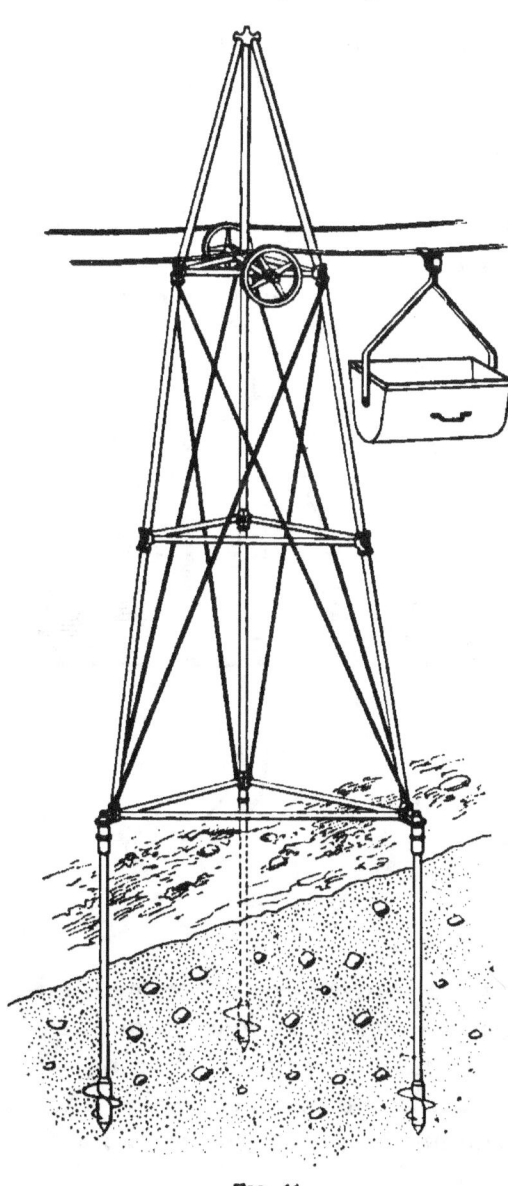

FIG. 11

tailings.   To overcome this, tripods carrying the conveyer may be supported on screw piles, which can be lifted as fast as the pile of material grows.   One of these adjustable supports for a conveyer is shown in Fig. 11.

62.   Differential Pulleys.—Differential pulleys are indispensable around mills for lifting heavy apparatus, such as stamps, mullers of amalgamating pans and settlers, etc.   By their use one man can raise easily, and with no risk of dropping and breaking, weights which several men could not move by main strength alone.

63.   Crawls.—Overhead crawls or tackle-block carriages are merely movable hangers or supports for the differential pulleys.   They are usually made entirely of iron, the

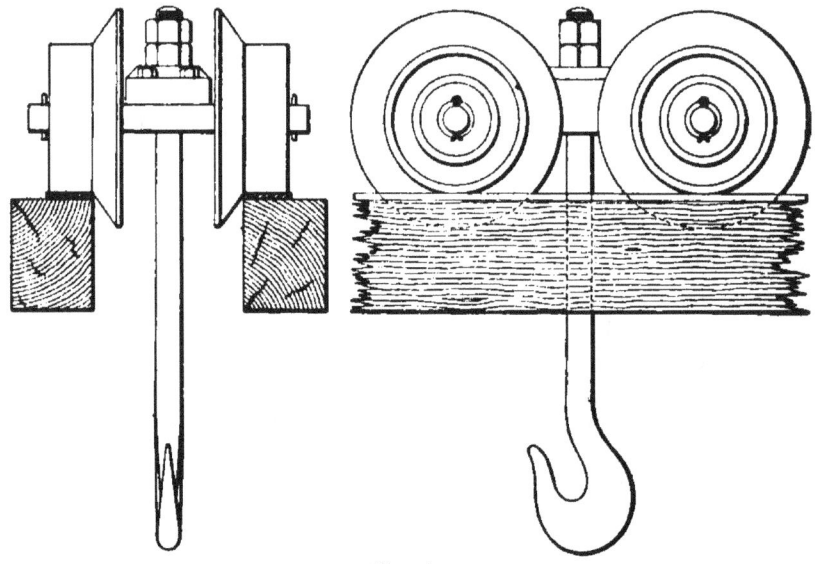

Fig. 12

most common form being the four-wheeled carriage shown in Fig. 12.   There are various other forms in use, however,—four-wheeled carriages, with flanges outside, running on a single horizontal timber, with iron or steel strips for tracks; two-wheeled carriages running on a single track; and a single wheel carrying a hook.

The tracks on which the crawls run are suspended from the timbers above the stamp batteries, pans, etc., so that the crawl with its pulley can be run back and forth to any point, and the stamp or muller lifted and swung out of the way. One crawl and one pulley are supplied for each row of apparatus. The tracks most commonly used are made by fastening flat steel or iron strips by screws to heavy horizontal timbers, which are suspended by wooden hangers from the overhead frame timbers. Flat iron bars, suspended edgewise by iron hangers, are also common.

64. Exhaust Fans.—Exhaust fans are used in all modern dry-crushing mills to draw off the dust, more or less of which will escape even from the most carefully housed machinery. The fans are placed at advantageous points in the upper part of the mill and keep up a draft through the mill, drawing in the dust and discharging it outside or into a settling room. They are indispensable to the health and comfort of the workmen in dry-crushing mills. In addition to these, fans are connected sometimes by pipes directly to the housing of the machinery, drawing off the dust before it can get out into the mill.

SPECIAL EXAMPLES OF CONCENTRATION

65. Concentration and Preparation of Copper Ores. The concentration of copper ores naturally divides itself into two distinct methods: (1) The concentration of those ores in which the copper occurs in metallic form, as, for instance, the ores of the Lake Superior region in the United States. (2) The concentration of ores in which the copper occurs as a copper mineral, usually one of the compounds with sulphur or oxygen.

66. Native Copper Ores.—In the first case, the copper is all in metallic form, and hence it is practically impossible to produce slimes from the metal itself; also the great difference in specific gravity between the copper and the associated gangue renders the separation much easier than

is the case with the copper minerals.   Owing to these facts,
a special system of concentration has been developed in the
Lake Superior region.   The old gravity stamps have been
abandoned and heavy steam stamps introduced.

The copper ore as it comes from the mine is, to a certain
extent, hand sorted in order to remove any large masses of
the metal.   The ore is all stamped through coarse screens
and is sized by means of hydraulic classifiers.   As a rule,
no screens or trommels are used for sizing the material,
which upon leaving the hydraulic classifiers passes to Collom jigs.   These jigs usually produce three classes of material: fine concentrates passing through the jig sieve and into
the hutch box under the jig (commonly called hutch work);
barren tailings over the end of the jig; and a bed of mineral
on the sieve, composed, in the case of the coarse jigs, mainly
of nuggets of metallic copper; and on the jigs farther down
in the series, a mixture of gangue and metallic copper, called
ragging, which requires further crushing.   The beds of the
jigs are cleaned out at intervals, the metallic copper being
placed with the concentrates, and the ragging being returned
to the stamps for further reduction.   The hutch work
passes to other jigs or settling boxes and is worked over
again.   The slimes passing through the screens of the finishing jigs are worked on buddles or slime tables and in tossing
tubs.   The slime-concentrating machinery at many of these
copper mills appears very complicated, and yet it is mostly
composed of such simple machines as buddles and keeves,
which have very few parts requiring renewal.

67.   Ores Containing Copper Minerals.—Ores of the
second class present a very different problem, for the copper mineral naturally crushes finer than a large portion
of the gangue rock.   This results in the production of a
great percentage of slimes.·   The engineers in charge of
concentration works in the West have followed two lines in
dealing with this class of material.   In some cases they have
introduced the steam stamp on account of its great capacity
and the low cost per ton for which it will crush the material.

Where steam stamps have been introduced, an attempt has been made to follow the Lake Superior system of concentration as closely as possible; but most mills have introduced some more expensive machinery for dealing with the slimes (such as Frue vanners and other special concentrating machinery), in addition to buddles or slime tables. They have also been forced to introduce large slime pits, in which an attempt is made to catch the valuable portions carried off in the fine slimes. The overflow from these pits is carried out of the mill and saved behind dams. The pits are cleaned and the material mixed with lime before it is fed to the furnaces. The material which settles behind the dams is cleaned out every few years and either allowed to dry in solid cakes or is mixed with lime and charged into the furnaces. This method of handling slimes is very expensive and leaves a large amount of copper locked up for months or years before it is recovered.

68. The other general method followed by engineers in charge of this class of concentrating works is that of successive reduction and separation, the crushing being accomplished by means of rock breakers and rolls and the material being sized by means of trommels or screens. Hydraulic classifiers are also used to make intermediate classifications. The material is passed over jigs, and any of the middle products (corresponding to the raggings of the Lake Superior ore) are recrushed by rolls or special machines (such as Huntington mills). The jigs used for the coarse concentration work are usually of the Hartz pattern, those for the finer work being of the Collom, Evans, or Slide pattern. The work is carried on very much as described under the heading of "Concentrating Mills."

69. Points to be Observed in Concentration.—In concentrating any ore of copper, the object is to produce a product suitable for the copper furnace, and on this account iron pyrites is not unwelcome, as it will assist in forming the matte, and the iron is useful in the subsequent processes of treating the matte. In concentrating ores of tin, lead, or zinc, it is of considerable importance that the different minerals be separated, for iron and lead will injure the retorts or furnaces in which the zinc is treated, while zinc in the lead furnace renders the smelting very difficult and tends to carry off both lead and silver as fumes or to carry them into the slag. When it becomes necessary to separate two minerals which have specific gravities varying but little, the material must be closely sized before it is passed to the jigs or other concentrating machines.

70. Before any ore can be concentrated, it should be crushed to such a size that the grains or crystals of the different minerals are set free, and the first crushing should be such that the average size of the product is the average size of the mineral grains. The result of concentrating such a product will be pure grains of the mineral and barren tailings; also a third or intermediate product, consisting of particles which contain both the mineral and the gangue and require further crushing before they can be concentrated. This rule applies equally to the methods of dry, magnetic, and wet concentration.

71. When gold and silver are present in the ore to be concentrated, they may have a decided effect upon the method pursued, for if the gold or silver occurs in one particular mineral, the concentration will be carried on with an idea of saving the greatest possible percentage of that mineral.

72. Concentration of Lead Ores.—Lead ores which occur pure, that is, free from other metals, are frequently hand-picked or washed on hand jigs to separate them from the gangue. Where the ores occur in somewhat harder formations, they require crushing and sizing previous to jigging, and if the mineral is finely disseminated through the ore, fine crushing and close sizing will be necessary, especially if zinc ores occur associated with those of lead.

Some concentrating mills also use buddles or other slime-working machines for concentrating the fine ore. Where the lead ores are associated with some zinc ores or with iron pyrites, a separation may be effected by closely sizing the ores and then carefully jigging them, but unless the zinc crystals are comparatively coarse, it becomes difficult to make a first-class separation by hydraulic means. There is only one mine in the United States worked for lead alone, and that is Mine LaMot in Missouri. Most lead is derived from silver lead ores.

73. Concentration of Zinc Ores.—Where the ore of zinc is blende, either associated with lead ores or occurring by itself, it may be separated to a certain extent under certain conditions only by picking and by crushing and jigging. When the zinc ores contain iron or manganese minerals, they may be magnetically concentrated. Where the objectionable material is franklinite, it may be separated from such minerals as willemite or zincite by means of magnetic concentrators, as is done at the New Jersey mines, the oxide being removed and employed for the manufacture of zinc pigment; the residue from this process being used in the manufacture of spiegeleisen. The non-magnetic portions, consisting mainly of willemite, are used for the manufacture of zinc. The non-magnetic portions carry more or less gangue material with them, and this has to be separated by means of ordinary wet concentration on jigs. Some ores of zinc have been freed from iron by roasting the ore until the iron is rendered magnetic and then separating it on magnetic concentrators.

74. Concentration of Tin Ores.—Tin ore should be free from other compounds before it is introduced into the smelting furnace. These facts render the concentration of such ores somewhat more difficult, but by taking advantage of the difference in their specific gravities, the problem is by no means impossible.

Ordinarily tin is concentrated by hand-sorting or crushing and then stamping the ore, after which it is sized and separated by means of jigs, buddles, and keeves. The concentrates consist of tinstone and minerals carrying iron associated with sulphur and arsenic. The sulphur and arsenic are driven off by roasting, after which the iron oxide is removed by further concentration. The fact that tinstone is already in the form of an oxide keeps it from being changed during an oxidizing roasting.

When tinstone is associated with large amounts of mica and close sizing is not desirable, most of the concentrates of the jigs are formed as hutch work; that is, the concentrates pass through the sieves of the jig and the tailings over the ends, while the material forming the ordinary concentrate, discharged over the bed, is middlings or raggings, which require further treatment.

75. Concentration of Mercury.—As a rule, the ores of mercury are separated entirely by hand picking and sorting, no concentrating machinery being employed.

76. Concentration and Preparation of Iron Ores. Most of the iron ores need no preparation before they are charged into the furnace, but there are great deposits of lean ores, or ores containing certain ingredients which it is desirable to remove, and various processes have been introduced to prepare these ores for smelting. These operations or processes may be described as follows:

1. Separation of the ore from barren rock or gangue by means of ordinary wet concentration.

2. The separation of ore from clay by washing.

3. The elimination of sulphur or carbonic acid.

4. Magnetic concentration.

77. Wet Concentration and Sorting.—Under the first operation can be considered all ores in which the iron mineral is fairly hard and the gangue consists of quartz or other worthless material. Such ores are commonly cobbed and separated by hand. For this purpose, special sorting floors are sometimes provided; the best or clean ore being picked out in the mine, goes directly to the cars or stock pile, while the ore which is mixed with more or less gangue is sorted into two or more classes of merchantable ore (the number of classes depending upon the percentage of ore in each), and a worthless class composed of the barren rock or gangue. The sorting or picking may be done on floors from tables, or picking belts. A great many plants have been introduced for the preparation of ore by ordinary wet concentration, using rock crushers and rolls to reduce the material, screens or hydraulic classifiers to size or sort the crushed material, and jigs or other concentrating machines to separate the ore from the worthless material. Owing to the extremely low price per ton which iron ore brings at the present time, it is impossible to concentrate most low-grade ores at a profit. This is especially true in cases where both the ore and the gangue are hard and cause excessive wear on the rolls, crushers, and other machinnery, besides requiring a large amount of power. The expense per ton is often more than the price of the ore warrants, and as a result nearly all of the plants which were operating upon the harder non-magnetic ores, by means of wet concentration, have been closed.

78. Iron Ores Containing Clay.—Many of the iron ores, especially those of the Eastern and Southern States, are associated with more or less clay. A number of machines have been devised to wash this material from the ore, but the primitive log washer has developed into a form which seems the best adapted for this purpose. Figs. 13 and 14 illustrate a log-washing plant that has been in successful operation for a number of years. Fig. 13 is a front and Fig. 14 a side elevation; the engine is not shown in the plant, but is an automatic Buckeye giving 25 H. P. with 60 pounds of steam when running at 285 revolutions per minute. The plant is driven by a 12-inch belt passing over a 3-foot pulley on the engine shaft. The washer proper is driven with 6-inch rubber belts running over the pulleys D and E, Fig. 14, F and G in the same figure being the loose pulleys on to which the belts can be shifted when

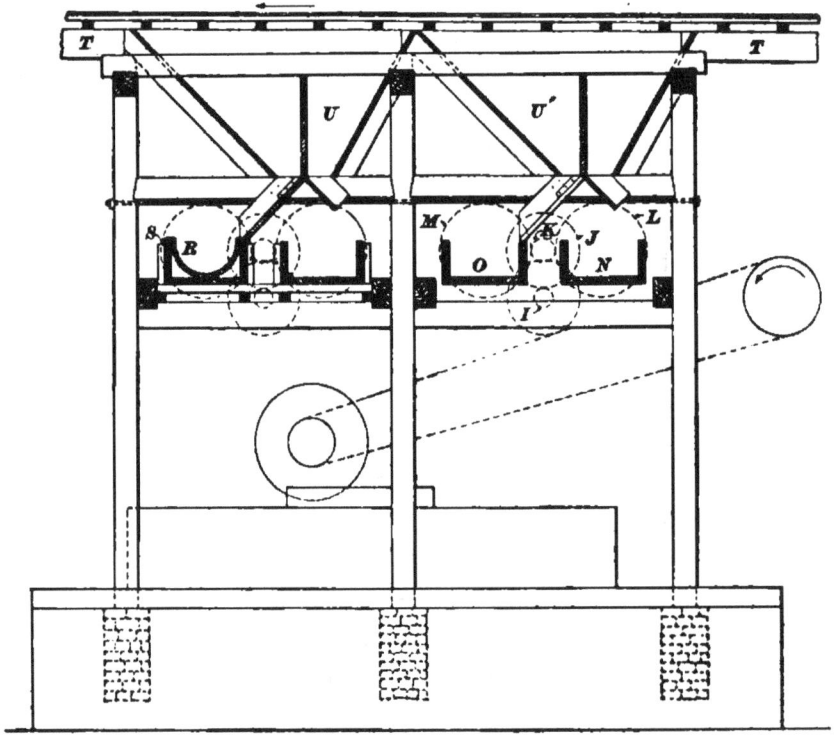

FIG. 18

it is desired to stop the engine. The logs are arranged in pairs, and as they are alike, a description of one pair will be sufficient.

To the end of the shaft H, Fig. 14, on which is fastened the pulley D, is keyed the small pinion I, which meshes into the spur wheel J. This drives another pinion at K, and this in turn gears into the spur wheels L and M, which drive the logs in the two washers N and O at the rate of 12 revolutions per minute.

The driving gear is connected to the logs, which are on a slope of $\frac{3}{4}$-inch per foot, by cast-iron clutches, one of which is shown at P, Fig. 14.

The rear bearing is 5¼ inches in diameter and is of cast iron. It is cast solid with a flange, on the face of which is turned a shoulder. This shoulder fits into a corresponding

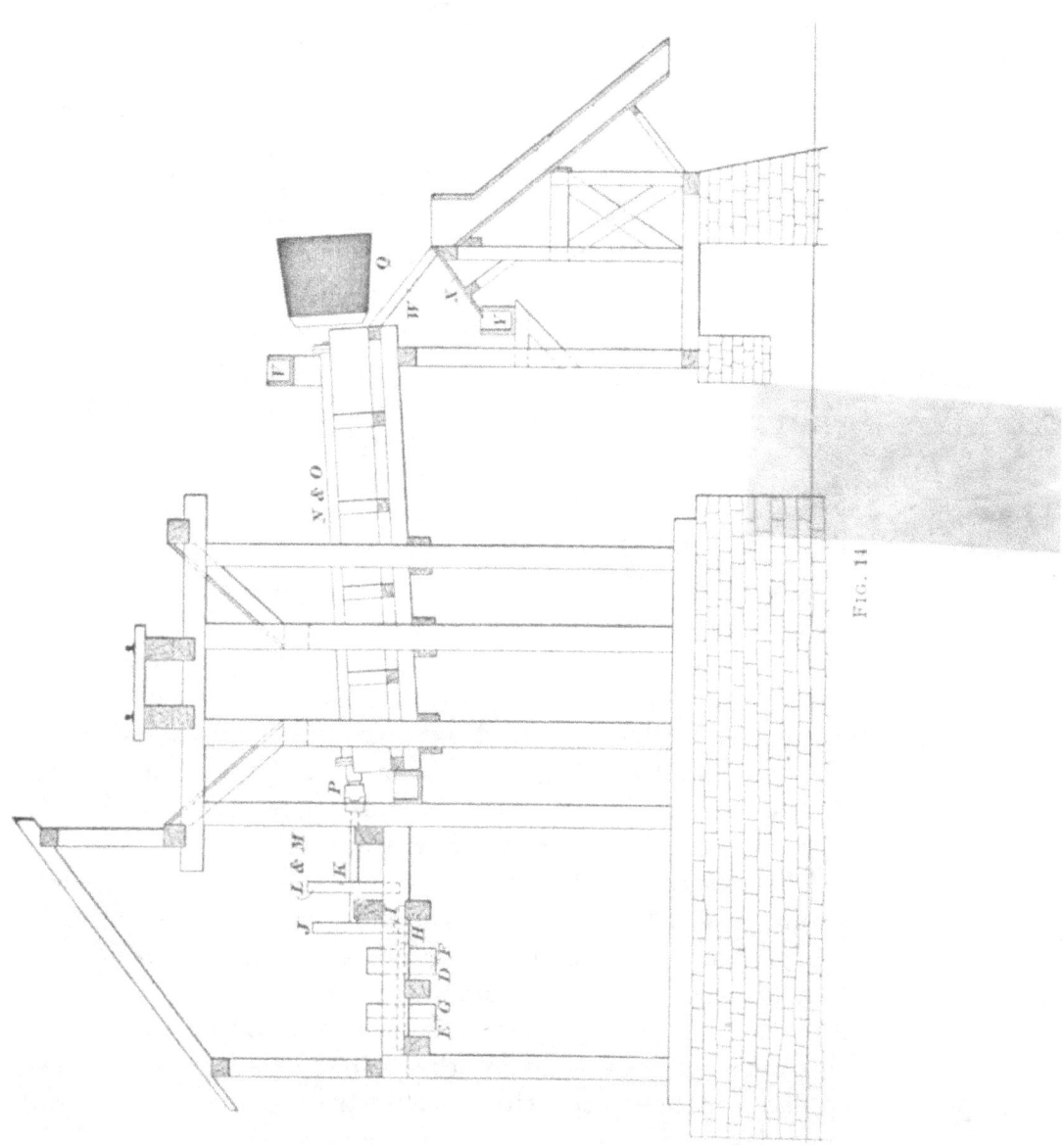

recess turned in the similarly flanged end of the log. The two flanges are bolted together and make a very stiff joint, as the shoulder prevents any lateral motion.

The logs are simply pieces of cast-iron pipe, 17 feet 5½ inches long, 11½ inches in diameter, with metal ¾ inch thick, and flanged at each end. This makes a splendid log —one that is stiff and wears well.

The method of attaching the spoons is shown in Fig. 15. They are put on in two spiral threads, 180° apart, and with

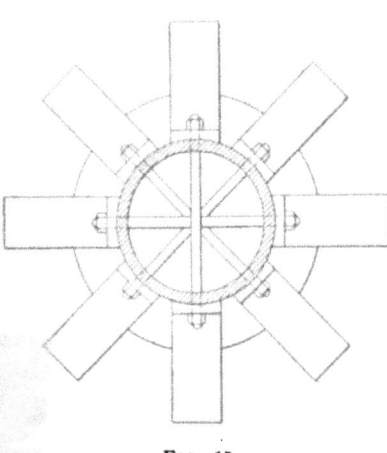

FIG. 15

a 5-foot pitch. They are set 45° apart on the circumference, thus making 8 spoons to each revolution, as shown in Fig. 15. By this method of laying out, there are, at every ⅛ of a revolution, two spoons opposite each other and 180° apart. If, now, holes be bored through the pipe, under the two holes with which the foot of each spoon is provided, two through bolts will fasten on two spoons. These bolts are ¾ inch in diameter and are made with nuts at each end as shown.

At the upper end of the log there is a gudgeon, similar to the one at the lower end, except that the bearing is only 4½ inches in diameter and extends 2 feet beyond the box. To this end the revolving screen Q, Fig. 14, is attached. The screens are made of $\frac{3}{16}$-inch steel plates perforated with $\frac{3}{16}$-inch holes, ½ inch from center to center.

The troughs in which the logs work are made of a wooden frame, in which are fastened the iron plates constituting the trough proper. The bottoms and sides of the frames are of 3-inch pine, thoroughly braced by the yokes shown at S, Fig. 13. Both bottom and sides are bolted to iron end pieces, in which are cast seats for the chilled-iron gearing boxes.

The iron plates constituting the trough proper rest upon the sides of the frames, to which they are attached by ⅝-inch lagscrews. As indicated at R, Fig. 13, they are of the usual semicircular pattern and are cast in sections only 15 inches long. This permits them to be made as open-sand castings.

In the operation of the plant, the ore is brought from the mines in side-dump cars, holding about 5½ tons each. The cars are pushed out past the washer on the trestle T, Fig. 13, which is built with a grade ascending in the direction of the arrow shown in the drawing. The cars are then allowed to drop down, two at a time, until they come over the chutes U and U', which are lined with 1-inch iron plates.

The ore, falling through the chutes to the logs, is caught by the spoons, which force it up against a descending current of water from the trough V, Fig. 14, until it reaches the revolving screens Q, into which a stream of water from the same trough V is flowing. There the ore is further washed and at the same time separated. All over $\frac{3}{16}$-inch diameter passes along the screen and falls into the "chute to cars," Fig. 14.

The fines, which drop through the perforations, fall on the 14-mesh wire-cloth screen W, Fig. 14, where they are further washed and screened, all over 14 mesh going to the cars, while the sludge falls on the apron X, and is thence carried away in the trough Y, which also conveys away the water from the rear end of the washers.

The current of water descending in the troughs is apt to carry off more or less ore through the rear end, and to prevent this loss, two perforated screen plates (not shown in the drawings) are used. The muddy water from the trough passes through a gate upon these screens, through which it falls and is carried away into Y, while the ore remains upon the screen. Only one screen is used at a time, and as soon as ore enough has accumulated upon it to stop the perforations, the water is shut off and turned into the other. The ore is shoveled back into the washers. This device saves a great deal of ore at a very low cost, as it requires the attention of one man for only part of his time, thus leaving him free to help at other points.

79. When clay is the only material removed from an iron ore, the percentage of impurities will not be much affected; that is, the phosphorus and similar impurities, as a rule, occur in the iron ore and not in the associated clay; hence the washing will not increase the grade of the ore materially, except in its percentage of iron.   Washing plants are sometimes introduced to clean ore from clay before it is put through some other form of concentrating machinery, such as jigs, etc., or before it is hand-picked.

80.   Ores Containing Sulphur or Carbonic Acid. These impurities have to be removed by roasting, and as this is usually done at the smelter, the apparatus need not be fully described in a work on Mining.   It will be sufficient to say that the roasting kilns now in use for this class of work are fired with gas and that the ore is fed to the kilns and the roasted material drawn from them continuously, much as in the case of a lime kiln or an iron blast furnace.   This continuous action greatly increases the capacity of the kilns, and the firing with gas results in a more uniform roast than was possible in the old style of intermittent kilns or roasters.

81.   Preparation of Salt.—As rock salt comes from the mine it usually carries more or less foreign matter, and if it were ground to a fine powder for table use, it would have a dark color, and hence would not find a ready market.   For this reason, the greater portion of the table salt of commerce is made by the evaporation of brine solutions obtained either from salt wells, salt lakes, or from the sea.   There are a few rock-salt mines which produce perfectly clear crystal salt, and from these table salt can be manufactured without the intermediate stages of dissolving and reevaporation.   The regular product of salt from any mine is treated as follows: The large lumps are laid under sheds to undergo a process called weathering, for the salt attracts moisture, and if the lumps are not properly weathered, they are liable to break up more or less during shipment.   The water which the salt absorbs from the air forms a brine on the surface which effectually cements all crevices and renders the masses solid pieces.   The portion of the salt intended to be treated in the mill is crushed in rock breakers and toothed rolls, after which it passes over shaking screens which separate the different grades. The coarsest material (up to ¼-inch cubes) is used for cappings in packing meat. These cappings are added to the brine on top of the meat in order to maintain the brine at its full strength. The finer grades of salt are used in the manufacture of ice cream, for preserving hides, and for similar commercial purposes. The large lumps of weathered salt are shipped as cattle salt. When salt is prepared for the table, it is ground fine and either separated by means of shaking screens or by a blowing machine, the different grades of salt being collected in various bins or chambers, owing to their various sizes, the lightest material being blown the greatest distances.

82. Points to Observe in Designing Concentrating Works.—When an engineer is called upon to design a concentration plant, he should be very careful that he is not deceived by new conditions. For instance, the condition of the mineral may be such that concentration is impossible, as in the case of a silver sulphide ore in a comparatively hard gangue. The silver sulphide would be pulverized so fine and form such bad slimes that it would be impossible to recover the greater part of the values from the ore.

83. Another case which might be mentioned is that of the hard iron ores banded with jasper. The jasper is frequently so intimately associated with the iron that it is impossible to separate them by crushing, and as the specific gravity of the jasper is frequently very high, a separation of the two would be practically impossible. On the other hand, the magnetic ores of iron being in the form of crystals can easily be separated from the gangue by crushing, and hence can be concentrated.

84. General Rule.—*As a general rule, it may be stated that in order to concentrate any substance, it must be of such a nature that by crushing the mineral it separates in the form of distinct crystals or distinct pieces.* Where minerals are practically of the same hardness and are intimately associated, it is rare that concentration by mechanical means is successful.

SAMPLING ORES

INTRODUCTORY

1. Object of Sampling.—The object of **ore sampling** is to obtain for chemical or mechanical tests a small quantity that will contain all the minerals in the same proportions as the original ore. If the sample is not correct, there will be a loss to either buyer or seller. In concentrating mills and leaching plants, samples are also taken of the tailings, and in smelters of the slag, in order to determine how much value is being lost. In concentrating mills, the different products of each machine are sampled, in order to know whether the machine is doing the work expected from it. In blast-furnace smelting, samples and analyses of the ores, fluxes, and fuel are necessary in order to calculate the proportion in the charge that will make the furnace run properly. Careful sampling is very often disregarded ; but no furnace or mill manager can afford to guess at values when the exact knowledge can be so accurately obtained by sampling and assaying. This is especially true in these days of close competition, when values are being profitably saved that could not be recovered by the old methods. It may be stated, as a rule, that the best extraction cannot be attained unless checked by careful sampling and assaying.

2. Obtaining a Correct Sample.—To obtain a correct sample, a systematic method must be used. Selecting lumps of ore haphazard here and there will not answer, for however honest the sampler may be, it is impossible to

§ 29

For notice of copyright, see page immediately following the title page.

judge the right proportion of rich and poor ore by the eye. The more thoroughly the ore is mixed and sized, the more certainly will a perfect sample be obtained.

3. Sampling is done either by hand or by machine; machine sampling, however, is seldom completed by machine, the final process being done by hand.

Although great improvements have been made in sampling machinery, metallurgical works usually do their own sampling by hand, while public sampling mills do their sampling by machinery, which seems to give satisfactory results to both buyer and seller. Public samplers really occupy the position of umpire between buyer and seller, the seller frequently believing that the buyer takes unfair advantage of him, especially if the latter is a public smelter or reduction mill. Miners have been known to send ore to smelters so lean that, if freight and smelting charges were added, the miner would be in debt to the smelter. The only way for a miner to be satisfied that he is getting full value for his ore is to sample his ore before he sends it to the smelter or else send it to a public sampling mill. Sampling is as important as assaying, and the hand sampler should have no interest whatever in the ore if he would obtain an average sample.

HAND SAMPLING

SAMPLING DUMPS

4. General Consideration.—The method of sampling dumps or any large piles of ore depends on the character of the ore, the amount to be sampled, and the disposition that is to be made of the ore. If all the ore is to be moved, the first sample may be obtained by taking a certain portion of the ore as it is being moved, as, for instance, every fifth shovelful or every fifth car or wheelbarrow load. This is called *fractional selection.*

If the main portion of the ore is not to be moved, the first sample may be taken by digging trenches or channels through the mass and either taking all the ore from the channels or a certain proportion of it by fractional selection. This method is called *channeling*.

The first sample from a large mass may also be obtained by sinking shafts into the pile or by driving tunnels through it. The sample is sometimes obtained by taking small portions of ore from various parts of the surface of the pile. A sample taken in this manner is called a *grab sample*.

5. Grab Sample.—When it is desired to get an approximate idea of the value of a large ore heap, a sample is sometimes obtained by taking a shovelful of the material from various points, equally spaced, all over the surface of the heap. This method should be used only for materials that are pretty uniform in composition and of low value, such as iron ores, fuels, and fluxes. Even with these care must be used to take coarse and fine pieces as they come and not to take all lumps, for the fines are quite certain to differ from the lumps in composition.

An improved modification of this method is sometimes used when unloading iron ore from a vessel. When enough ore has been removed to expose a face of ore reaching to the bottom of the vessel's hold, small quantities of ore are taken from all over the face, the samples being taken in regular order from side to side and from top to bottom. When considerable more ore has been taken out, samples are taken from the new face and so on. This procedure has the advantage of taking portions from all parts of the heap instead of merely from the surface. The sample may be further reduced by fractional selection, by quartering (which will be described later), or by a machine.

6. Fractional Selection.—When a large lot of ore is being shoveled from cars or elsewhere, a sample may be obtained by throwing aside every third, fifth, tenth, or twelfth shovelful. The richer the ore or the more unevenly the minerals are distributed through it, the oftener is a sample shovelful taken. Each shovelful should be taken from the floor and from the bottom of the pile. When the ore comes in sacks and is of a fairly uniform character, every fifth or tenth sack may be taken for a sample. Fractional selection is probably the most accurate method of obtaining a sample from a large amount of ore.

7. Channeling.—Channeling, as applied to dumps or large piles of ore or other material, consists in shoveling channels through the mass and taking all or a portion of the ore from the channels as a sample. When only a portion of the ore is taken as a sample, the reduction is generally made by fractional selection ; that is, by throwing every third, fifth, or tenth shovelful of ore from the channel aside as a sample. When sampling dumps by channeling, care must be taken to see that the ore from the sides of the cuts does not fall into the channel to such an extent as to give an unduly large proportion of ore from the upper part of the pile.

8. Size of Sample.—The size of the sample taken from a dump or any other large quantity of ore should depend on the manner in which the values are distributed through the ore. In the case of ores in which the values are uniformly distributed, such as iron ores, a grab sample may be all that is required or every twentieth shovelful may be taken; but if the values are not uniformly distributed, as in the case of ores carrying free gold or valuable minerals, the first sample must be larger and the ore must be crushed finer if correct results are to be obtained. In some cases the first sample must be at least one-third of the ore. After the first sample is taken it is reduced by fractional selection, channeling, quartering, or by a machine.

SAMPLING SMALL LOTS OF ORE

9. General Consideration.—The sampling of small lots of ore does not differ greatly from the sampling of dumps or large lots. In dealing with a carload or other small lot of ore, it is generally necessary to handle all the ore, hence a grab sample is rarely taken. If the ore is being unloaded from cars, the first sample is generally taken by fractional selection and then reduced by the same method or by quartering or channeling.

10. Fractional Selection.—When this method is used for sampling fairly large lots of ore, the first sample is taken in wheelbarrows or cars and dumped in a pile. When this pile is completed, it is removed by shoveling and a certain proportion of the material thrown into a new pile, care being taken to throw all the ore on the top of the pile, so as to thoroughly mix the sample. It is not often that the second sample is so large that it has to be taken in wheelbarrows or cars and piled in another place. When shoveling the ore, care must be taken to see that the place where the sample is piled is swept clean and that each shovelful of ore is taken from the bottom of the pile. Fractional selection is probably the most accurate method for obtaining a sample of 100 or 200 pounds from a lot of ore in which the values are not regularly distributed through the ore. The more irregularly the values are distributed through the ore, the larger should be the sample taken at each handling of the ore and the finer should the ore be crushed. After the sample has been reduced to 100 or 200 pounds by fractional selection, it is generally still further reduced by quartering or channeling.

11. Channeling.—For channeling, the ore is spread out in a flat heap or layer a few inches thick and the sample taken by shoveling out two or more parallel channels, like paths through a snow bank. All or a part of the ore from these channels constitutes the sample. Sometimes two sets of channels are made, one set at right angles to the other. Channeling is fairly accurate if carefully done, but is little used, since it requires a large floor space. In channeling, the ore does not require as much mixing as in quartering, but there is sometimes considerable difficulty experienced in making the channels without knocking down some ore, either coarse or fine, from the sides of the channels.

12. Quartering.—When ore is sampled by quartering, also called the **Cornish method** of sampling, the ore is first thoroughly mixed and then divided into four parts or quarters and two of these parts taken as a sample, while two are discarded. In order to thoroughly mix the ore, the work may be done as follows:

The ore is dumped in a large circle, as shown at *a*, Fig. 1, and the samplers move slowly round the ring, shoveling the

FIG. 1

ore into a pile in its center. It is common for two men to work together, always keeping diametrically opposite each other. They drop each shovelful exactly on the apex of the resulting cone, as shown at *b*. This is done by holding the shovelful of ore above the apex of the cone and then suddenly pulling the shovel away from the ore in the direction of the arrow. This distributes the ore on all sides of the cone and gives a pretty thorough mixture. As another aid to complete mixing, the samplers do not shovel all the ore in walking once round the circle, but make at least two trips. Care must be taken to sweep up all the fine ore and place it on the top of the pile. It will not do to simply sweep the fine ore up to the edge of the pile, as this is apt to be the richest part of the ore.

In some cases, to insure thorough mixing, the ore is shoveled from a ring to a pile and back to a ring again several times or it may be shoveled from one pile to another. After the ore has been thoroughly mixed and piled up in a cone, the samplers walk round the cone continuously in one direction and with their shovels draw the ore into a wide, flat pile or heap. The manner of doing this is illustrated in Fig. 2, which shows the cone partly spread out. It may sometimes be necessary to shovel the ore from this flattened cone into a second cone in order to make the mixing more complete.

FIG. 2

13. It is hard to get a thorough mixture when very coarse and very fine material occur together, but this difficulty is lessened by moistening the ore a little. Water should not be added in sufficient quantities, however, to make the ore cake. Very wet ore cannot be properly sampled by quartering, because the water washes the fine material away from the coarse and so prevents satisfactory mixing.

Having spread out the ore flat, it is divided into four quarters, as shown in Fig. 3. This may be done by pressing the edge of a board down through the heap on two lines that pass through the center at right angles to each other. If a comparatively large mass of ore is being sampled, the quarters may be separated by shoveling two channels across the pile at right angles to each other. When shoveling these channels, the ore removed is thrown alternately to the right and left. After the quarters have

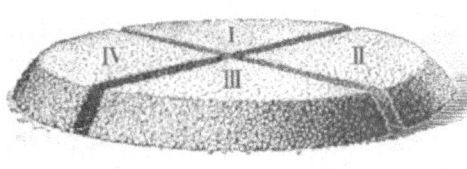

FIG. 3

been marked off for separating, two opposite quarters, as, for instance, *I* and *III*, are shoveled away, taking care to remove all the fine particles belonging to them. The remaining two quarters are shoveled into a new cone and the process repeated. This time, quarters *II* and *II'* are shoveled away instead of *I* and *II*. The object of this is to overcome any possible tendency of the shovelers to make one side of the heap richer than the other. The reduction of quantity must not proceed too far before the ore is crushed to a smaller size. Even after fine crushing the sample is seldom reduced in size below 8 pounds by quartering.

14. Sampling Floor.—The floor on which the quartering is done should be smooth and free from cracks, and for this reason it is best to have it covered with iron. It

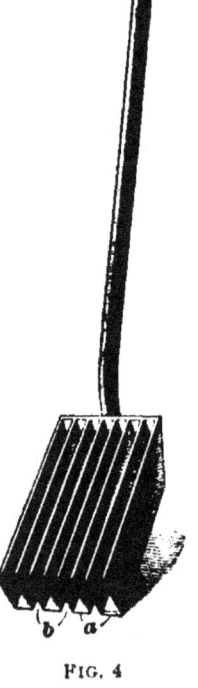

FIG. 4

should be carefully swept before it is used, to prevent rich dust from previous samplings being mixed with the ore in hand. When a good floor is not available, a large piece of strong canvas may be spread out.

15. Split Shovel.—With lots weighing not more than half a ton and when the largest lumps are not more than about ¼ inch in diameter, the split shovel, Fig. 4, is sometimes used. This shovel consists of several long scoops *a*, with open spaces *b* between them, the spaces being the same width as the scoops. The sampler takes the ore on an ordinary shovel and spreads it over the split shovel, moving his shovel back and forth across the scoops as the ore slides off. When the scoops are full, the split shovel is lifted and the ore in the scoops is put one side as the sample, the rest of the ore remaining on the floor. The action of the split shovel is very much like that of the sample riffle or the Jones ore sampler, both of which are described under the head of "Finishing the Sample,"

SAMPLING TAILINGS

16. A slotted-pipe sampler is sometimes used to sample fine material that is of fairly uniform grade, such as concentrates. The best form of this tool is shown in Fig. 5. A 1-inch iron pipe *a* has a slot *b* cut lengthwise in it from one end to within 3 or 4 inches of the other. A **T** is screwed to the unslotted end, for the insertion of a handle *c*. A second pipe *d*, just large enough for the first to easily slip inside, is also slotted, but this slot *e* does not extend quite to either end of the pipe. The lower end of this larger pipe is forged to a point, so that it can be easily pushed into a pile of fine ore.

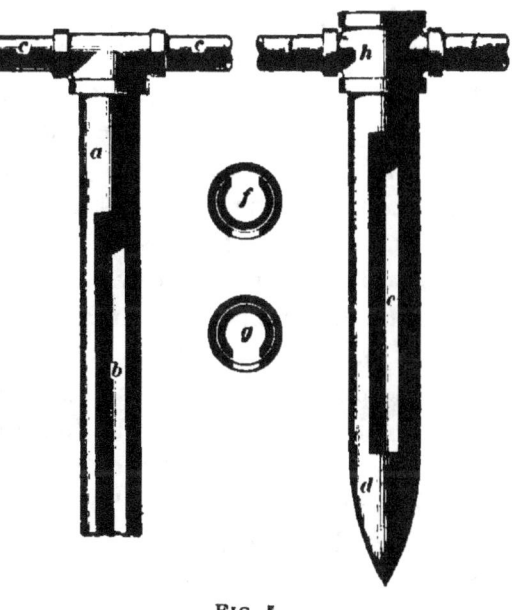

FIG. 5

The tool, while held in the position shown by the cross-section *f*, is forced into the material to be sampled; the inner pipe is then turned into the position *g* and is twisted back and forth until it fills with ore. It is then returned to the position *f* and the sampler is withdrawn. To pull out the driven pipe *d*, a cast- or wrought-iron cross *h* is screwed to the pipe and a lifting handle *i* inserted through *h*.

17. Dipper or Bucket Sample.—To sample any running stream of material, such as the different products in a concentrating mill, a dipper or bucket may be passed through the stream at regular intervals, once an hour, for example. The dipper must pass across the whole stream, because one side may be richer than the other. It must not be allowed to fill and run over, especially in the case of products carried in a stream of water, for this would cause a loss of some of the finer part of the sample.

18. In this connection, it is necessary to mention an important point that is often disregarded. When fine material carried in water is to be sampled, it is common to let a bucketful settle for 10 or 15 minutes, then to pour off most of the water and dry the residue. This is thought by some to be a safe plan because, although the water poured off is in some cases quite turbid, the quantity of solid matter suspended in it is apparently insignificant. The trouble with this reasoning is that this extremely fine material is often very rich. With most ores of silver, gold, copper, lead, etc., the valuable minerals are pulverized more than the waste during the crushing, so that a small quantity of very fine material may contain more value than a large quantity of the coarse part. To show the importance of catching these very fine slimes, a case can be mentioned where material to the value of about $1,000 a day was going to waste in this form in the mill tailings without the manager's knowledge. Sometimes, however, the very fine slimes will not settle completely in several days or even weeks; but right here an interesting scientific fact comes to our aid. Small quantities of common salt, alum, and various other substances dissolved in the water cause the finest slimes to settle much more rapidly than otherwise, as is shown by the following experiments.

19. The tailings from a gold stamp mill were run through a settling tank to remove the coarse portion and the overflow was allowed to fill a small tank. After the water had stood undisturbed in the latter for $\frac{1}{2}$ hour, a large sample was taken from the top of the tank and, of course, it contained only extremely fine slime. Numerous tests showed that 95 to 100 per cent. of all the suspended slimes settled out in $\frac{1}{2}$ hour when as much as 1 per cent. of salt or alum was dissolved; while only 22 per cent. settled out in the same time where nothing was added; in fact, 20 hours were necessary for 95 per cent. to settle out when nothing was added.   Lime was still more effective; the equivalent of .05 of 1 per cent. or less of burned lime CaO added in the condition of clear lime water caused nearly all the slimes to settle out in less than 5 minutes.   It should be noted that slimes from different ores act differently.   In some cases neither lime nor common salt gives good results; therefore, the best substance to use in any case must be found by experiment.   Fairly vigorous but not too violent stirring sometimes makes the particles that have been coagulated by lime or any other substance come together in large flakes and so settle more rapidly.

FINISHING THE SAMPLE

20. General Remarks.—When the sample has been cut down to a few pounds of finely crushed ore by successive crushings and reductions of quantity, the final sample is prepared.   In order to do this, the sample is generally put through a sample grinder and then reduced to 2 pounds or less.   This small sample is then ground on a bucking board and passed through a fine screen, after which it is divided into three or four portions, one for the seller, one for the buyer, and one for the umpire.   In some cases, a fourth sample is kept by the sampler for reference in case of need.

Some ores contain metallic particles that cannot be ground and will not pass through the screen.   These "metallics" and the ore passing through the screens must be weighed and assayed separately.   *If this point is not carefully attended to, the value of the ore cannot be correctly determined.*

21. Sample Grinder.—When the sample is ready for finishing, it generally consists of from 8 to 12 pounds of ore that has been crushed in some form of fine crusher or that has been passed through finishing rolls.   This sample is then passed through some form of sample grinder before it is still further reduced in bulk.

A good form of sample grinder is shown in Fig. 6. The sample is introduced into the hopper *a* and is ground between

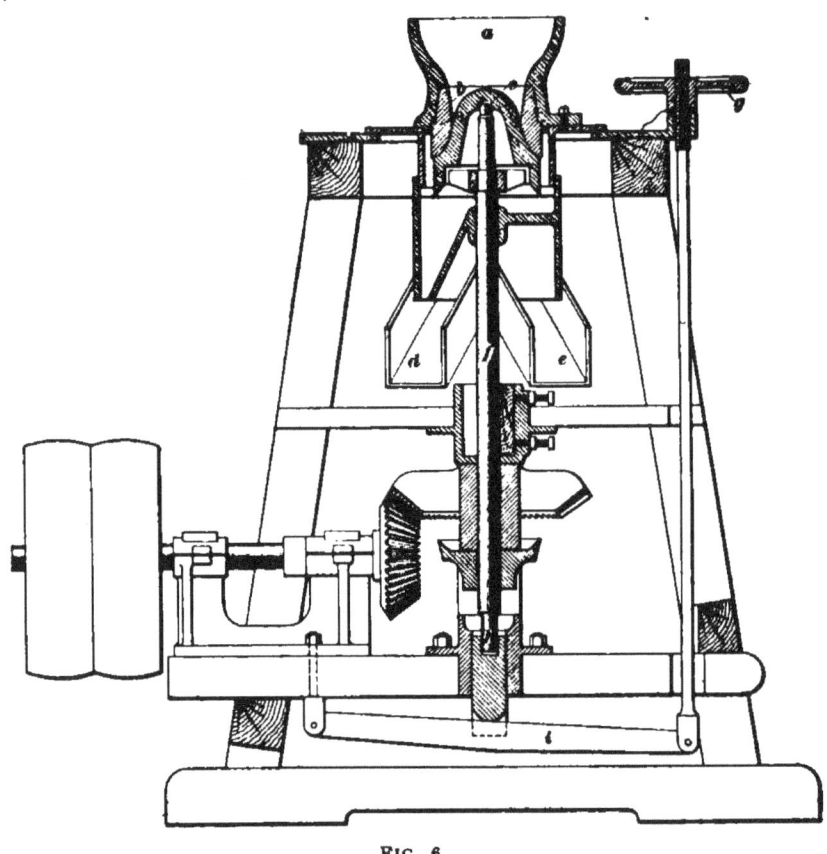

FIG. 6

the head *c* and the ring *b*. After grinding the material is discharged through the spouts *d* and *e* and collected in pans. The head *c* can be raised or lowered by means of the hand wheel *g*, which controls the thrust bearing *h* at the foot of the shaft *f* by means of the lever *i*. If the head is raised, the particles of the material will be ground fine; on the other hand, if lowered, the particles will be ground coarse.

22. Cutting Down the Sample.—After the sample has been ground it is cut down to about 2 pounds or less. This may be done by quartering on the bucking plate or on any other hard smooth surface or by using some form of sampler, such as the riffle or the Jones sampler.

23. Riffle, or Tin Sampler.—This consists of a series of troughs arranged side by side with open spaces between

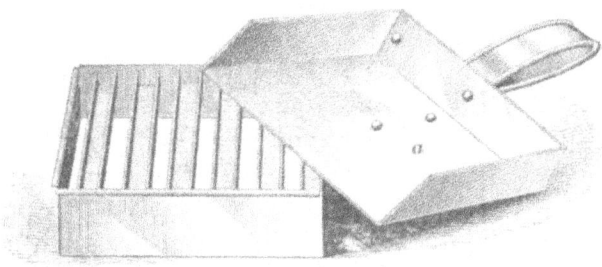

FIG. 7

them, as shown in Fig. 7. The width and number of the spaces and troughs can be so arranged that the sampler will take out any desired portion of a crushed sample that has been spread evenly over it. The riffles are commonly made to cut the sample into two equal parts, half remaining in the troughs and half falling through. The ore is taken up on a scoop a and spread over the troughs, care being taken not to heap the ore above the top of the troughs; the sampler is then lifted and the portion in the troughs or that which remains on the plate taken as the sample.

24. Jones Ore Sampler.—In the case of the ordinary riffle, it is necessary to lift the sampler out of the ore to separate the samples, and care must be taken not to let the ore pile up above the top of the troughs. The Jones sampler overcomes these objections and makes provision for taking duplicate samples from much larger amounts of material than can be handled on the riffle at one operation. This sampler is illustrated in Fig. 8. It is really a riffle sampler in which all the

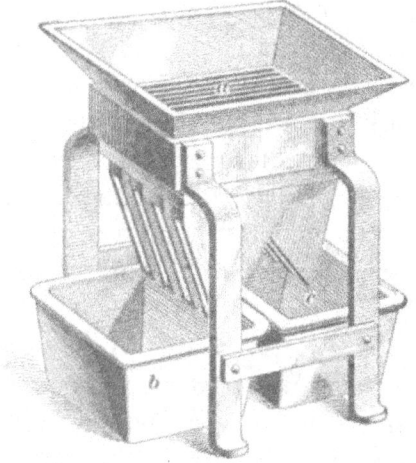

FIG. 8

spaces are connected to spouts that discharge alternately to the right and left. The ore is distributed over the riffles at *a* and two samples result, one being received by the tray *b* and the other by the tray *c*.

25. Final Grinding.—After the sample has received its final cutting down, it is placed upon a *bucking plate*

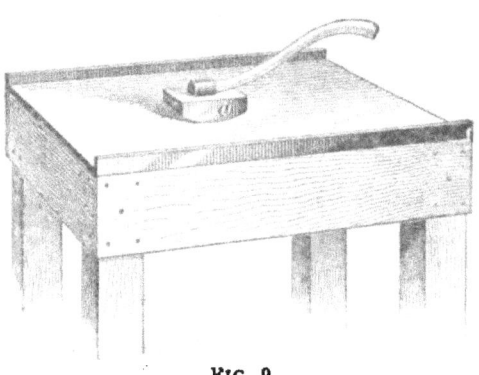

FIG. 9

(also called a *bucking board*) and ground with a *muller* until it will pass through a sieve having the desired mesh, which is usually one that has 80 or 100 meshes to the linear inch. One form of bucking plate, or rubbing plate, is shown in Fig. 9. The muller with which the grinding is done is shown at *a*. The plate shown has two raised edges, though some rectangular plates have three raised edges. This form of plate is best if the ore comes on to the plate in rather large pieces, so that they must be broken with a hammer or with a very heavy muller, for the raised edges tend to keep the ore from flying off the plate. For reducing samples that are already comparatively fine, some persons prefer a circular bucking plate. Such a plate is usually about 3½ feet in diameter, and when in use is placed on a low table or bench so situated that the operator can walk around the table as he draws the muller back and forth. These circular plates wear more evenly than the rectangular ones, because the rubbing takes place in different directions across the plate, and there is no tendency to wear grooves in the surface of the plate.

26. Separation of the Sample.—After the sample has been passed through the sieve, it must be divided into three or four portions, so that the buyer, seller, and umpire may each have one. A common method of separating the sample is to place the ore on a piece of oilcloth or glazed paper and to mix it thoroughly by drawing up first one corner and then another of the oilcloth or paper in order to roll the material over.　After the ore is mixed, it is spread out in a thin layer by means of a spatula and then small portions of the material are taken from various parts of the mass, as shown at *a*, Fig. 10, and placed in the dish *b*.　After

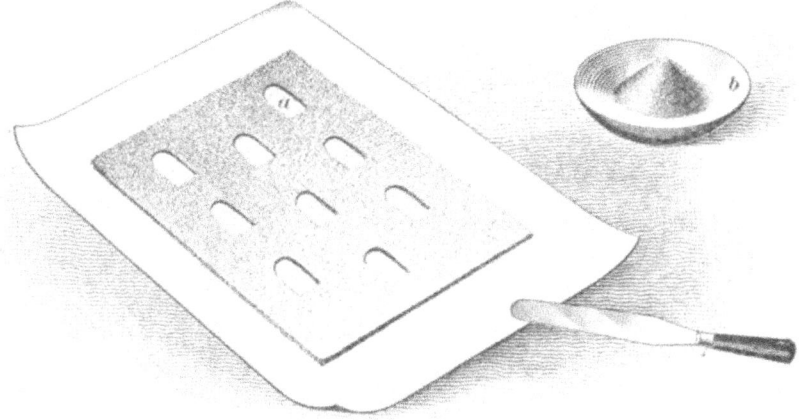

FIG. 10

the first sample is taken, the process is repeated to obtain the second and third samples.　There is some danger that this method will not give accurate results on account of the difficulty of getting equal portions from the top and bottom of the pile.　Some persons prefer to thoroughly mix the sample and then to divide it by quartering.

27.　Probably one of the most accurate methods of dividing a sample is by means of the Bridgman mixer and divider, which is shown in Fig. 11.　In using this device, 2 or 3 pounds of finely ground ore is mixed in the funnel, or mixer *a*, the opening at the bottom being closed by one finger.　The finger is then removed and the mixer passed back and forth over the divider *b*, which discharges into the sample bottles as shown.　All the ore that is put into the mixer must be run into the bottles, or there may not be a proper distribution of the heavy and lighter portions.　A cover, which is not shown in the illustration, is placed upon the mixer when the ore is being shaken.

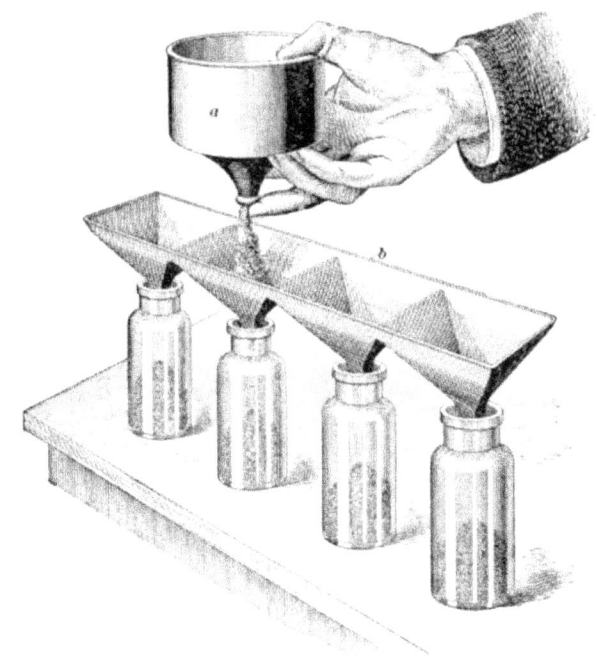

FIG. 11

MOISTURE SAMPLE

28. Nearly all ores are more or less wet; and, since the moisture gradually evaporates, the weight does not remain constant. Hence, the only way to find the real value of any lot of ore is to have the assayer's sample perfectly dry and also determine what the entire lot would weigh when perfectly dry. For the latter purpose, a special sample should be taken when the ore is weighed. This sample must be obtained as rapidly as possible and put into a closely covered iron pail or box to prevent evaporation. It must not be taken simply from the top of the ore pile, because that will generally be drier than the average. The best way, if possible, is to dump the ore immediately after weighing and take a sample by rapid fractional selection.

The moisture sample may be very much smaller than the regular sample. As soon as the moisture sample is all taken, it should be weighed and dried; or, if it is too bulky, it may be rapidly mixed and a portion taken for drying, but it must be handled as little as possible before drying. *The difference between the wet and dry weights divided by the wet weight and multiplied by 100 gives the percentage of moisture in the wet ore.*

EXAMPLE.—A sample of 20 pounds of ore after drying weighed 19.6 pounds. What percentage of moisture was in the ore?

SOLUTION.— $20 - 19.6 = .4$ and $\dfrac{.4 \times 100}{20} = 2$ per cent. moisture.

Ans.

MECHANICAL SAMPLING

29. General Consideration.—Sampling ores by hand is tedious, slow, and expensive, besides the operator's judgment may be warped unintentionally in regard to the proper proportion of coarse and fine ore in the sampling. It has been stated with considerable truth that a man who samples a mine should have no conscience, as he may err almost as badly by trying to be fair as by trying to be unfair. This does not apply to ore sampling at metallurgical works to as great an extent as at mines; nevertheless it does in some measure, unless the operator acts in a mechanical manner and merely directs the sampling according to some fixed method. A machine has no judgment, still it has method due to its construction, and while, if properly planned and constructed, it may prove accurate, on the other hand it may not be accurate, and then must be discarded. The main objection to the most accurate mechanical sampler is the difficulty experienced in cleaning it after each sampling operation.

30. Limiting Conditions.—Mechanical sampling can be used wherever the ore runs in a fairly steady stream. A mechanical sampler has some form of diverting spout that delivers the sample in one direction while the bulk of the ore goes in another. To give accurate results, *it must take the whole stream part of the time and not a part of the stream all the time;* that is, it must cut out a portion of ore across the whole stream at regular intervals. The reason for this is that the large lumps of ore are almost sure to roll to one side, away from the fine, and the heavy to roll away from the light, so that one part of the stream does not represent the whole. The stationary devices, which take one part of the stream all the time, are at present not used by careful operators. If the diverting spout of a sampler swings back and forth through an ore stream, it must move completely out of the stream in each direction or it will take too much from one part of the stream and not enough from other parts. The machine should be simple in construction so that it can be cleaned easily, thus preventing rich particles of ore from one sampling getting into the next sample.

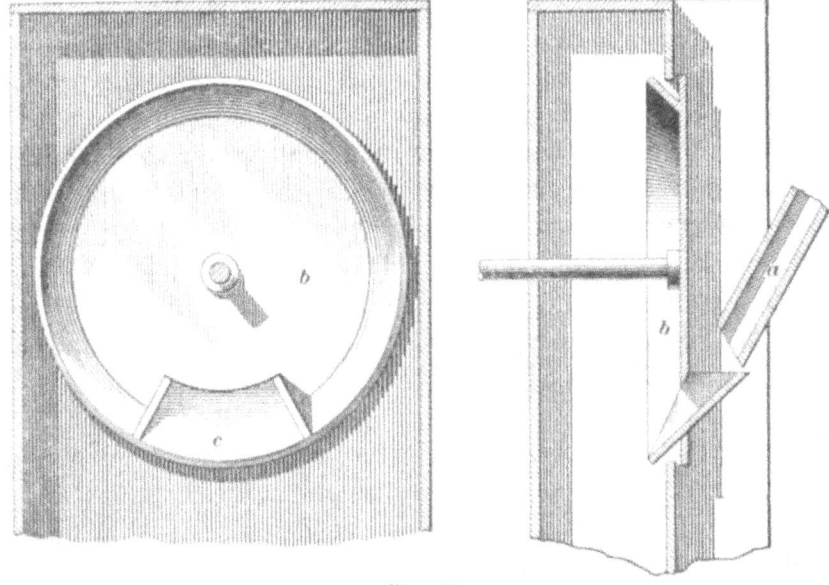

Fig. 12

31. Topham's and **Snyder's samplers** are shown in Figs. 12 and 13. Each of these resembles a pan with flaring sides, turned on edge and fastened to a revolving horizontal shaft. Near the circumference of the pan is an opening through which the sample passes to a bin, but the main portion of the ore is diverted into a separate bin by the flange. In the Topham sampler, which is illustrated in Fig. 12, the ore comes down a sloping chute *a* placed at the back of the pan *b*, while in Snyder's, Fig. 13, the ore is fed to the front side. Topham's sampler is not on the market, but Snyder's is made by the Allis-Chalmers Company of Chicago. It will be noticed that the two sides of the opening for the sample to pass through converge towards the axis of

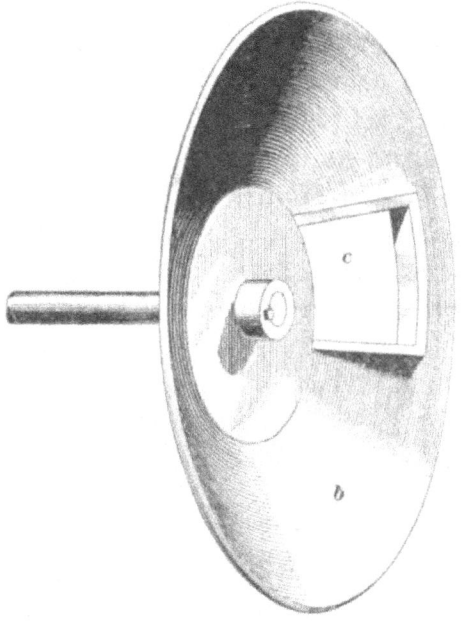

FIG. 13

rotation, so that the part of the opening near the center of the pan and the part near the circumference pass through the

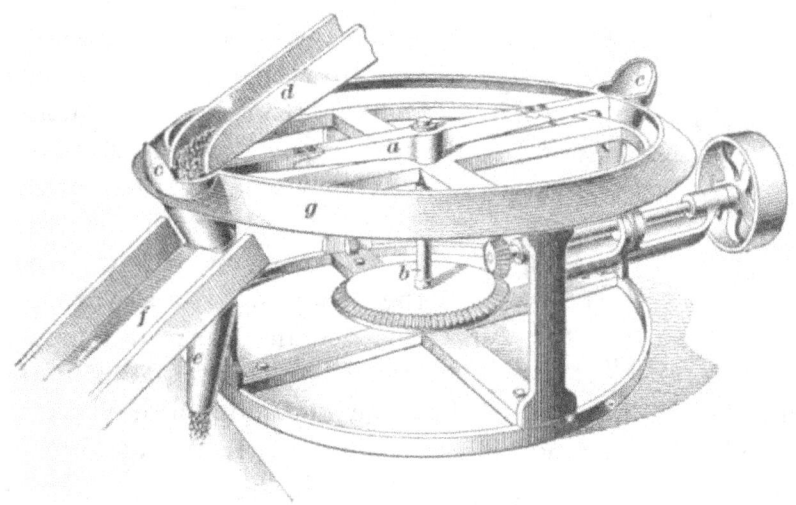

FIG. 14

stream in exactly the same time and take equal proportions

of ore from all parts of the stream.   This point is provided
for in all the mechanical samplers that will be described.

32.   The **Collom sampler** shown in Fig. 14 has a horizontal arm a fastened at its middle to a revolving vertical
shaft b; on each end of the arm is a diverting scoop c.
When these scoops come under the end of the spout d, the
ore falling from d is directed into the spout e, which leads
to a sample box or bin.   At other times the stream of ore
passes through the spout f to the shipping or storage bin.
The sloping flange g prevents any ore falling into the
sample spout e except when
the sample scoops come round.

33.   The **Vezin sampler**
is similar to the Collom in its
essential feature, which is a
diverting scoop revolving horizontally, as shown in Fig. 15,
but it differs in construction.   Two truncated cones
made of heavy sheet iron are
joined at their bases, where
they are attached to a set of
spider arms extending horizontally from the vertical
shaft.   From the upper cone,
one or two radial scoops a
project beyond the base of the
cone.   The stream of ore falls
from the spout b, and when
the scoop a passes through
this stream a sample passes
inside the cones and is deliv-

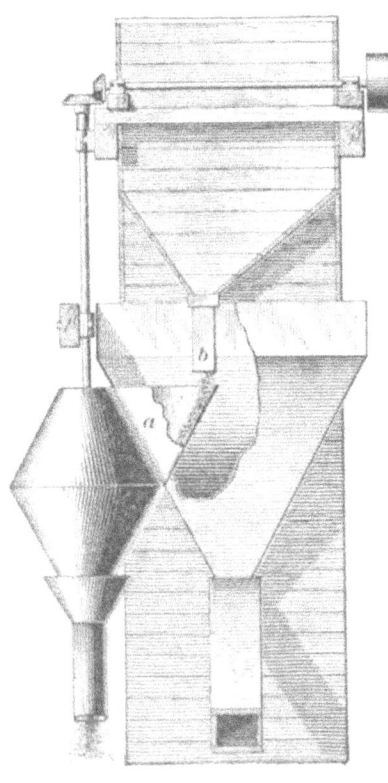

FIG. 15

ered through the small end of the lower one into a box or
bin.   The main portion of the ore passes to a storage bin.

34.   **Brunton's sampler,** shown in Fig. 16, consists of
an oscillating divider b b that swings back and forth beneath
the end of the feed spout g.   The divider is fastened to

the horizontal shaft c, which receives an oscillating motion from the wheel f by means of the arm e and the lever d. The scoop a cuts completely across the stream of falling ore at

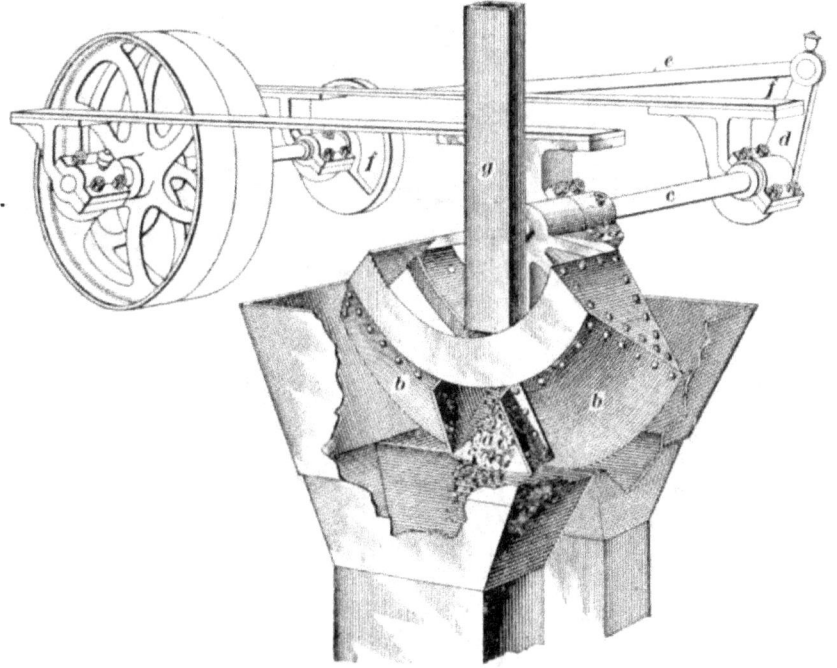

FIG. 16

each oscillation and diverts a sample to a sample bin; but the rest of the ore is deflected in the opposite direction to the stock bin by the sloping faces b. There are two Brunton samplers, but the one described is the better.

35. Constant's sampler, shown in Fig. 17, is a hollow cylinder fastened on a revolving horizontal shaft and having openings b in the sides. When these openings pass under the chute a, samples of the falling ore run inside the cylinder and are delivered from the ends into the hoppers c. The rest of the ore rolls over the surface of the cylinder into the hopper d. This machine is arranged to take duplicate samples.

36. Self-Acting Samplers. — In some cases where tailings are to be sampled and there is no shafting from which to drive any of the machines described above, devices are used that are operated either by the force of the running stream of tailings or by an auxiliary stream of water. The

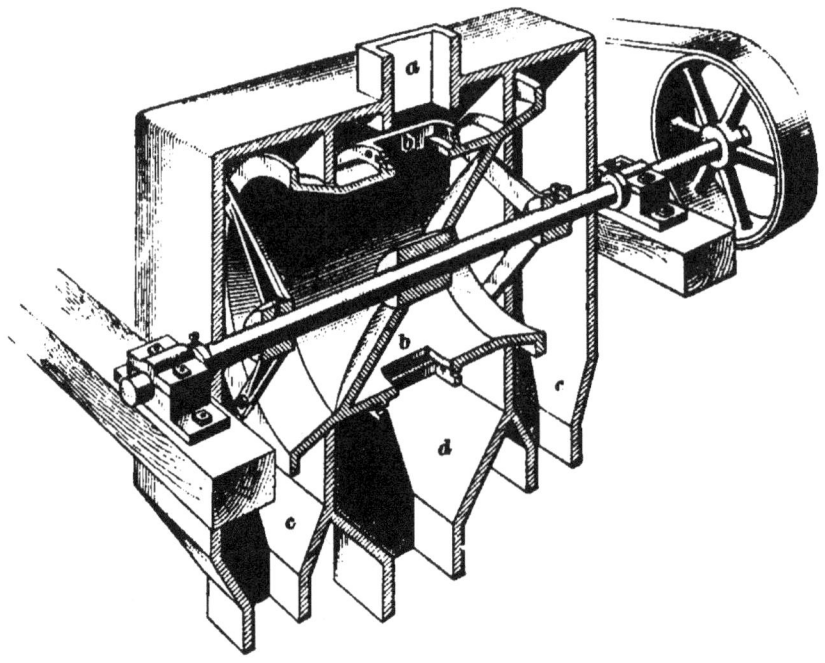

FIG. 17

Richards sampler, Fig. 18, is a hollow cylinder made of sheet iron and having internal curved blades against which the tailings and water strike, causing the machine to revolve. The axis of the cylinder is inclined so that the tailings pass between the blades and out the lower end of the cylinder. A hole is cut through the cylinder between two of the blades and a tube is soldered on the outside to deliver the sample that passes through the hole into the channel *a* of the casing. The space between the two blades mentioned should be covered on top and at the lower end, so that whatever gets into this space can get out again only through the hole in the side of the cylinder. The sample delivered into the channel *a* passes through a tube, at the lowest point of the channel, into a bucket or box. Professor Richards, of the Massachusetts Institute of

Technology has also designed an overshot waterwheel to be driven by the stream of tailings, one of the buckets having

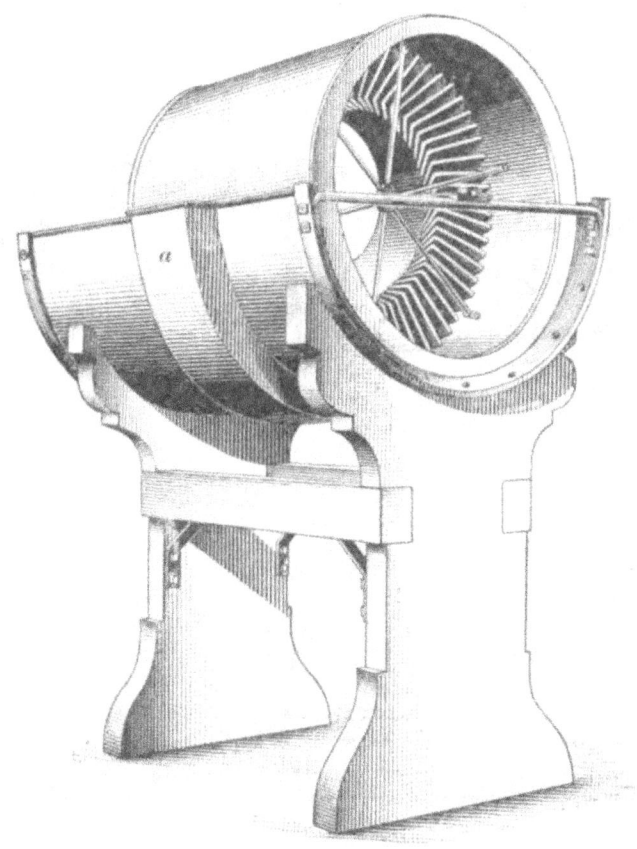

Fig. 18

an opening towards the center of the wheel, through which a sample passes.

37. The apparatus shown in Fig. 19 is used to sample tailings at some of the gold mills of the Transvaal. It is not strictly self-acting, but it does not have to be connected to shafting. The pipe a, in which there is a longitudinal slot, is attached to the shaft c by means of the arms b. The box d is also attached to the shaft c. A small stream of water from the pipe e runs steadily into d. When d is filled, the weight of the water makes it tip over and the pipe a is thereby moved downwards through the stream falling from the launder g. The water then runs out of d and the apparatus is returned to its original position by the weight f; in the act of returning, the pipe a

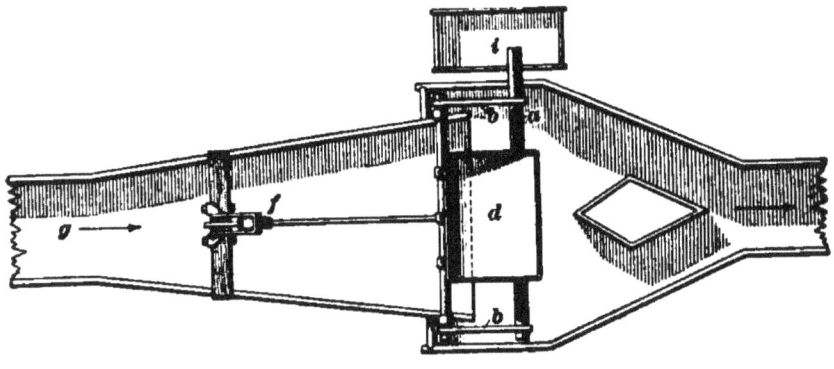

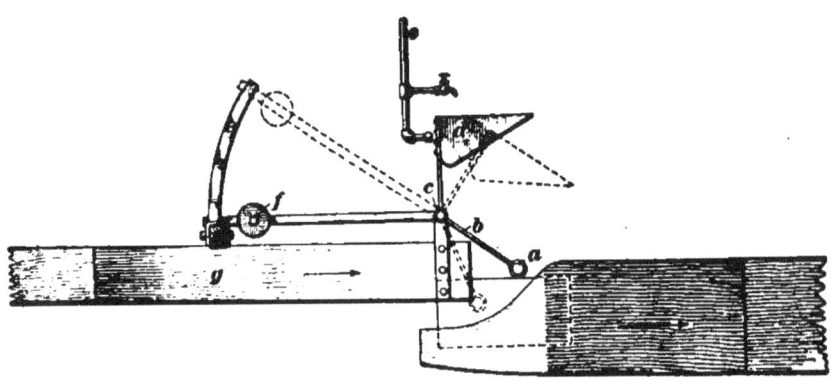

FIG. 19

again passes through the stream. One end of this pipe is closed, but the other end is open and delivers the sample into the box i.

38. The **Lamb tailings sampler,** made by the Allis-Chalmers Company, is shown in Fig. 20. The box b, which is arranged to rock endwise, has a central partition c. When enough water has run into one side of the partition, that end of the box tips down and discharges the water at the end. This allows the suspended spout c to run down to the lower end of the guide rods and thereby cut through the stream of tailings that falls from the launder *a* and deliver a sample into a box or barrel. When one end of the box *b* tips down, the pipe *d* delivers its water into the other end of the box, which, in turn, gradually fills and is tipped

FIG. 20

down, allowing the spout *e* to cut across the stream of tailings again, but in the opposite direction. The frequency with which samples are taken depends on how fast the water is allowed to flow from the pipe *d* and to escape from the box *b*.

39. Revolving Tailings Sampler.—One form of sampler that is operated by the flow of the tailings is shown in Fig. 21. It consists of an undershot waterwheel *c*, driven by the flowing tailings, and which, in turn, drives the gear *k* on the vertical pipe *h*. The pipe *h* is carried in bearings in the flanges g and f, and one end of the water-wheel shaft is carried in the vertical flange d. The vertical pipe h is connected to the horizontal trough i, so that whatever is caught in i will flow down through h to the launder j. The wheel c causes the pipe h to revolve and the trough i to cut through the tailings as they flow from the launder a.

FIG. 21

The sample that passes through j must be caught in a large tank, so that none of the water will escape, for the water must be separated from the solid matter by means of a settling tank, but in all cases any water that is allowed to flow away must be clear, so that it will not carry any values with it.

40. Comparisons and General Remarks.— Hand sampling is generally used for small lots that do not justify the outlay for a mechanical plant. The latter, however, generally costs less for operation and saves time. Moreover, a properly constructed mechanical sampler always gives correct results, but in hand sampling the operator may take an unfair sample if he is so inclined.

In regard to different methods of hand sampling, it may be said that quartering is probably more accurate than channeling and requires no more labor. For very large lots of ore, neither quartering nor channeling can well be used, because with such lots they would require too much floor space and too much labor. Fractional selection, however, can be used for a lot of any size. The limitations of the grab sample are stated in Art. **8.**

41. Some machines are arranged to take duplicate samples, but there is no advantage in this, for it has been shown in practice that with a properly constructed machine the duplicates always check. Indeed, if they did not check, that would be conclusive proof that the machine did not take fair samples. In some cases these duplicates might serve to detect any "salting,"* but it would generally be a simple matter for any one who tampered with the ore to add equal proportions of "salt" to both samples. The final small samples for assaying are made up in duplicate where ore is being sold, so that both buyer and seller can have a sample. Indeed, a third sample is commonly prepared, so that any discrepancies in assaying may be settled by an assayer who is independent of both buyer and seller.

PRINCIPLES OF SAMPLING

42. General Consideration.—Since ores are never perfectly uniform in composition, a single shovelful taken from a lot is not at all likely to be a correct sample. But if enough shovelfuls are taken from regularly apportioned parts of the lot, as in fractional selection, the rich portions will balance the poor and give a correct result In the method of quartering, instead of many small portions, a

---

* "Salting" is the unscrupulous addition of rich material, with a view to making the ore appear richer than it really is.

few large portions are taken; but this requires careful mixing before cutting out the sample. Without this mixing, the chances of making the sample too rich will not be so well balanced against the chances of making it too poor as they are in fractional selection. By considering this balancing of the rich against the poor, it will be seen that the accuracy of a sample depends not on the ratio of its weight to the weight of the whole lot, but simply on the actual weight of the sample, whether the lot is large or small.

43. Effect of Size of Lumps.—With any given ore, it is clear that a greater weight must be taken when the lumps are large than when they are small, because in the former case the different grades of ore cannot be as uniformly mixed as in the latter case. The greater the weight for any size, the greater the accuracy of the sample; but for every case there is a weight that comes so near to perfect accuracy that there is no practical advantage in taking more. In this connection, it may be stated that ore in any sized lumps can be correctly sampled if a large enough sample is taken; but if the lumps are too large, the sample will be inconveniently bulky.

44. Effect of Character of Ore.—Larger samples must be taken of rich than of poor ores, because the former must be more accurate in order to prevent losses. When the rich minerals occur irregularly, the sample must be larger than when those minerals are uniformly distributed; otherwise there will not be the proper balance of rich and poor.

45. Size of Sample.—If a 100-ton lot of ore was crushed fine enough and was perfectly mixed, a correct sample would be obtained by taking a few ounces from any part of the lot; but this would be very expensive, both for the labor of mixing and for the power used in crushing, and would generally leave the ore too fine for any following treatment. Moreover, we should never feel sure of getting a perfect mixture of such a large lot. The method used is to crush the entire lot, leaving it as coarse as is required for concentrating, smelting, or other subsequent process, and take a suitable weight for a sample. This is crushed finer and a smaller portion taken for the sample.

46. Rules for Size of Sample. — To decide what weight to take for lumps of a particular size, an approximate estimate of the quality of the ore is made by judging from its appearance or from previous knowledge of the mine from which it comes. Having decided this for lumps of one size, the weight to be taken for lumps of any size may be found by the following rule: *Take such a weight from each size as will be equal to a fixed number of the particles;* that is, if the weight taken after crushing to 1 inch is equal to 50,000 1-inch lumps, then after crushing to ¼ inch, the quantity taken should be equal to 50,000 ¼-inch lumps. Since the weight of a lump of ore is proportional to the cube of its diameter, this rule may be stated as follows: *For a given ore, the weight taken for a sample should be proportional to the cube of the diameter of the largest particle of the ore.*

47. This rule would be all right if the different minerals were entirely detached from one another and if all the particles of any given mineral were equally rich. But most ores have to be crushed finely before the minerals will be wholly detached from one another. Now, if the entire lot of ore was in one large lump, that single lump, of course, would be a perfect sample. If, on the other hand, the ore was crushed so fine that all the minerals were perfectly detached from one another, a single particle could not be a correct sample, for it would contain only one of the minerals. Thus when the particles are large, fewer of them can be taken than when they are small. The following rule conforms to this last statement and gives results that agree with good practice: *For any given ore, the weight taken for a sample should be proportional to the* SQUARE *of the diameter of the largest particle.*

48. This rule was worked out by Professor R. H. Richards, of the Massachusetts Institute of Technology, after studying the practice of several careful managers, and is given here, together with the accompanying table, by his special permission.

The accompanying table embodies this rule, and each column in the table is based on figures taken from practice. Column 1 applies to such materials as iron ore, very low-grade lead or copper ores, or even to low-grade gold ores when the gold is contained in pyrite and the pyrite is evenly distributed through the ore. Column 2 applies to ordinary low-grade copper, lead, and zinc ores, or to any ores of not too high a grade in which the valuable minerals are uniformly distributed. Columns 3 and 4 apply to richer ores in which the valuable minerals do not occur in irregularly distributed spots. Column 5 applies to native gold ores with fine nuggets of gold (these nuggets occurring in spots), to ordinary telluride ores, and to certain ores containing sulphide of silver. Column 6 applies to native gold ores in which the gold occurs as large nuggets, also to rich telluride ores, and to rich ores of silver sulphide (silver glance) and silver chloride (horn silver).

49. Such a native gold ore as that just mentioned cannot be sampled in the sense of gradually reducing the size of the particles and at the same time reducing the quantity of ore. Suppose such an ore contained all its gold in nuggets no smaller than $\frac{1}{8}$ inch in diameter and there was an average of five such nuggets in a ton of ore. When the quantity of sample was reduced to 20 pounds, a single nugget in the sample would make the latter too rich; and if this 20 pounds contained no nugget, it would be too poor. The only way to find the value of such an ore is to extract the gold from a 10- or 20-ton sample.

50. In using the accompanying table, it is not necessary to pass successively through all the sizes shown in any one of the columns; but before taking any of the quantities given in the weight column, the ore must be reduced to the corresponding size shown in one of the other columns.

WEIGHTS TO BE TAKEN IN SAMPLING ORE

Weight. Pounds	Diameters of Largest Particles					
	1	2	3	4	5	6
	Very Low-Grade or Very Uniform Ores.	Low-Grade or Uniform Ores.	Medium Ores		Rich or "Spotted" Ores.	Very Rich or Excessively "Spotted" Ores.
	Milli-meters *	Milli-meters *	Milli-meters *	Milli-meters *	Milli-meters*	Milli-meters *
20,000.000	207.00	114.00	76.20	50.80	31.60	5.40
10,000.000	147.00	80.30	53.90	35.90	22.40	3.80
5,000.000	107.00	56.80	38.10	25.40	15.80	2.70
2,000.000	65.60	35.90	24.10	16.10	10.00	1.70
1,000.000	46.40	25.40	17.00	11.40	7.10	1.20
500.000	32.80	18.00	12.00	8.00	5.00	.85
200.000	20.70	11.40	7.60	5.10	3.20	.54
100.000	14.70	8.00	5.40	3.60	2.20	.38
50.000	10.70	5.70	3.80	2.50	1.60	.27
20.000	6.60	3.60	2.40	1.60	1.00	.17
10.000	4.60	2.50	1.70	1.10	.71	.12
5.000	3.30	1.80	1.20	.80	.50	
2.000	2.10	1.10	.76	.51	.32	
1.000	1.50	.80	.54	.36	.22	
.500	1.00	.57	.38	.25	.16	
.200	.66	.36	.24	.16	.10	
.100	.46	.25	.17	.11		
.050	.33	.18	.12			
.020	.21	.11				
.010	.15					
.005	.10					

* 25.4 mm. (millimeters) = 1 inch.

51. The following figures, based on practical work, show what are considered proper sizes and quantities for certain kinds of ore; but they are not to be taken as a universal guide. In some cases smaller quantities may be sufficient, and in other cases larger quantities will be needed.

52. Sizes of lumps and their corresponding quantities for iron ores and other low-grade material:

With 10-inch lumps, take 10 tons for a sample.
With 6-inch lumps, take 5 tons for a sample.
With 3-inch lumps, take 1 ton for a sample.
With 1-inch lumps, take 200 pounds for a sample.
With $\frac{1}{4}$-inch lumps, take 10 pounds for a sample.
With $\frac{1}{25}$-inch* lumps, take $\frac{1}{2}$ to 1 pound for a sample.
With $\frac{1}{80}$-inch † lumps, take 4 to 6 ounces for a sample.

53. Sizes of lumps and their corresponding quantities for copper or lead sulphide ores of a value not exceeding \$50 or \$75 a ton (including silver and gold):

With 6-inch lumps, take 15 to 20 tons for a sample.
With 3-inch lumps, take 4 to 5 tons for a sample.
With 1-inch lumps, take 1,000 to 1,500 pounds for a sample.
With $\frac{1}{4}$-inch lumps, take 200 to 400 pounds for a sample.
With $\frac{1}{12}$-inch ‡ lumps, take 2 to 4 pounds for a sample.
With $\frac{1}{100}$-inch § lumps, take 4 to 6 ounces for a sample.

SAMPLING MILLS

54. Introductory.—The arrangement of a complete mill depends on the opinions of the designer and on the character of the ore. Some managers prefer all hand sampling; but in most cases mechanical sampling is better, at least for part of the work, unless the ore comes in very

---

* Through 16-mesh screen.   † Through 80-mesh screen.   ‡ Through 12-mesh screen.   § Through 100-mesh screen.

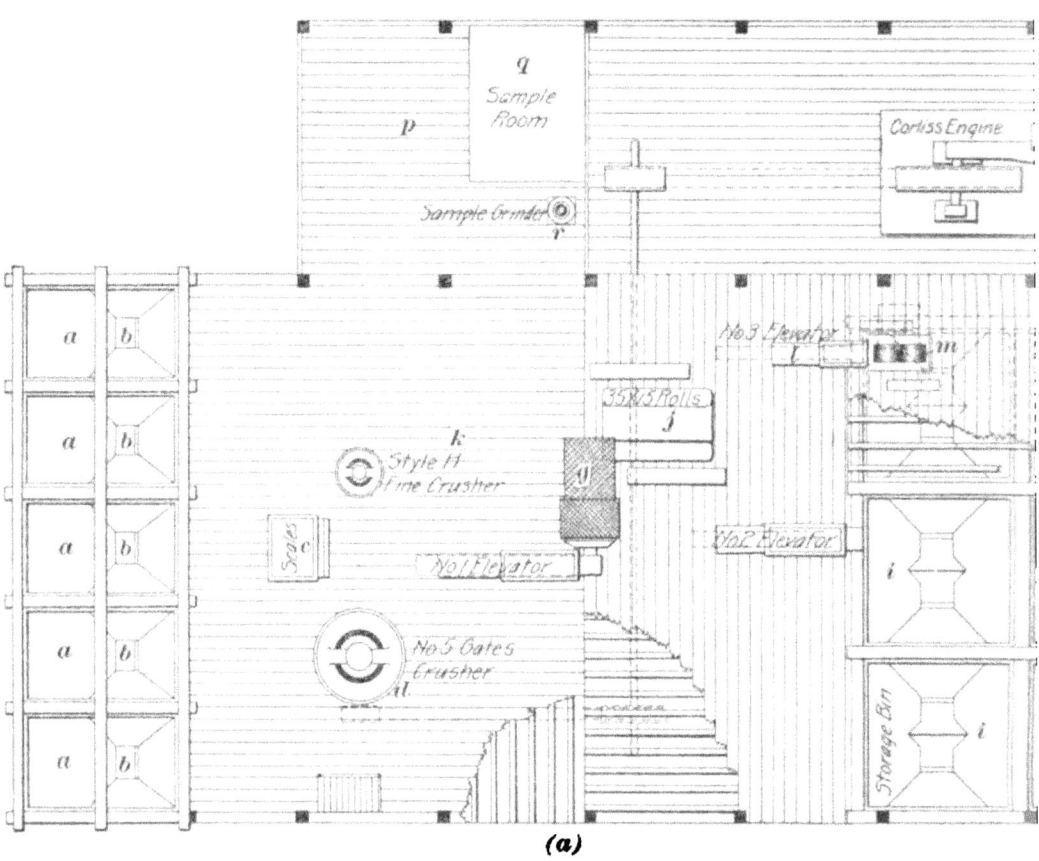

(a)

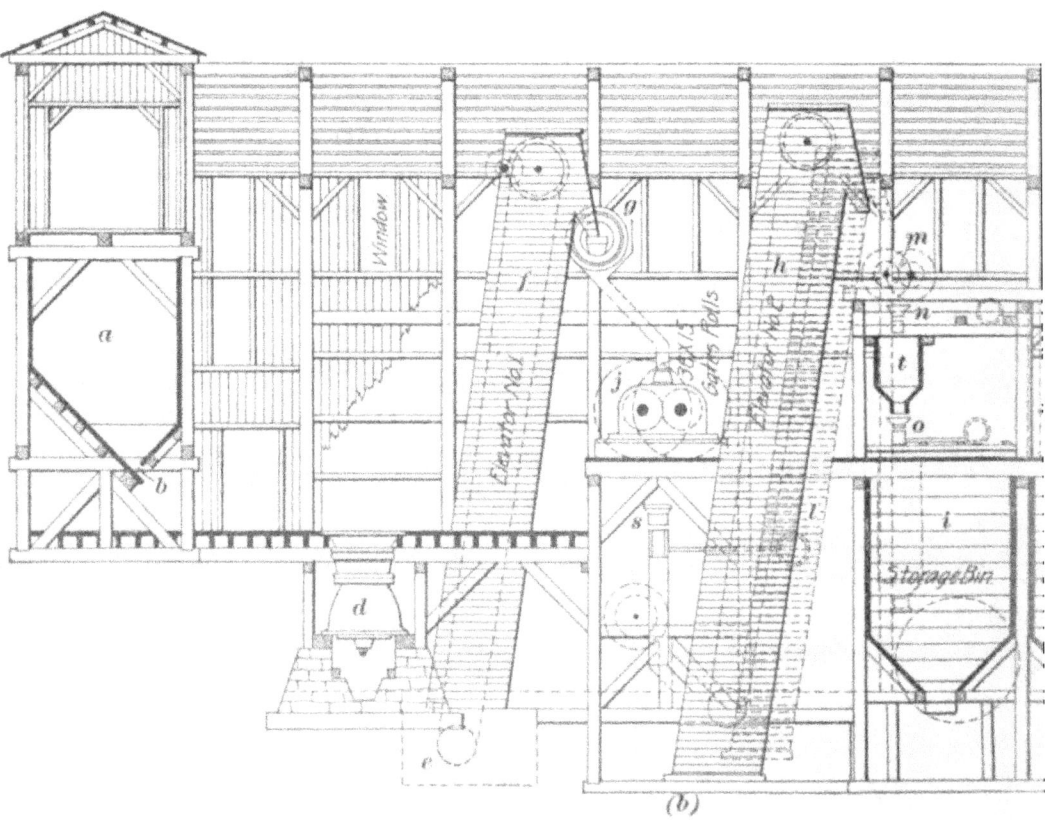

(b)

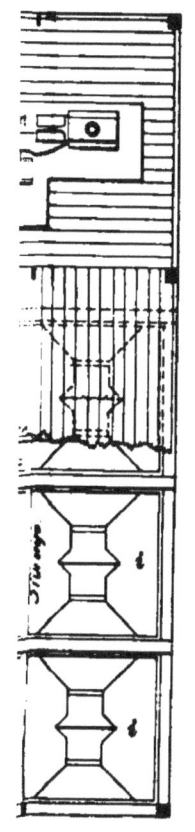

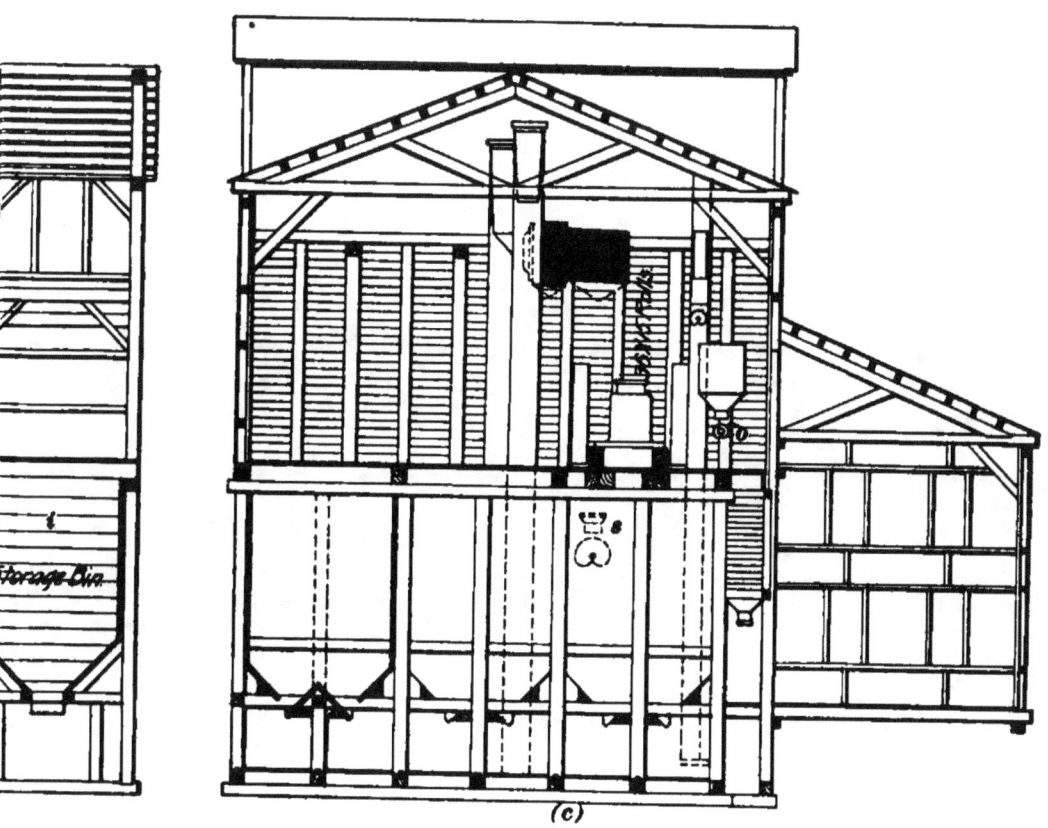

(c)

small lots. The two methods are compared in Art. **29.** When the bulk of the ore is to be left coarse for smelting, the first sample is taken by fractional selection and crushed and the further sampling is done either by hand or by machine. When all the ore is to be crushed fine for treatment by some leaching process or when lead ores are to be roasted, mechanical sampling is most convenient, except for the final steps in which the quantity is small.

SAMPLING MILL FOR A GOLD MINE

55. General Description.—In Fig. 22, (*a*) is a plan, (*b*) a side elevation, (*c*) an end elevation of a sampling mill used by a gold-mining company. The ore comes to the mill in cars and is dumped into ore bins *a*, shown in views (*a*) and (*b*). There are several of these ore bins, so that each lot of ore can be kept separate. The ore is drawn from the bins through gates *b* into ore barrows and is then wheeled to the platform scales *c* to be weighed. If a moisture sample is needed, a shovelful of ore may be taken from every fifth barrow after weighing. The ore is then dumped into the coarse-rock breaker *d* and is raised from the boot *e* by No. 1 elevator *f* to the revolving compartment screen *g*. The ore that sifts through the fine screen passes directly to No. 1 sampler *s*, where a sample is taken and the discard carried by No. 2 elevator *h* directly to the storage bins *i*.

56. Sizing the Ore and Taking the First Sample. The revolving screen gives three products: the fine, which has just been disposed of, the intermediate, and the coarse. The ore that sifts through the intermediate screen passes to crushing rolls *j*, 36 inches in diameter, and then to No. 1 sampler *s*, where a sample is taken. The discard is taken to No. 2 elevator *h* and by it to the storage bins *i*. The third, or coarse, product of the screen passes through a rock crusher *k*, termed the "fine crusher." The ore is taken from this crusher by a scraper line to the boot of elevator No. 1, is raised by the elevator, sifted in the revolving screen, and the product disposed of as described for fine and intermediate ore.

57. The object of using the screen instead of passing all ore directly through the three crushers in succession is to remove the fine ore from the coarse as soon as possible and thus prevent its being reduced to dust. There are some cases where so much crushing as is described above would not answer, for the reason that there would be too much very fine ore for the subsequent operations. Coarse sampling is, therefore, used for the product passing through the screen to the fine-rock crusher mentioned.

58. Reduction of the First Sample.—The ore that is being sampled passes through the screen and the No. 1 sampler, where a rather large sample is taken. This sample is raised by No. 3 elevator l to sample rolls m, and passed from them to No. 2 sampler n, where a sample is taken and the remainder of the ore discarded. The sample is next carried to an intermediate bin t, which is arranged between No. 2 and No. 3 samplers o, in order to give No. 3 sampler a continuous feed. No crushing is done between No. 2 and No. 3 samplers, but it is safer to take, say, a 20-per-cent. sample and then to take 20 per cent. of this, which makes 4 per cent. of the original sample, than to take 4 per cent. by a single operation. In doing this, however, the intermediate bin is necessary, otherwise the sample from the No. 2 sampling machine might all pass to the sample side or to the discard side of the No. 3 sampling machine. The plant is so arranged that all discarded ore can be delivered from No. 2 elevator h to any one of the six storage bins i. The final machine sample is carried by car to the sample room p, where it is cut down by quartering or fractional sampling on the iron-covered floor q. A sample grinder r is provided so that the ore may be pulverized to reach the final sampling for the assayers.

59. Diagram of the Course of the Ore.—Fig. 23 illustrates the progress of the ore through the sampling mill into the sample bin. From the sample bin it is taken to the sample room for further reduction by hand.

60. Drying the Sample.—Some form of drying table should be provided to dry the samples, for, as stated in Art. **28,** the ore should contain no moisture when assayed.

61. Sampling Room. The door to the sampling room should have a lock, so that only the head sampler can enter when the final samples of purchased ore are being prepared. The room

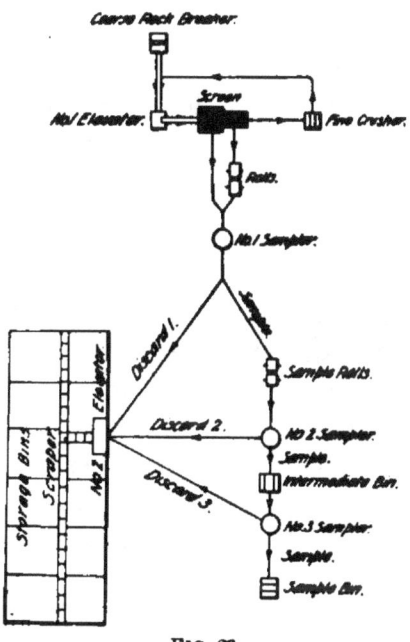

FIG. 23

should be well lighted, and the windows should be so placed that interested persons can watch the work from the outside.

THE TAYLOR-BRUNTON SAMPLING MILL

62. General Description.—In Fig. 24 is shown a section of the Taylor and Brunton sampling system, which is about as accurate and complete as any present automatic sampling mill, although some prefer the Vezin sampler. In the figure, a broad-gauge box car is shown in a shed. The ore is shoveled out of this car into a steel hopper *a*. The gate *b* of the hopper is arranged with a notched lever, so that it can be opened to suit the size of the ore passing on to the shaking grizzly *c*. The grizzly is not placed at a steep angle, because it is desired to feed the crusher *d* slowly and it is further desirable to riddle the ore and not crowd the crusher with ore that is fine enough.   The ore that passes through the grizzly falls into a steel hopper c, where it meets the ore that passes through the $20'' \times 10''$ crusher d.   The ore slides down the steel chute and hopper into the elevator boot f, from which place it is raised by the elevator buckets g to the top of the sampling mill.   The elevator buckets discharge into a complete bucket discharge at h, which, however, is not shown in the figure.

63. Taking the Sample.—From the complete discharge, the ore passes down a chute, where it meets the No. 1 sampler i, which takes out 20 per cent. of the ore, the remaining 80 per cent. is discarded through the chute j into the ore bin k.   From the sampler i, the 20-per-cent. sample passes through a set of $16'' \times 36''$ crushing rolls l to the No. 2 sampler m, which takes a 4-per-cent. sample of the original ore and discards 16 per cent.; in other words, 20 per cent. of the No. 1 sample is retained and 80 per cent. discarded.

The No. 2 sample now passes through $14'' \times 27''$ rolls n and then through No. 3 sampler o, where .8 per cent. of the original ore is taken and 3.2 per cent. discarded; that is, 20 per cent. of No. 2 sample is retained and 80 per cent. discarded.

The No. 3 sample next falls to a rotating steam drier p, from which it is fed by the feeder q to a pair of $12'' \times 20''$ crushing rolls r.   From the crushing rolls the ore passes to the No. 4 sampler s, which retains .16 per cent. and discards .64 per cent. of the original ore; i. e., 20 per cent. of No. 3 sample is retained and 80 per cent. discarded.

The No. 4 sample passes to the sample safe t, from which it is removed and sampled by hand, so as to still further reduce the size of the sample.

64. Size of Final Mechanical Sample.—If the car contained 20,000 pounds of ore, the size of the mechanically obtained sample would be 32 pounds, which can be further reduced by hand.   Thus it is demonstrated that mechanical ore sampling is not entirely mechanical, but depends for its final stages on hand sampling.   The

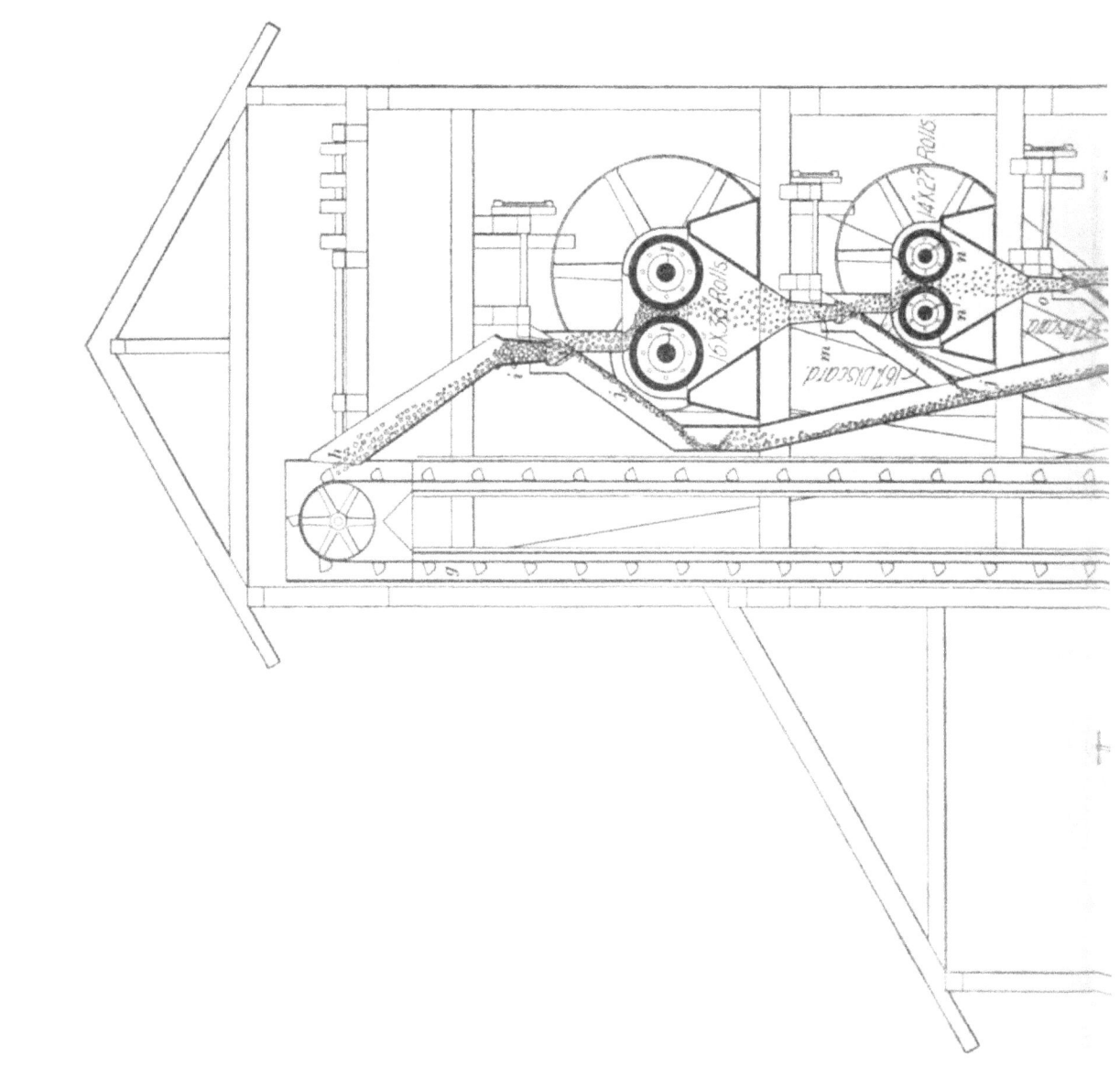

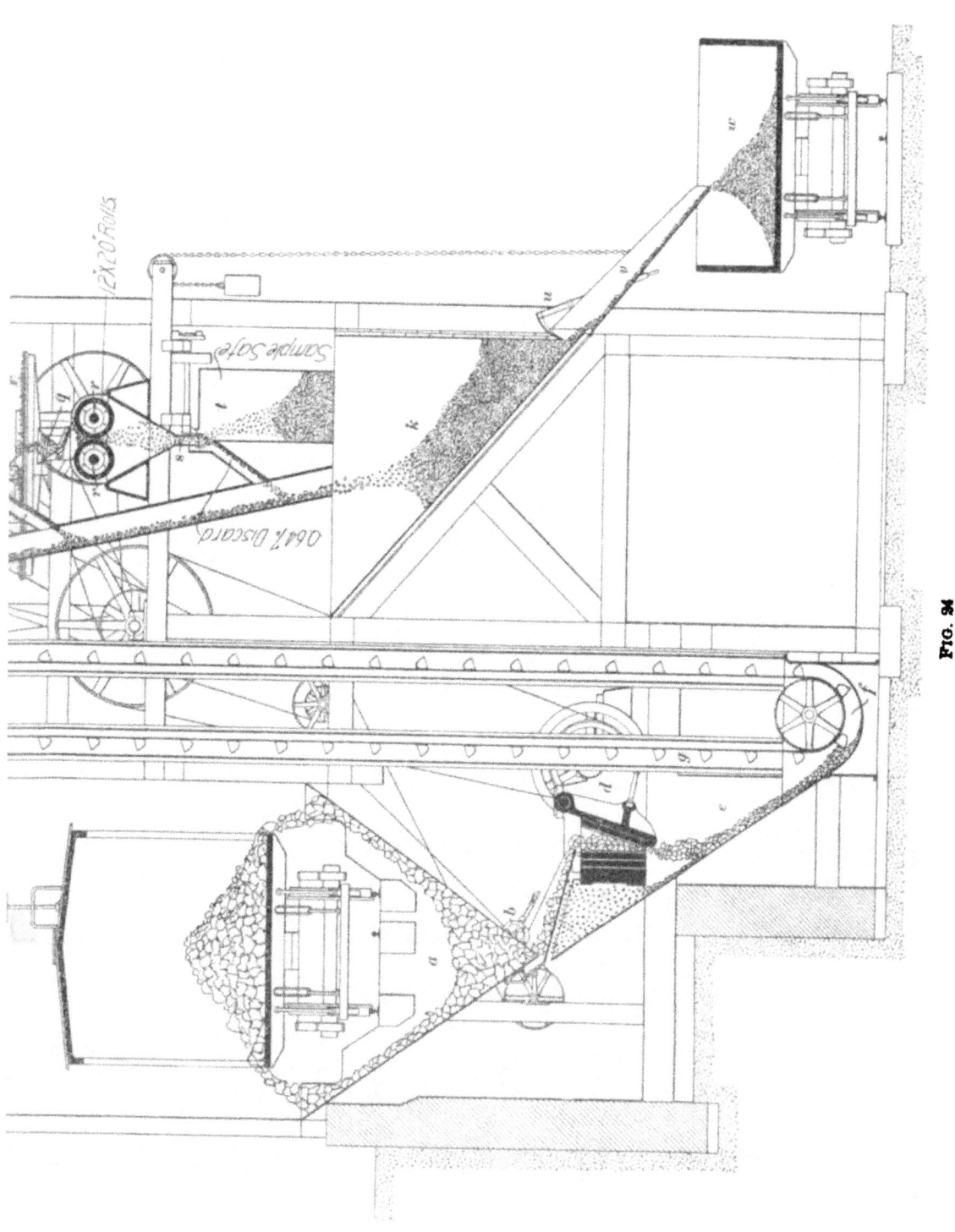

FIG. 94

sampled ore has now reached the ore bin *k* and is drawn off through the gate *u* and chute *v* into the car *w*, which removes it to mill or furnace, as the case may be. If 1 ton only were sampled, the first sample would be 400 pounds, the second 80 pounds, the third 16 pounds, and the fourth 3.2 pounds from 1 ton. In this case the ore could be finally divided by the Bridgman apparatus.

65. Advantage of Height in a Mill.—It will be noted that the Taylor and Brunton sampling mill is higher than that shown in Fig. 23. This permits the ore to be raised high enough to descend by gravity from one machine to another, thus economizing in power, as the second and third elevators of Fig. 23 are not required. In this plant, little bin capacity is needed, but by using a scraper line under the spout as the ore comes from the discard a series of bins can be filled.

66. Use of the Drier.—The drier is used only for ores that are quite moist, such ores forming lumps when ground fine. Soft damp ores when passing through rolls simply flatten out in sheets, thus making it necessary to dry such ores before attempting to sample them by hand or machinery.

ROASTING AND CALCINING ORES

INTRODUCTION

DEFINITION

1. Ore roasting is the process by which certain chemical changes are produced by the aid of heat, but at a temperature so comparatively low that the ore does not fuse. It is one of the most important operations in metallurgy, because the quality of the product from this process controls the results in the various processes that follow.

2. The most common example is the **oxidizing roast**, by means of which sulphides are converted into oxides by the oxygen of the air, the sulphur passing off principally as sulphur dioxide SO_2. Arsenic and antimony are also oxidized, the object being to form as much arsenic trioxide As_2O_3 or antimony trioxide Sb_2O_3 as possible, because these compounds pass off by volatilization. Other compounds of sulphur, arsenic, and antimony which are not volatile are also produced by roasting; these, however, can generally be changed to a condition in which they can be volatilized by the **reducing roast,** which consists in the admixture of fine coal to take away oxygen, atmospheric oxygen being excluded during this operation. When the sulphur, arsenic, and antimony have practically all been removed by a series of alternate oxidations and reductions, the ore is said to be **dead-roasted** or **sweet-roasted.** For the **sulphatizing roast** the operation is conducted at an especially low

§ 30

For notice of copyright, see page immediately following the title page.

temperature with a small supply of air and with the ore bed somewhat deeper than usual; by this means considerable sulphur is converted into sulphur trioxide SO_3, which combines with the metals to form sulphates.

3. In the **chloridizing roast,** certain metals, especially silver, are changed to chlorides by mixing their ores with common salt $NaCl$ after practically all the sulphur, arsenic, antimony, etc. have been expelled.

PURPOSES OF ROASTING

4. The object of roasting the sulphides of lead and zinc is to convert them into oxides, which are afterwards reduced to metal in smelting furnaces. In the case of copper sulphide, the purpose generally is to burn off the excess of sulphur, leaving enough so that when smelted it will combine with all the copper to form an artificial sulphide called **matte,** which always contains iron. Copper-sulphide ores with 4 per cent. or less of copper are often roasted to sulphates, or sometimes to chlorides, the copper being subsequently dissolved by some liquid.

5. Silver ores are often chloridized so as to prepare them for amalgamation. Gold ores that are to be treated by the chlorination process generally contain sulphides or arsenides, and as these minerals are detrimental to chlorination, such ores are dead-roasted. If oxides of copper, calcium, or magnesium remain after this roasting, common salt is added to the ore towards the end of the roast to convert them into chlorides; otherwise they will absorb chlorine gas during the process of chlorination. It often happens that with finely pulverized ores and slimes which are to be treated by the cyanide process, the solutions will not easily pass through the ore. Again, some ores, for instance tellurides, are so compact that the solutions cannot enter and attack the gold. In such cases roasting will often remedy the trouble by producing physical changes in the ore.

6. Iron ores are sometimes roasted to remove carbon dioxide CO_2 and water and also to remove sulphur. Zinc carbonate is also roasted to remove carbon dioxide. This kind of roasting is termed **calcining.** Lump ores are sometimes heated in order to eliminate moisture and allow them to crush more readily. The heat in drying sometimes expands the rock and produces small cracks. Ores that are to be crushed fine and screened generally have the moisture driven from them by heat to render them more friable.

7. Roasting is usually preliminary to some metallurgical process, but in some cases it is preliminary to mechanical concentrating. Mixtures of blende and pyrite are roasted for the purpose of concentration by magnetic concentrators.

Experiments have been made with hematite and limonite iron to convert the ferric oxide Fe_2O_3 into magnetic oxide Fe_3O_4, but they have not proved commercially successful. On the other hand, magnetic iron ore containing sulphur has been successfully converted into hematite by roasting to eliminate the sulphur.

METHODS OF ROASTING

8. The methods adapted for roasting fine ore are not suitable for coarse ore, and the methods used for coarse ore are still less suitable for fine. Fine ores are generally treated in beds only a few inches deep, which are turned over and over to expose all parts to the air. With coarse ores this would be troublesome in various ways; they are, therefore, roasted in masses from 6 to 30 feet deep, the necessary air passing up through the mass between the lumps. Fine ore cannot be treated in this way, because it would pack so solidly that the air could not pass through it. A diameter of $\frac{1}{2}$ inch is probably the limit or boundary line between coarse and fine ore for roasting purposes, and in most cases the largest particles in a lot of fine ore are much

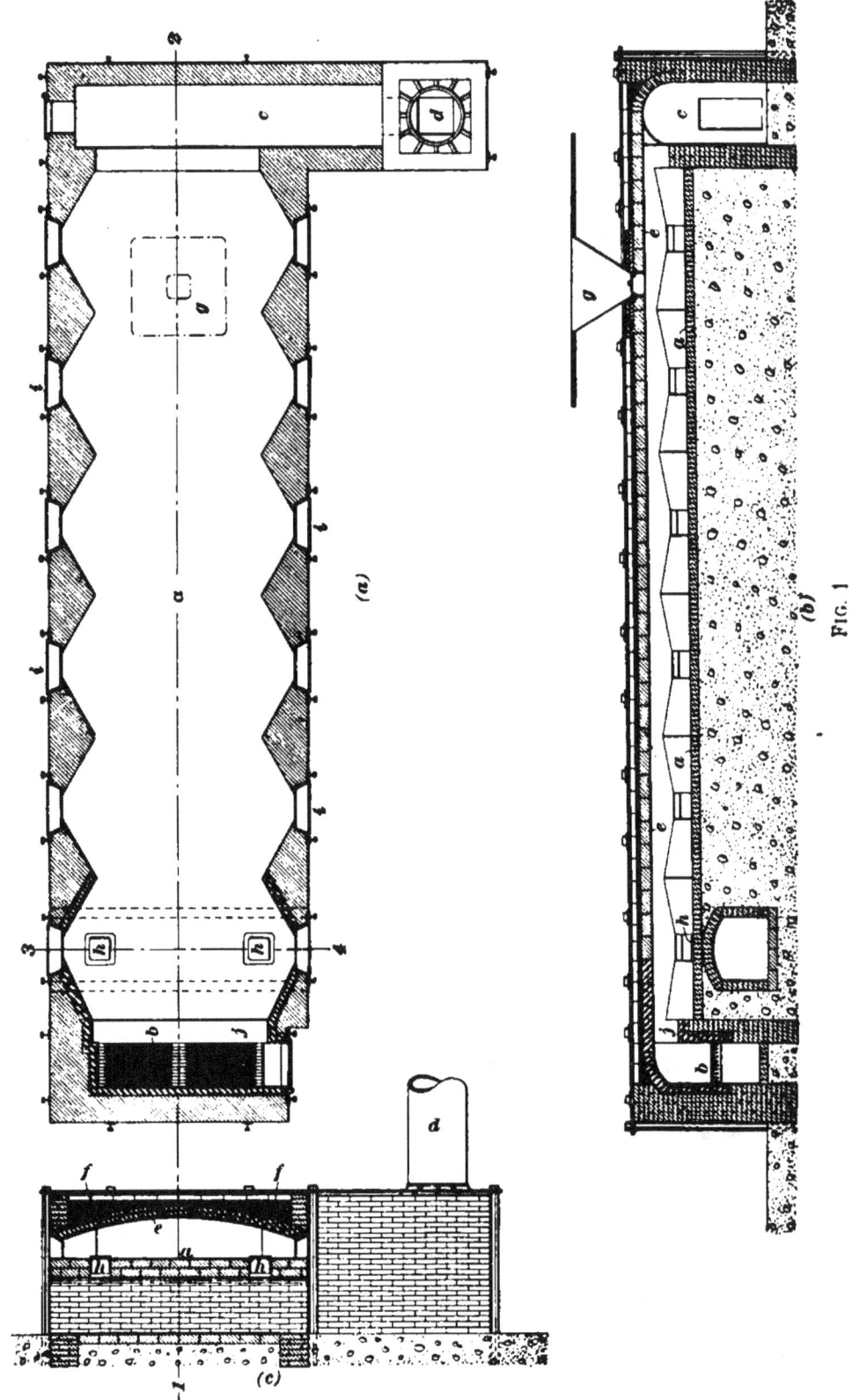

FIG. 1

smaller. Since in roasting one must deal with fine ore more often than with lump, the furnaces for the former will be described first and the methods for roasting lump ore afterwards.

ROASTING FINE ORE

9. Furnaces for fine ore may be classified as *hand-rabbled reverberatories, mechanically rabbled reverberatories, revolving cylinders,* and *shaft furnaces.*

HAND-RABBLED REVERBERATORY FURNACES

10. A hand-rabbled reverberatory furnace is shown in Fig. 1. In this figure, (*a*) is a horizontal section, (*b*) a vertical section through the line *1–2*, and (*c*) a vertical section through the line *3–4*. The furnace consists of a long hearth *a* (on which the ore is spread from 2 to 6 inches deep) with a fire-grate *b* at one end and a flue *c* at the other, leading to the chimney *d*. The fireplace and hearth are covered by the arched roof *c*, shown in Fig. 1 (*c*). This arch is made as flat as is possible without danger of collapse, so that the flame and heat will spread to the sides of the furnace. The top of arch *c* is commonly covered with sand *f* to prevent loss of heat by radiation.

11. The ore is dumped from the hopper *g*, Fig. 1 (*b*), on to the hearth *a*, and is occasionally *rabbled*, or stirred, to expose fresh ore to the heated air and to prevent caking. The ore is gradually moved by a paddle towards the fire end, where it is discharged through the openings *h* into cars. The openings are closed when ore is not being discharged through them. In order to expose as much surface as possible to the oxidizing action of the air, the workmen make furrows in the ore instead of leaving the surface flat. The rabbles and paddles for working the ore are inserted through doors *i* on both sides of the furnace. The bottoms of these doors are level with the hearth.

The rabble for stirring the ore is shaped like a hoe, the blade being about 4 inches by 9 inches. The paddle for pushing the ore forwards is similar but larger. As these tools are long and heavy, an iron bar is placed across each doorway, a few inches higher than the hearth, to serve as a rest or fulcrum on which to swing the tools from one position to another. Rollers are also used, but they are apt to get out of shape and not roll.

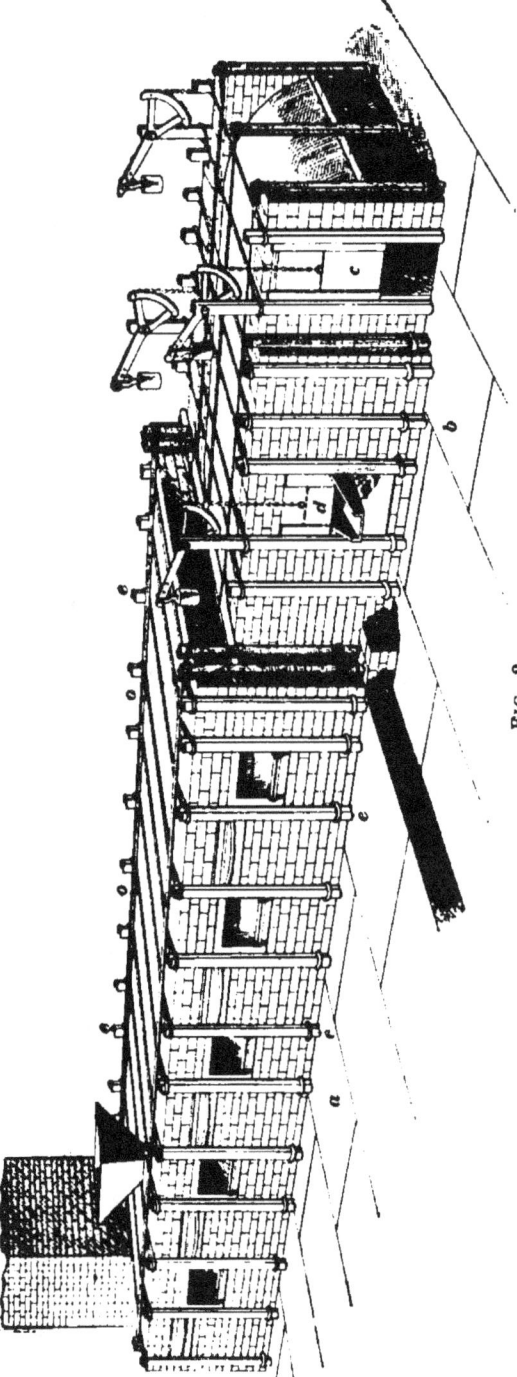

Fig. 2

12. In order to exclude cold air, the working doors are kept closed when the men are not rabbling the ore or moving it forwards. All the air for roasting should be heated. Sometimes the whole air supply comes through the fire; but for an oxidizing roast, it is better to obtain a special supply of fresh air through openings in the fire-bridge *j*. For the latter purpose, the bridge is made hollow from end to end; and the air becoming heated while passing through this hollow space, enters the furnace below the flames from the fire through several openings on the side towards the hearth.

13. For a reducing roast, this fresh air supply must not be used. Indeed, the bed of fuel on the grate should for the reducing roast be so deep that no free oxygen will pass through it into the furnace. On the contrary, the gases from the fire should be incompletely oxidized, in order to assist in the reduction of the ore.

14. Lead ores are often partially fused after roasting, so that they will be in better mechanical condition for smelting in the blast furnace. Fine ore makes the blast furnace work irregularly and creates losses. For the purpose of fusing, modifications to the furnace shown in Fig. 1 are made. The roasted ore is dropped about 2 feet from the roasting hearth *a*, Fig. 2, to the **slagging hearth** *b*. (This name is given to the hearth *b* because the ore is *slagged* or well fused on this hearth.) It will be noticed that the roasting hearth is considerably wider than the slagging hearth. By making it this way, the gases from the firebox *c*, which are hot enough to fuse the ore in *b*, expand when they enter *a* and are thus cooled to a suitable temperature for roasting. When the ore is properly fused on the hearth *b*, it is drawn out through the side doors *d* into slag pots, cooled, and broken up.

15. The slagging volatilizes a considerable portion of the lead, so that it cannot be recovered; therefore, this furnace should be used only for low-grade ores. For rich ores, the hearth *b*, Fig. 3, is made wide, so that the heat from the fire is sufficient only to **sinter** or slightly fuse the ore; in this case *b* is called the **sinter hearth**. At some works for treating ores that are easily fused, the hearth *b* is omitted and the ore sintered at the fire end of the hearth *a*. In these cases a depression is made in the hearth near the fire-bridge,

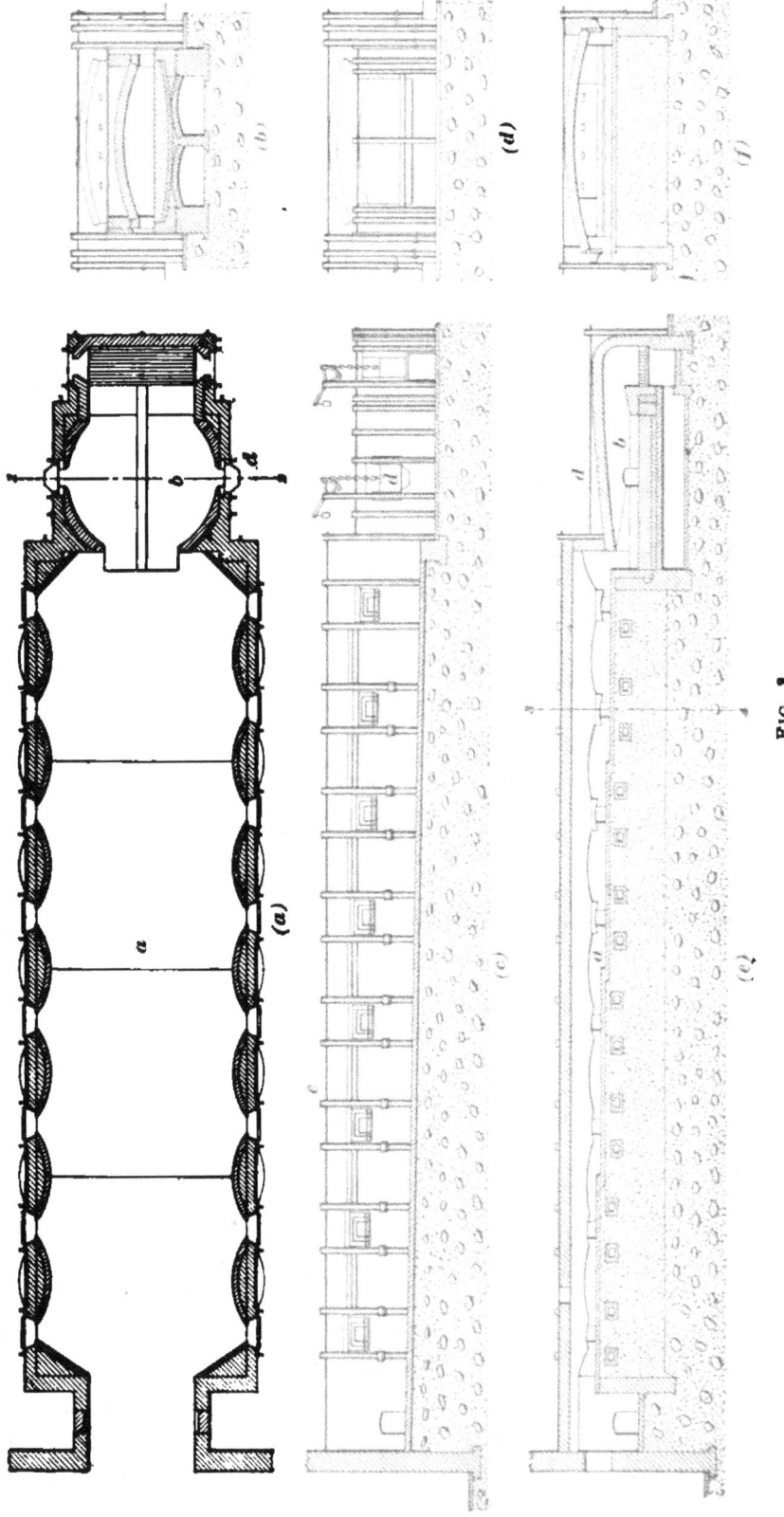

(a)

(b)

(c)

(d)

(e)

(f)

FIG. 8

in which the sintered ore is collected.   When enough sintered ore has collected here, it is scraped through the side doors into iron pots or barrows and is pounded in order to solidify it.   When cold, it is broken up.   If the sintering is done on the same hearth as the roasting, there is no distinct line between the ore that is roasting and the ore that is sintering; consequently, some of the galena may not be roasted before it becomes sintered, and in the latter condition roasting is very imperfectly done, if at all.

16.   Unless the ore contains a good deal of silica, some highly silicious ore is spread on the slagging hearth before drawing the charge on to it from the preceding hearth. This is to prevent the corrosion of the hearth by the hot ore. Some of this silicious ore is also thrown on top of the charge on the slagging hearth if necessary and is well stirred in. When this has been done, the fire is made hot for several hours and the ore thoroughly rabbled at stated intervals.

17.   When a furnace is heated the brickwork expands, and therefore it must be bound with iron in order to prevent its becoming distorted and falling to pieces.   One method of tying the walls of a furnace is shown in Fig. 2, which is a perspective view of a hand-rabbled reverberatory with a slagging hearth.   Steel rails or **I** beams c, termed **buckstaves,** are placed in vertical positions at intervals along the sides and ends of the furnace, their upper ends extending a few inches above the top of the furnace.   These buckstaves are held together by iron tie-rods o, shown across the top and through the foundation near the bottom of the furnace.   The tie-rods are loosened a little when the brickwork expands, to prevent their breaking, and are again tightened when the brickwork contracts on cooling.

18.   The size of a furnace depends on the ore to be roasted therein.   For an ore containing much pyrite, the length of the hearth is commonly about 65 feet, but for an ore rich in galena or blende it does not generally exceed 40 feet.   The reason for this difference is that pyrite generates so much heat in roasting that it is less dependent on the heat from the grate fire than are the other minerals. The width of the hearth is usually about 15 feet. It is difficult for the men to work the ore if the width is much greater than this. For roasting zinc ores, it is quite common to have working doors on only one side of the furnace, in order to admit as little cold air as possible. The width of the hearth is in that case reduced to 7 or 8 feet, since a greater width would be inconvenient for the workmen.

19. When ores are to be dead-roasted, more satisfactory results can be obtained with a hand-rabbled reverberatory than with any of the mechanical furnaces, because the ore is more under control. In most mechanical furnaces the ore is rabbled with the same frequency at all stages of the operation; but with a hand furnace this can be varied to suit the condition of the ore at different times. Fig. 3 shows a plan, an elevation, longitudinal cross-section, also a cross-section through the line *1–2*, a front elevation, and a cross-section through the line *3–4*.

20. If galena that is roasting on the surface is stirred into the ore bed before it is properly oxidized, it is apt to form lumps, and these do not roast well. Therefore, rabbling should not be done too often during the first part of the roast; but towards the end of the roast it should be done more frequently, in order to expose all parts of the charge to the heat from the grate fire. As both galena and the oxide that results from roasting galena are quite readily fusible, an ore that contains any considerable percentage of this mineral is apt to cake and stick to the hearth. It requires special care to prevent or correct this in mechanically rabbled furnaces, but it is much more easily prevented in hand-rabbled furnaces. For these reasons most metallurgists prefer the long-hearth, hand-rabbled furnace for lead ores. Mechanical furnaces are, however, used in some cases.

The fuel used for roasting should be one that burns with a long flame. With a long flame extending some distance into the roasting hearth, the heat is generated in the hearth, where it is wanted. With hard coal, which burns without any flame, the heat is all generated on the fire-grate. Although the products of combustion from such a fire pass through the hearth and heat it, the heating is not so well done as when the fuel gas burns in the hearth itself.

MECHANICALLY RABBLED REVERBERATORY FURNACES

21. In Fig. 4 (*a*) is shown a horizontal section of a mechanically rabbled reverberatory furnace known as the **Ropp straight-line furnace.** It has a long hearth *a* divided by a narrow slot *b* that extends from end to end. Below the hearth of this furnace is a pit *c*, which is shown in section in Fig. 4 (*b*). In this pit is a narrow track *d*, upon which several carriages *e* are moved by a steel rope *f*, to which they are attached. From each carriage, a strong steel plate *g* extends upwards through the slot *b* for the purpose of carrying the rabbles *h*, which work the ore. The rakes *h* reach the width of the hearth and have their blades pitched at an angle, so that when the rake is moved along, the ore is not only pushed forwards, but is stirred and turned to one side, exposing new surfaces. The blades on one rake are set in the opposite direction from those on the preceding rake; thus the ore is scraped first to one side and then back. Each rake has the blades on one side of the hearth set in the opposite direction from those on the other side, in order to balance the rake; if all the blades were set alike, the resistance of the ore to the inclined blades would force the rake and its carriage to one side instead of allowing it to run true on the track.

22. The steel rope to which the carriages are attached passes around large horizontal sheave wheels *i* and *i'*, one being placed at each end of the furnace. The power to move the rope and rakes is delivered to the sheave *i'* by bevel gears, moved by the horizontal shaft *j*. When the rakes come out of the furnace, they raise the two iron doors *k* and *k'*, which are hinged at their upper edges. The door *k* closes before *k'* opens, so that no cold air

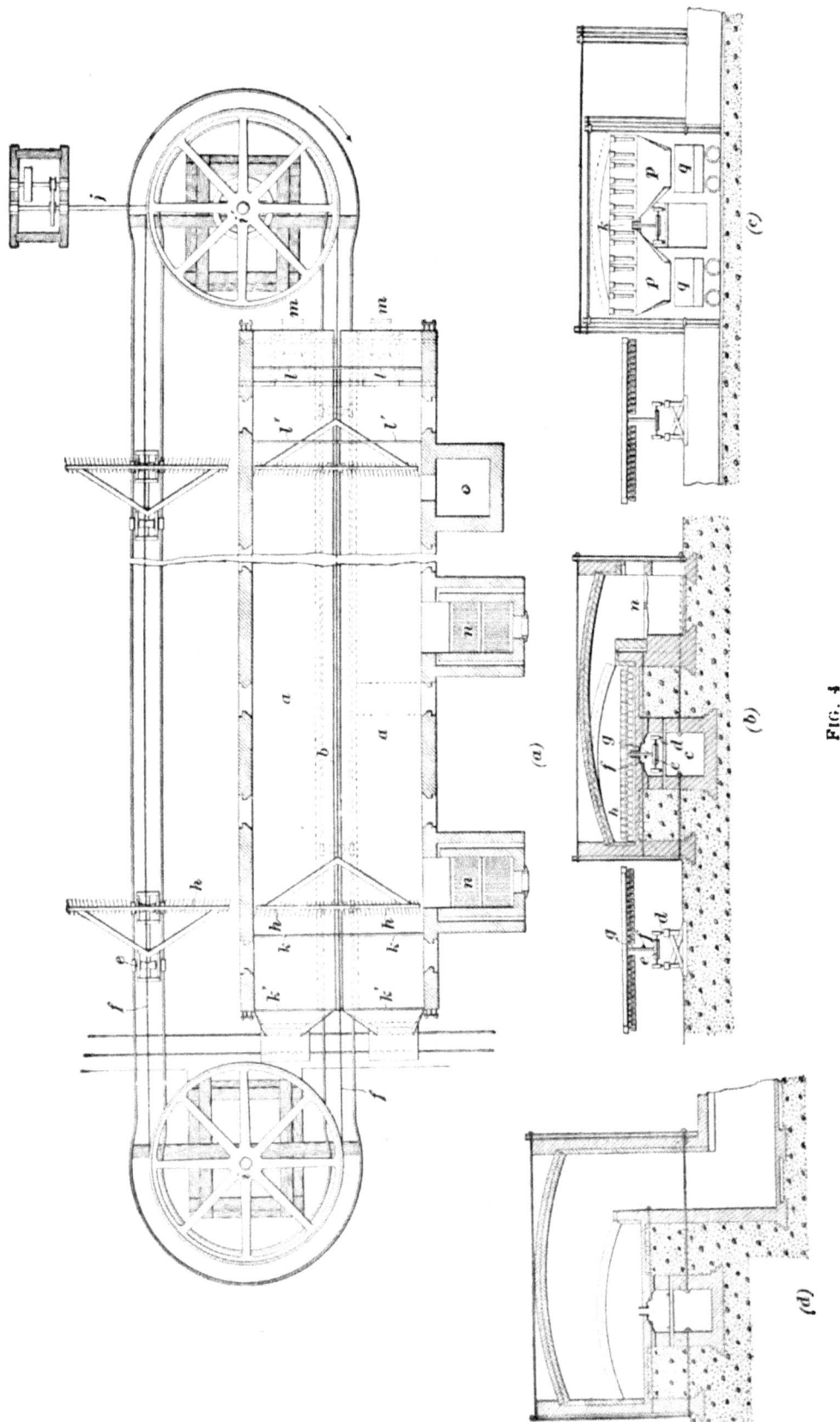

FIG. 4

can enter the furnace. The carriage and rake then pass round the sheave *i* and return to the other end of the furnace in the open air, on that part of the track that is built outside of the furnace. This gives the rakes plenty of time to cool and is an effective method of causing them to wear longer and lessening the expense for repairs. On entering the furnace again, the rakes raise the doors *l* and *l'*.

23. The ore is fed into the furnace by two automatic feeders *m* ; heat is supplied from one or more fireplaces *n* and the fumes made in roasting pass off through the flue *o*. The discharge end of this furnace is shown in elevation at Fig. 4 (*c*). The rake pushes the ore under the doors *k'* to aprons *p*, from which it slides into cars *q*. This furnace is made in different sizes, varying from 100 to 180 feet long and from 11 to 16 feet wide. It is used with good results for copper sulphide ores, for mattes (artificial sulphides produced in smelting) containing copper and lead, and for dead-roasting either pyrite or blende.

24. **Brown's "horseshoe" furnace** is built in the form of a ring, as shown in Fig. 5 (*a*). The hearth occupies about five-sixths of the circumference, while the remaining sixth is an open platform extending from *a* to *b*. The rabble arm *c* is carried by trucks running in the side passages *d*, shown in the cross-section in (*b*), which is a section through the line *1–2*. The rabbles are operated by a steel rope *e*, which is guided by the pulleys *f* set in that passage *d*, which is towards the center of the ring. The rope is driven by a suitable mechanism at *g*. Ore is discharged on the open platform by the automatic feeder at *h* and is moved into the furnace in the direction of the arrow by the rabbles. A door at *i* is automatically opened for the entrance of the rabbles; there is also a similar door at *i'*, where the rabbles leave the furnace and the ore discharged into buggies, which are not shown. Heat is supplied from the three fireplaces *j*, while the gases pass off by the stack *k*.

25. When one of the rabble arms comes out of the furnace on to the open platform, it is automatically detached from the driving rope; at the same time another carriage, which has been standing on the platform, is automatically

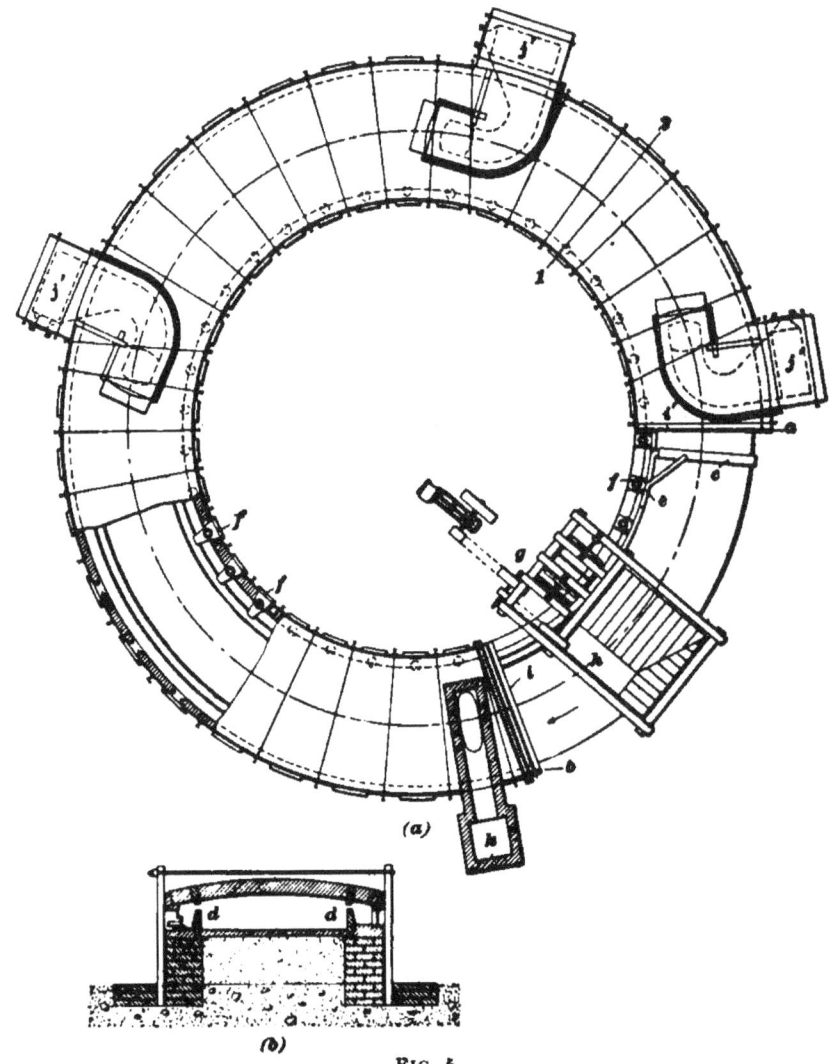

FIG. 5

attached to the rope. This gives the rabbles a chance to cool off, and they last much longer than they would without the cooling. The furnace is successfully used for pyritic ores, for copper matte (sulphatizing roast), and for dead-roasting zinc blende.

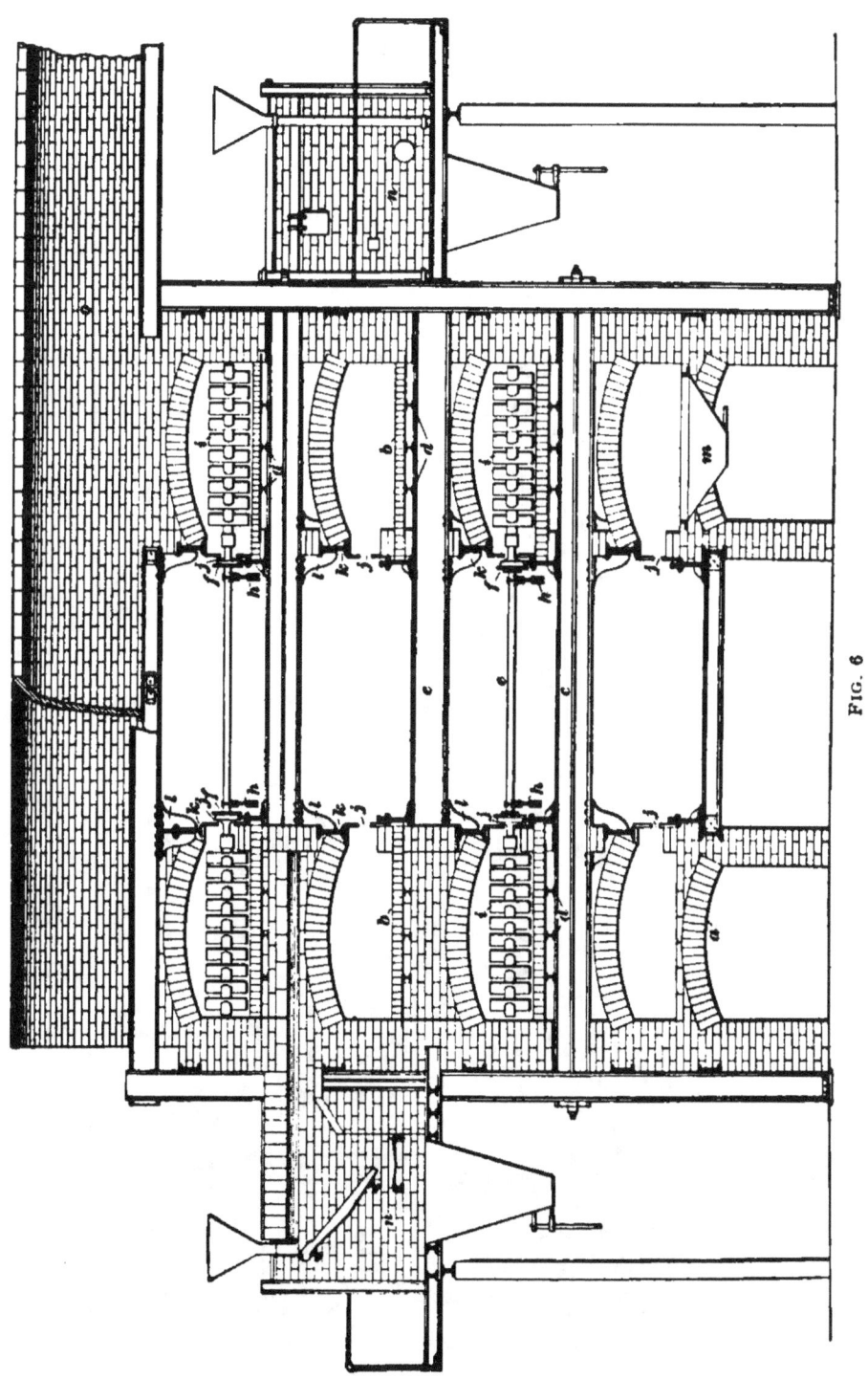

FIG. 6

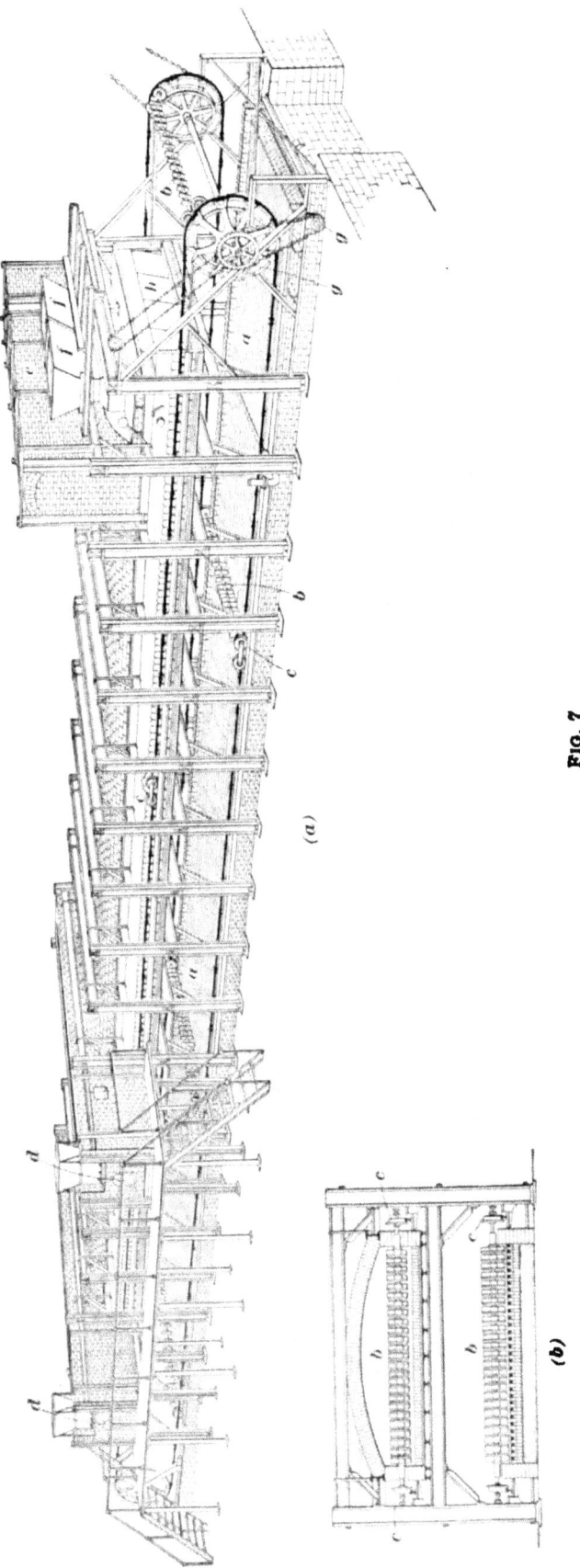

FIG. 7

26. The **Wethey multiple-hearth furnace** is made either single or double. Fig. 6 shows a cross-section of the double form, that is, two furnaces in one structure. The bottom hearths are supported either on solid foundations or on brick arches *a*. The other hearths *b* are supported by means of strong cross-beams *c* of steel, which carry smaller beams *d* running longitudinally. The brick hearths *b* are built on thick steel plates, which are laid on the beams *d*, each hearth being 5 feet wide and 50 feet long. The rabble arms *e* are carried by trucks *f*, which are drawn by chains *h* that pass over sheaves outside the furnace in a manner similar to that shown in Fig. 7. The rabbles *i* pass through and out of the top hearth, over the sheaves, then through and out of the second hearth, over a second set of sheaves at the other end, and back into the top hearth. The two lower hearths are operated in the same way.

Each rabble arm extends from the truck through a slot *j* running the whole length of the furnace. This leaves one side of the arched roof without any support from below; but it is supported by a longitudinal **I** beam *k* hung on brackets *l*. To prevent large quantities of cold air rushing in through the slots just mentioned, they are closed by a series of overlapping sheet-steel plates pivoted above, but not shown in the section given. Each rabble momentarily lifts one of these plates, which, when the rabble has passed, swings into place again. The hearths are closed at the ends by doors (which are not shown) that are hinged at their tops and which are automatically raised and lowered when a rabble enters or leaves a hearth. The ore is fed to the upper hearth by an automatic feeder (which is not shown) and is gradually moved to the opposite end, where it drops through a slot to the second hearth. In the same way it traverses the other hearths and is finally discharged through the hopper *m*.

For pyritic ores, each single furnace has a fireplace *n* leading into the upper hearth near the feed end. The hot gases rapidly dry the ore and soon heat it to a temperature at which it begins to burn. The gases pass off to the flue o at the other end of this hearth. The combustion of the sulphur furnishes enough heat to continue the roasting on the other hearths without any other fireplace. For ores that do not contain so much sulphur as pyrite, a fireplace is used for the bottom hearth also, especially if the ores are to be dead-roasted.

27. Fig. 7 (a) and (b) shows a modification of the Wethey furnace designed by *Holthoff* for the chloridizing roast of silver ores or for dead-roasting gold ores. The principal feature to be noticed is the cooling hearth a, beneath the roasting hearth. To insure thorough cooling, pipes are laid in the hearth and cold water passed through them. Sometimes, in place of a brick cooling hearth containing water pipes, a special hearth made of corrugated iron plates, supported on **I** beams, is used. By this arrangement effective radiation is obtained from the top and bottom of the hearth. In common with the single form of the ordinary Wethey furnace, the rabble arms b extend through both side walls, as shown in section in Fig. 7 (b), each end being supported by a truck c. There are two fireplaces d, and the gases pass off through the flue e. The hoppers f feed the ore automatically to the furnace, power being transmitted to them by sprocket wheels g and chains h.

28. The **McDougall furnace**, shown in section in Fig. 8, is designed to roast pyrite FeS_2 without using any heat except what is produced by the roasting ore itself. Pyrite generates more heat in roasting than any other mineral; indeed, it is the only mineral that can be roasted without the use of heat from some outside source. To be self-roasting, the ore should contain at least 50 or 60 per cent. of pyrite. Gangue minerals absorb heat, but do not contribute any. This furnace is so constructed that a large part of the heat generated by the oxidation of the ore is absorbed by the brickwork and is given out again to the ore, to continue the oxidation after the sulphur in the ore has been so nearly consumed that it can no longer supply the necessary heat.

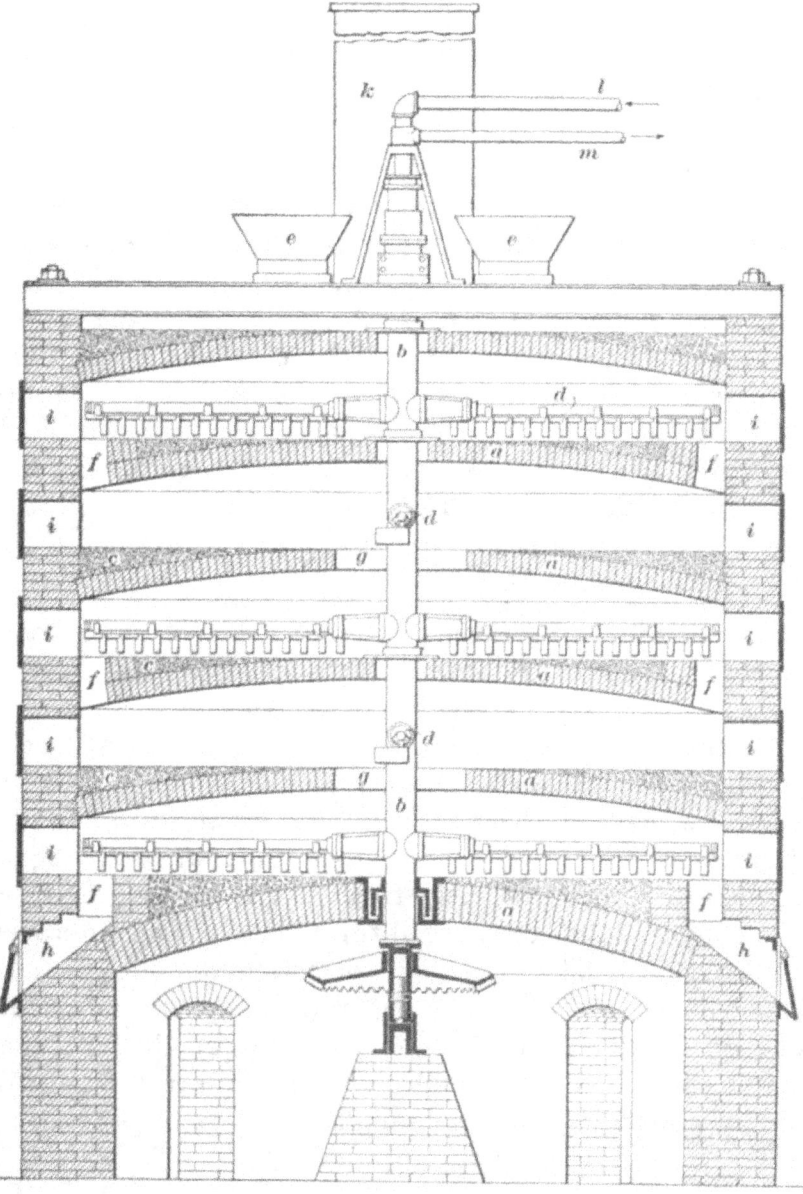

FIG. 8

The furnace consists of a number of circular hearths *a*, one above another, with a vertical revolving shaft *b* extending through the center. The hearths are made level by letting the fine ore *c* fill up on top of the brick arches *a*. The diameter of such furnaces varies from 9 to 18 feet. For each hearth there are two rabble arms *d* fastened to the shaft diametrically opposite each other.

29. To start the furnace, a wood fire is made on the hearths and kept burning until the brickwork is well heated. No more wood is used after that, but ore is then automatically fed from the hoppers *e* on to the upper hearth, where the rabbles gradually move it towards the circumference and drop it through holes *f* to the next hearth. Here the rabble blades are set so as to move the ore towards the center, where it drops through a hole *g* to the third hearth. The movement continues thus alternately to the circumference and to the center until the ore is finally discharged through the chutes *h*. The necessary air enters through the openings *i* in the side walls. The gases move in the opposite direction to the ore and are finally discharged from the upper hearth into the stack *k*.

30. It will be observed that the rabble arms are constantly exposed to the heat. This made a great deal of repairing necessary in the original styles of this furnace; but in the more recent forms this trouble is overcome by circulating water through the arms and the vertical shaft, which are made of heavy tubes with small tubes inside. Water flows in through the inner tube and out again between the inner and outer tubes. The water is supplied from the pipe *l* and discharged through *m*. This improved form of McDougall furnace has displaced the Brückner in certain large works for roasting pyritic copper ores. The capacity is two or three times as great as the Brückner roaster and the moving rabbles give it a still further advantage by preventing the ore balling.

BLAKE'S REVOLVING-HEARTH FURNACE

31. **Blake's revolving-hearth furnace** is shown in Fig. 9 (*a*) and (*b*). It consists of a circular, revolving, firebrick hearth *a* about 16 feet in diameter, made in step form and resting on cast-iron balls *b* arranged in a circular track *c*. These balls serve as an anti-friction bearing on which the hearth revolves. The ore is fed from the hopper *d*, through the center of the roof on to the upper step of the hearth; and as the latter revolves the ore meets one of the plows *e*, set at an angle of 45°, which scrapes it to the next

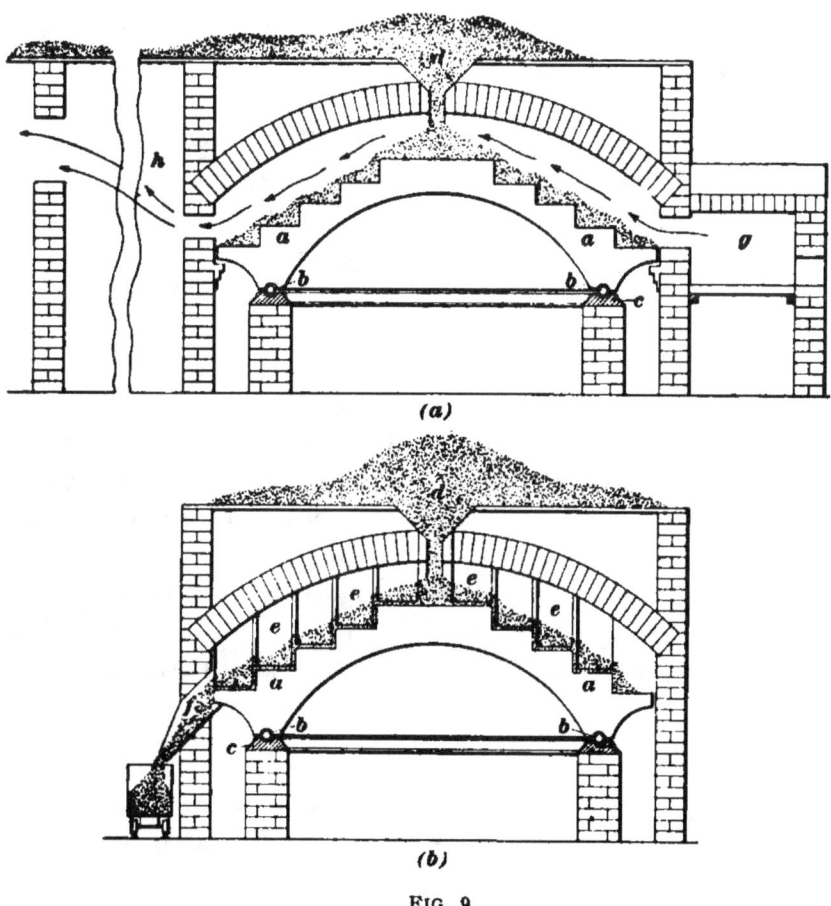

(a)

(b)

FIG. 9

step. The ore is thus gradually moved to the lowest step and discharged into the spout *f*. The drop from one step to another is 6 inches or more; and the purpose of this drop is to make the ore fall in a shower. This gives it a more thorough exposure to the oxidizing action of the hot air than it would have if the hearth was one continuous surface. The

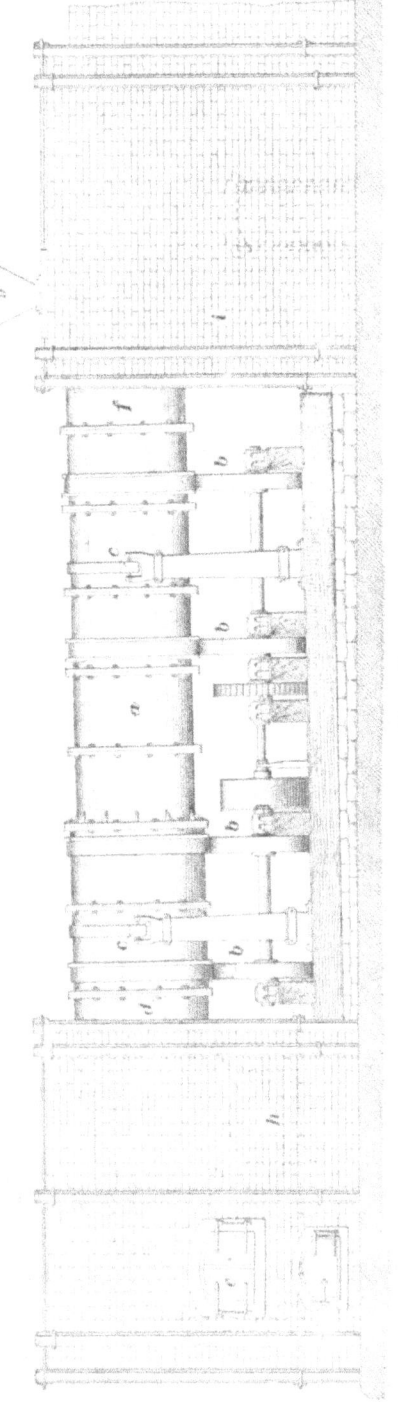

FIG. 10

ore is heated from a fire-box g on one side and a supply of heated fresh air enters through a number of holes (which are not shown) arranged around half the circumference. The waste gases escape into the dust chamber h.

This furnace was especially designed to roast a mixture of marcasite FeS_2, blende, and galena, so as to oxidize the first without changing the other two, and has given good results.* To do this, care must be used not to let the temperature get too high. Above a certain temperature the blende and galena will oxidize. The purpose of this roasting is stated in Art. **7.**

REVOLVING ROASTING CYLINDERS

32. The **Howell-White furnace,** shown in Fig. 10, is a long,

---

* The good results were not due to the special form of the furnace, but to the fact that it was arranged so as to permit accurate control of the draft and temperature.

slightly inclined, hollow cylinder *a*, supported by and rolling on several large rollers *b*. On each side of the cylinder are two guide rollers *c*. The furnace is made of cast-iron sections bolted together. The lower end *d*, where the hot gases enter from the firebox *e*, is lined with firebrick, but the upper part *f* is not lined. The shell of the lower part is just enough larger than the upper part so that, when lined, the internal diameter of the two parts is the same. Ore fed into the cylinder at the upper end from the hopper *g* is gradually worked to the lower end by the revolution of the cylinder and the slight inclination that is given it and falls into the chamber *h*. The furnace is from 20 to 30 feet long and from $2\frac{1}{2}$ to 5 feet in diameter. On the inside there are projecting ledges that raise the ore and allow it to fall in a shower through the hot gases. This makes the roasting more effective because it exposes a greater ore surface. Any dust that is carried out of the upper end of the furnace by the draft has a chance to settle in the dust chamber *i*. This furnace is applied principally to the chloridizing roast of silver ores; but similar furnaces are in some cases used for the oxidizing roast of copper ores and matte.

33. The **Brückner roasting cylinder** is shown in Fig. 11. Unlike the other mechanical roasting furnaces that are described, this one is not fed and discharged continuously. It is a hollow cylinder *a* made of heavy iron plates and lined with bricks. It is sometimes as large as $8\frac{1}{2}$ feet in diameter and 28 feet long, but $8\frac{1}{2}$ by $18\frac{1}{2}$ is a more common size. Each end is slightly conical and has an opening in the center to admit the hot gases from the firebox *b* and to discharge them into the flue *c*. The cylinder is surrounded by two heavy bands *d* that rest on friction rollers *e*. These rollers are driven, by gearing not shown in the figure, from the pulley *f*, thus giving the furnace a slow revolution.

The openings *g'* being closed and the furnace not revolving, ore is fed from the hoppers *h* into the openings *g*, the covers of which are then fastened.

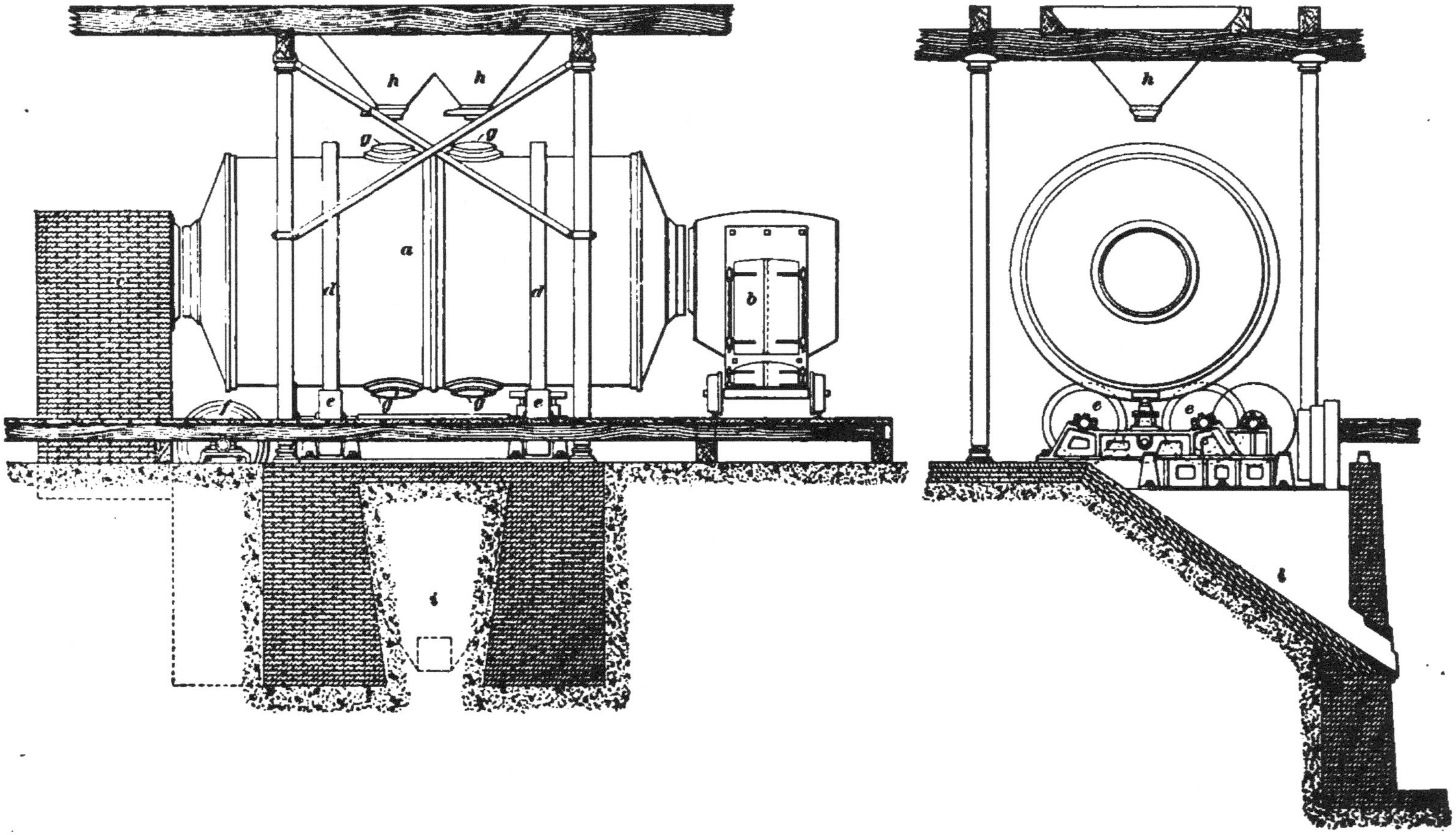

FIG. 11

34. The furnace is now revolved and the ore is carried up a certain distance on the inside perimeter of the cylinder until it forms a sloping surface, and then the grains tumble down the slope so that fresh ore is constantly exposed. When the roast is finished the ore is discharged into the bin *i*.

The Brückner cylinder has been extensively used to roast pyritic copper ores. It has also been used with special precautions for certain lead ores. It is not generally successful for lead ores, however, because it is difficult to prevent the ore becoming pasty and forming balls. The ore inside such balls escapes roasting. The furnace has been used in a few cases to chloridize silver ores.

SHAFT FURNACE

35. The Stetefeldt roaster is a shaft furnace, from 25 to 50 feet high, used for the chloridizing roast of silver ores. Hot gases from the fireplaces *a* of the Stetefeldt roaster shown in Fig. 12, mixed with air from the passages *b* and *c*, enter the shaft through *e*. Pulverized ore, mixed with the proper percentage of common salt, is fed into the stack by the automatic feeder *f* and falls in a shower through these hot gases into the hopper *g*. The finest part of the ore is carried over by the draft through *h* into the settling chamber *i* and is largely caught in the hoppers *j*. To assist the chlorination of this material, an extra fireplace *k* is provided. When cleaning is necessary, tools can be inserted through the openings *l*, *m*, and *n*. Although the ore remains in the roasting atmosphere only a short time, the operation is successful because each small particle is separated from all the others and is, therefore, thoroughly exposed to the hot gases.

This furnace gives excellent results in chloridizing silver ores, and has a much larger capacity than any other furnace for this purpose. It has never been used for the oxidizing roast of highly sulphureted ores, and probably would not do good work with them, because there is not long enough exposure to the heat and air to sufficiently oxidize the sulphur.

ROASTING LUMP ORES

36. Lump ores are usually roasted in heaps, stalls, or shaft furnaces. If the ores are high in sulphur and liable

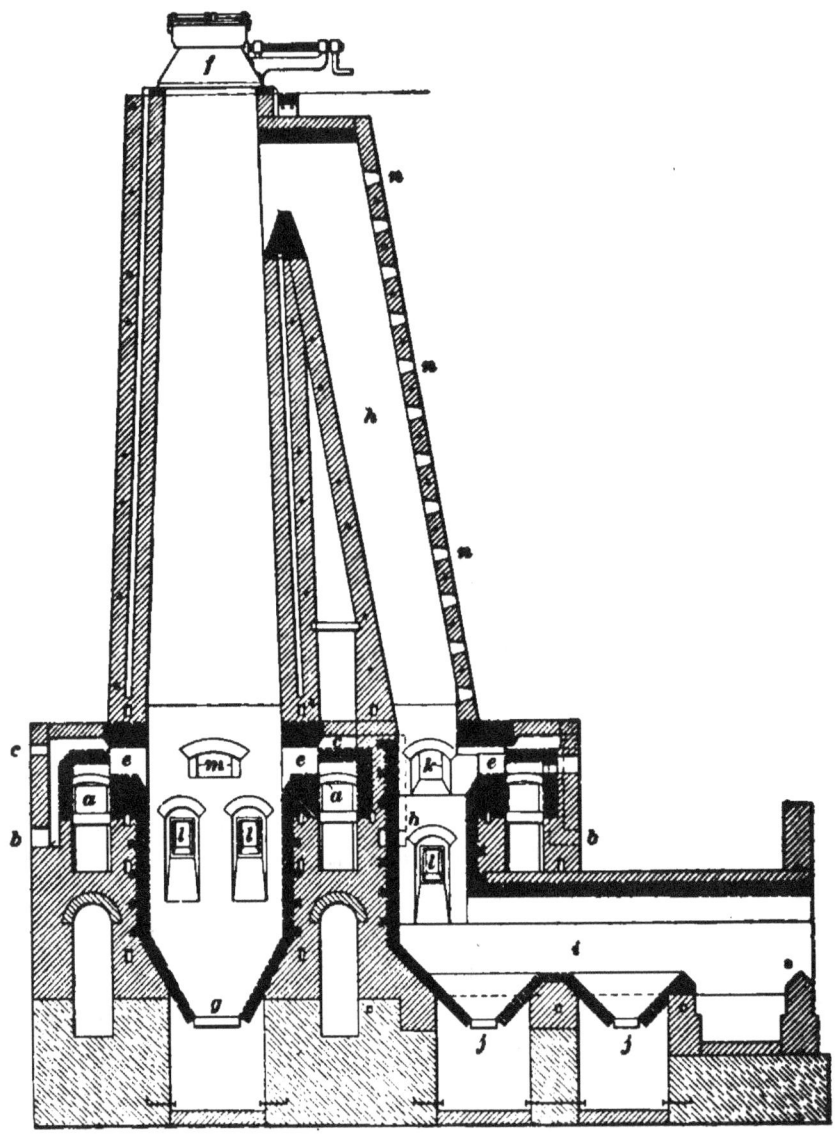

FIG. 12

to fuse, heap or stall roasting is to be preferred; on the other hand, if the ore is an iron sulphide low in sulphur or an iron or zinc carbonate ore, the shaft furnace is preferable. Many kilns have been invented for roasting pyritic lump ore. Very few of them, however, have produced as satisfactory a product at as low a cost as heaps or stalls.

- - -

HEAP ROASTING

37. Heap roasting requires very little outlay for apparatus; but it is expensive for labor, requires careful superintendence, and takes a long time. If proper care is taken, however, the results are excellent with suitable ores. The most suitable ores are such as contain a large percentage of pyrite, because the oxidation of the sulphur in the latter furnishes all the heat necessary after the roasting is well started. Pyrrhotite Fe_7S_8 and chalcopyrite $CuFeS_2$ also are sometimes roasted in heaps. As a matter of fact, heap roasting is mostly confined to pyritic copper ores. Lead ores are too apt to fuse. Zinc ores require too much fuel.

38. Heaps for roasting are built on open ground, but this should be carefully prepared. The surface earth is removed and a foundation of broken stone or slag (the waste material from the smelting furnaces) is put in. This is covered with gravel or loam and is well rolled until the whole surface is solid and smooth. On this is spread a thick layer of low-grade fine ore, the purpose of which is to prevent dirt being shoveled up with the bottom layer of ore after the roasting is finished. The surface should be made to slope slightly from what is to be the central longitudinal line of the heap to the sides, in order to shed water in case of rain. With the ground thus made ready, the area to be occupied by the heap is marked off. This should not be more than 20 or 25 feet wide, and a convenient length is 40 to 50 feet, though very much longer heaps are sometimes made. Cord wood in 4-foot lengths is placed all around the sides with the sticks pointing towards the middle of the area. Inside of this border, the space is filled with wood laid parallel to the

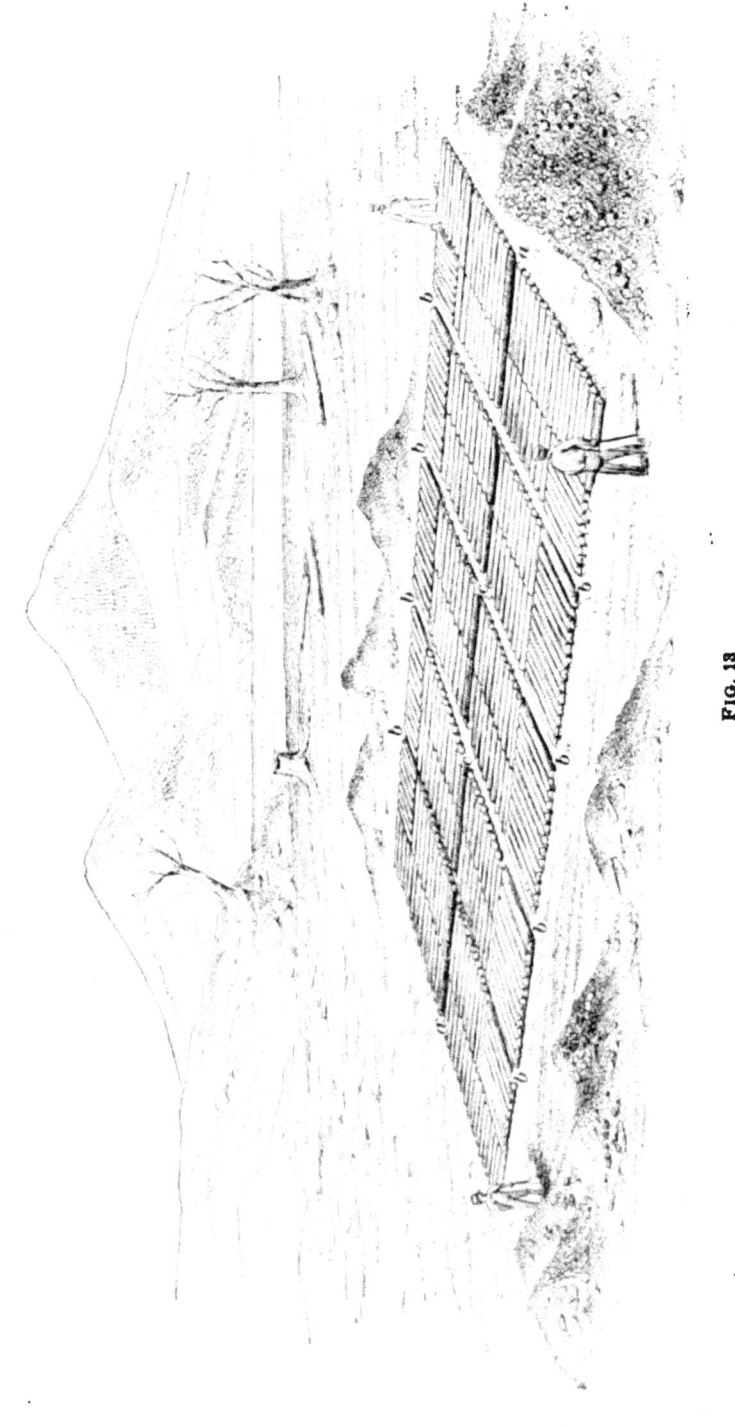

FIG. 13

long side of the area, as shown in Fig. 13. In order to have the heap ignite satisfactorily when lighted, good, sound wood must be used on the outside; but for the inside this is usually not important, waste ends, etc. being considerably used there. A second layer of wood is commonly laid on top of the first, with the sticks at right angles to those in the first layer, and still other layers are put on if necessary. Small brushwood or sawmill slabs should be used for the top to fill up the cracks and so prevent the ore (that is to be piled on later) from falling in among the wood. The amount of wood used depends in each case on the percentage of sulphur in the ore; but it should be as little as will satisfactorily ignite the heap, not only to save cost, but also because too much heat will fuse the ore and so hinder the roast.

39. Along the longitudinal axis of the area a space a 6 inches wide is left in the bottom layer of wood and is covered by the second layer. At intervals of about 8 feet, similar spaces b are left leading from this central space or flue to both sides. These flues are filled with small sticks, which are to serve for kindling. Where the branch flues unite with the longitudinal flue at c, vertical chimneys are erected. They are made simply by nailing together four old boards and are tall enough to come a little above the top of the completed heap.

40. The size of the ore is important, but it must be decided for each case. It should generally be broken so that there will be no lumps larger than 2 or 3 inches in diameter. After breaking, it is to be screened into three sizes, because, if all sizes were mixed together the air would not pass through the heap properly. The sizes might be coarser than 1 inch, from 1 inch to $\frac{1}{4}$ inch, and finer than $\frac{1}{4}$ inch. The coarse ore is piled on the wood, forming a truncated pyramid 5 to 6 feet deep, as shown at a, in the section of a pile, Fig. 14. On top of this the medium-size ore b is placed a foot or more deep. It is common and most convenient to distribute the ore from tracks elevated above the heaps.

41. If the ore contains only a small amount of pyrite, small wood or soft coal is mixed through the pile to furnish the necessary fuel to keep up the heat after the bottom bed of wood has burned out. Such an ore should be piled deeper than one that is higher in sulphur, but the total depth of ore is seldom more than 9 feet. Coal or wood scattered through the heap is useful also when the ore contains any considerable quantity of arsenic or antimony. In such a case, the coal reduces arsenates and antimonates, which are not volatile, to a condition in which the arsenic and antimony can be volatilized (see Arts. **59** and **60**).

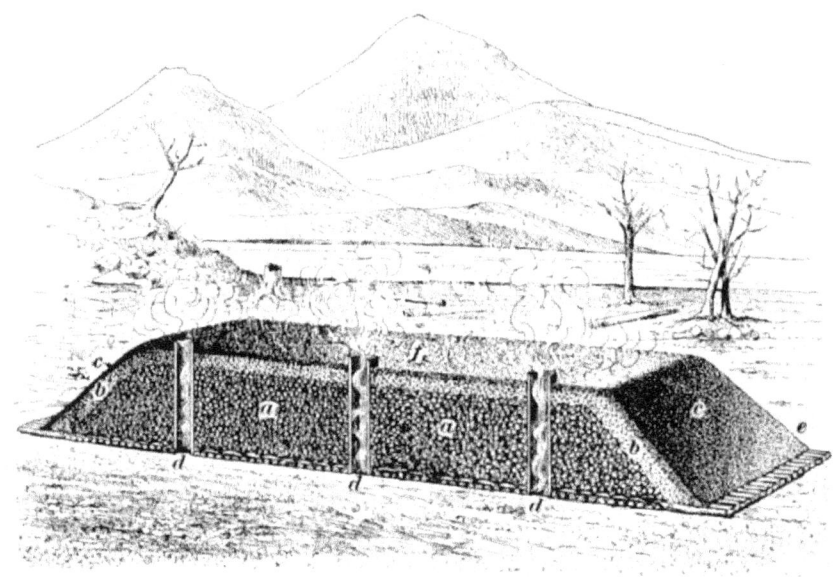

FIG. 14

42. The ore should be so arranged that about a foot of the wood bed projects on all sides of the heap. The sides of the heap are now covered with a layer of fines c, say $\frac{1}{2}$ inch deep, except near the bottom; and the heap is then ready to be fired. The reason that fine ore is not spread near the bottom nor on top of the heap at this stage is that the small air spaces between the large lumps must be left open in order to get the fire well started. The central wooden chimneys d are also important for this purpose. The fire is started at the outer end of each of the flues mentioned above; and after a period of some hours the heat from the wood will be sufficient to start the ore roasting near the edges of the heap. Fine ore is then spread on these parts *e* to check the draft, so that the roasting will not proceed so rapidly as to fuse the ore. As the roasting gradually extends to the middle of the heap, fine ore is spread over the entire surface, as shown at *f*, Fig. 14.

43. Fumes are now given off from all over the surface. For several days the heap has to be closely watched, and wherever the burning is too vigorous more fines are added to check the draft. Much experience and judgment are needed at this period. As the bed of wood burns away the heap settles, causing vent holes through which there will be strong drafts, if they are not covered with fine ore. These strong drafts would cause such rapid roasting as to fuse the ore. Any large cracks that appear around the sides of the heap, particularly at the bottom, should be immediately filled with fine ore. The completion of the roast usually takes 2 or 3 months and sometimes longer. The outside layer of fines is only partially roasted; but if proper care is taken, the coarse and medium ore is very thoroughly roasted.

44. In places where there is likely to be much rain, it may be best to have rough sheds built over the heap. Rain water percolating through the ore will dissolve the sulphate of copper that is formed during the roasting of copper ores. If the sheds are not provided, care should be taken to have all water drained away from the heaps and passed through tanks containing scrap iron. The copper dissolved by this water will be precipitated by the iron. This precaution has sometimes prevented large losses, since a large part of the total copper is converted into sulphate in heap roasting.

***

STALL ROASTING

45. A **stall** is an area enclosed by walls in which ore is roasted in small heaps. The advantages of stalls over open heaps are that they require less time, fuel, and labor.

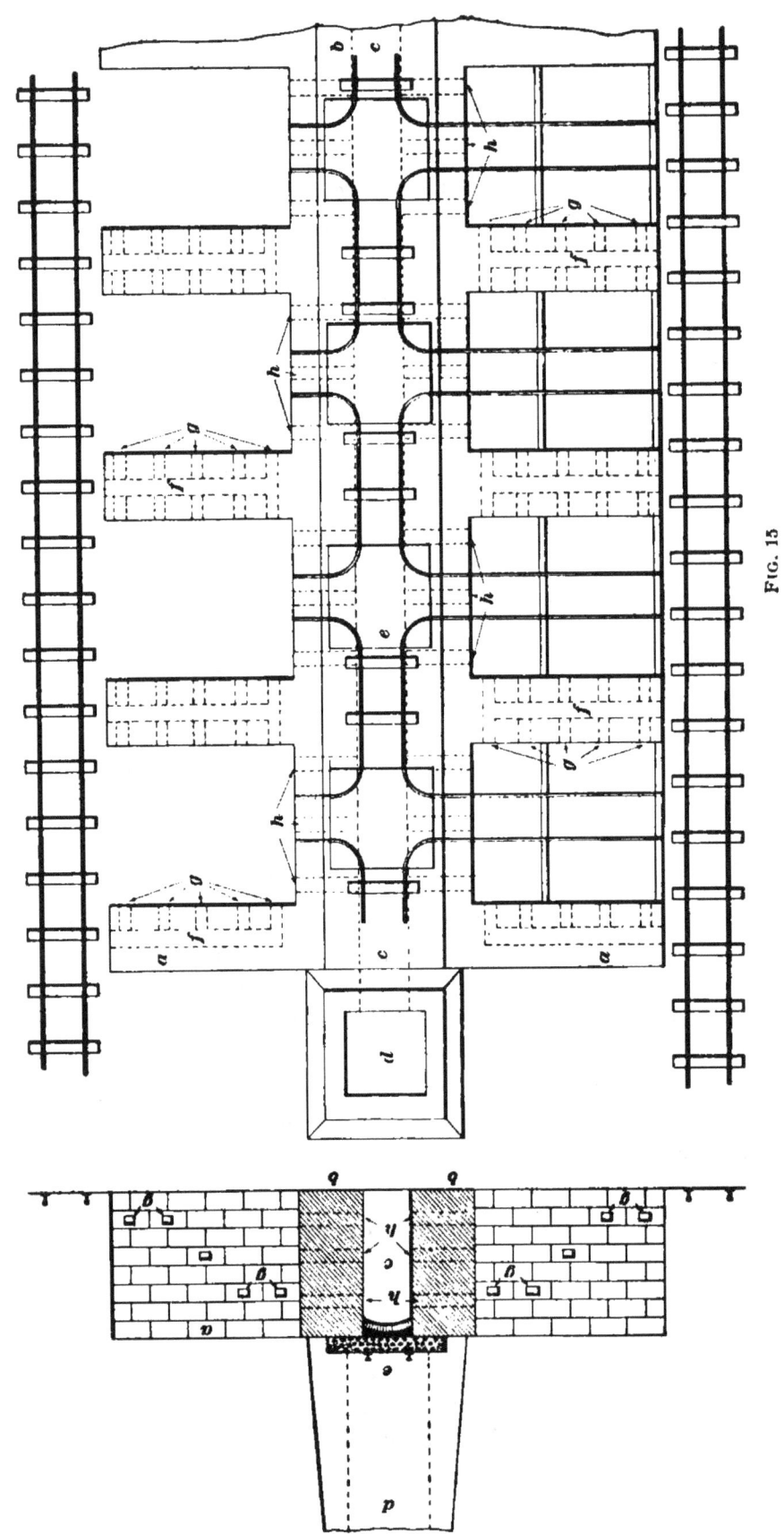

FIG. 15

One of the best forms of stall is shown in Fig. 15.   Each
stall is made with two side walls *a* and a back wall *b*, the
front and top being open; the bottom is paved.   Two rows
of stalls are built back to back with a large flue *c* between
them to carry the fumes into a tall chimney *d*.   In preparing for a roast, large lumps of ore are piled on the bottom of the stall, so as to form a small flue extending centrally from the open front to the back wall, and with branch
flues leading to the middle of each side wall.

46.   Small wood is placed inside these flues and a layer
of wood is placed between and on top of them.   The stall is
then pretty well filled with ore ranging from 2 or 3 inches
down to about $\frac{1}{4}$ inch in diameter, and this is covered with
a layer containing lumps from $\frac{1}{2}$ to 1 inch in diameter.   This
ore is now covered with small wood and shavings and on top
of all is spread a layer of fine ore.   It is well to place some
sticks of wood along the walls as the stall is gradually filled.
The front of the stall is built up either with lumps of coarse
ore or with bricks loosely piled up.   The top may or may
not be covered with a piece of sheet iron; some stalls are
permanently covered by a brick arch.   The ore is most conveniently brought to the stalls in cars on the track *e* running
above the large flue *c* between the two rows of stalls.   The
cars are run out over any stall by means of a turntable and
branch track, which are movable and can be placed wherever
desired.

The ore is fired by means of the wood in the bottom, as
in heap roasting.   The air for roasting enters partly through
the small flues made with lump ore on the bottom of the
stall, partly through the loose front wall, and partly through
the passages *f* and *g* in the side walls.   The waste gases
pass through the openings *h* in the back wall into the flue *c*.

47.   Stalls must be built with thick, substantial walls or
they will soon fall apart.   The reason for this is that sulphide ores expand considerably in roasting and will force an
unsubstantial wall out of place.   The walls in Fig. 15 are
about $2\frac{1}{4}$ feet thick.   The loose, temporary front wall should

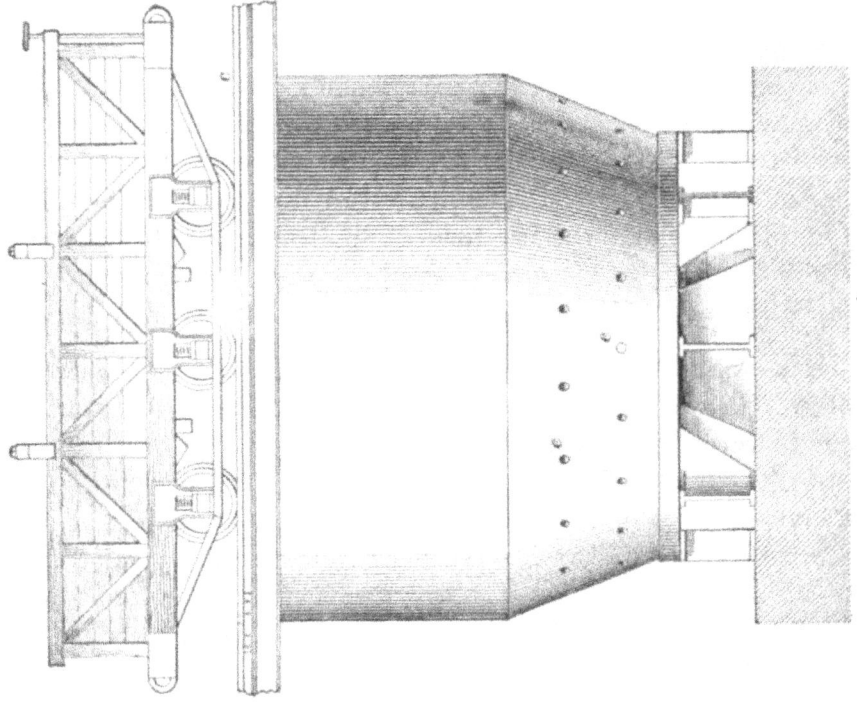

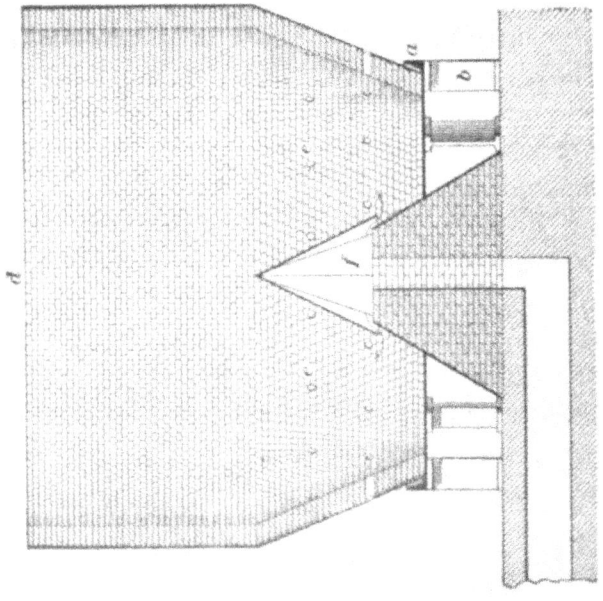

Fig. 16

be braced with timbers.   With these precautions, the expansion of the ore will take place upwards.

The walls of a stall should be plastered over with clay and this coat of clay should be patched where necessary before each roast.   The purpose of this is to prevent the ore sticking to the walls in case of any slight fusion.

48. In stall roasting, pyritic ores that contain more than 25 per cent. sulphur should be supplied with only sufficient wood to start the roast.   Too much wood will make too much heat and fuse some of the ore.   For this reason, it is best to use partly decayed wood rather than sound hardwood.   More wood is needed for ores that contain a good deal of gangue and so little pyrite that the oxidation of the ore itself does not produce enough heat to maintain a roasting temperature.   The size of the stalls is an important matter, and must be decided for each ore by experienced judgment.   The more heat the ore will generate in roasting, the smaller should the stalls be, since large heaps are not so readily controlled and the ore may fuse, thereby causing future trouble.   The management of the draft is also important and requires experienced judgment.

49. As previously stated, stalls have the advantage over open heaps that they save fuel, time, and labor.   An ore that would require 2 or 3 months to roast in heaps can be roasted in as many weeks or even less in stalls.   Sulphur is about equally well removed by either method; but stalls have the advantage of forming less ferric oxide Fe_2O_3, because the process is not so long continued as in heaps (see Art. **57**).   Stalls, however, require the investment of considerable capital for their construction, while heaps do not.   The former also require more careful superintendence.   For the lack of this care, stalls have in a number of cases been given up in favor of heaps.

GJERS KILN

50. The **Gjers kiln** is a shaft furnace and is shown in elevation and in vertical section in Fig. 16.   It is used to roast coarse iron ore.   A similar furnace is used to remove CO_2

from calamine $ZnCO_3$, leaving ZnO.   It is from 10 to 24 feet in diameter at the widest part and from 12 to 30 feet high. The walls are built of firebrick held in position by iron plates. They rest on a cast-iron ring a, which is supported by a number of cast-iron legs b.   The ore is brought in wagons or buggies on the tracks c and dumped into the open top d; 3 to 5 per cent. as much coal as ore is also used. As the charge descends, it is forced out in all directions from the center by the cast-iron cone f and is drawn away between the legs b.   The air to burn the coal and to oxidize the sulphur in the ore enters partly through the spaces from which the ore is discharged, partly between the cap and base of the central cone, and, if necessary, partly through the doors e. The charge can be poked through these doors if it happens to clog.   Very little if any air would reach the central part of the charge if it were not admitted at the cone f.

The combustion of the fuel and the actual roasting of the

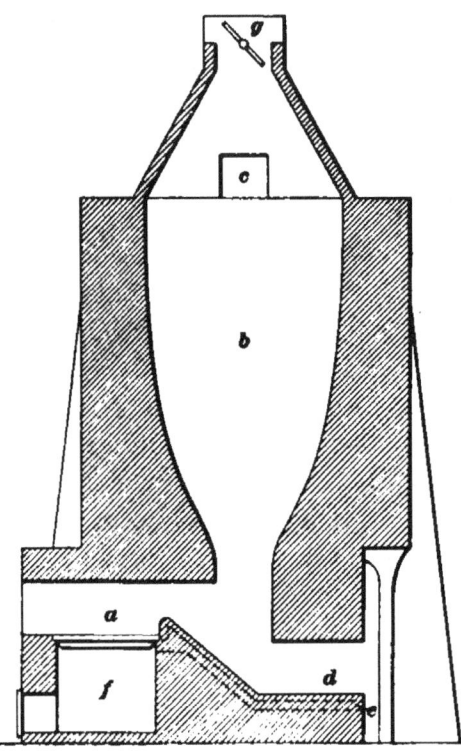

ore take place mainly in the central part (vertically) of the furnace. In the lower part, the cold air takes up heat from the roasted ore; and in the upper part, the cold ore is dried and heated by the hot gases.

51. Fig. 17 shows a kiln in which the fuel is burned on an exterior grate instead of being mixed with the ore. The flames from the grate fire a enter the ore chamber b and thus supply the necessary heat. The object of

FIG. 17

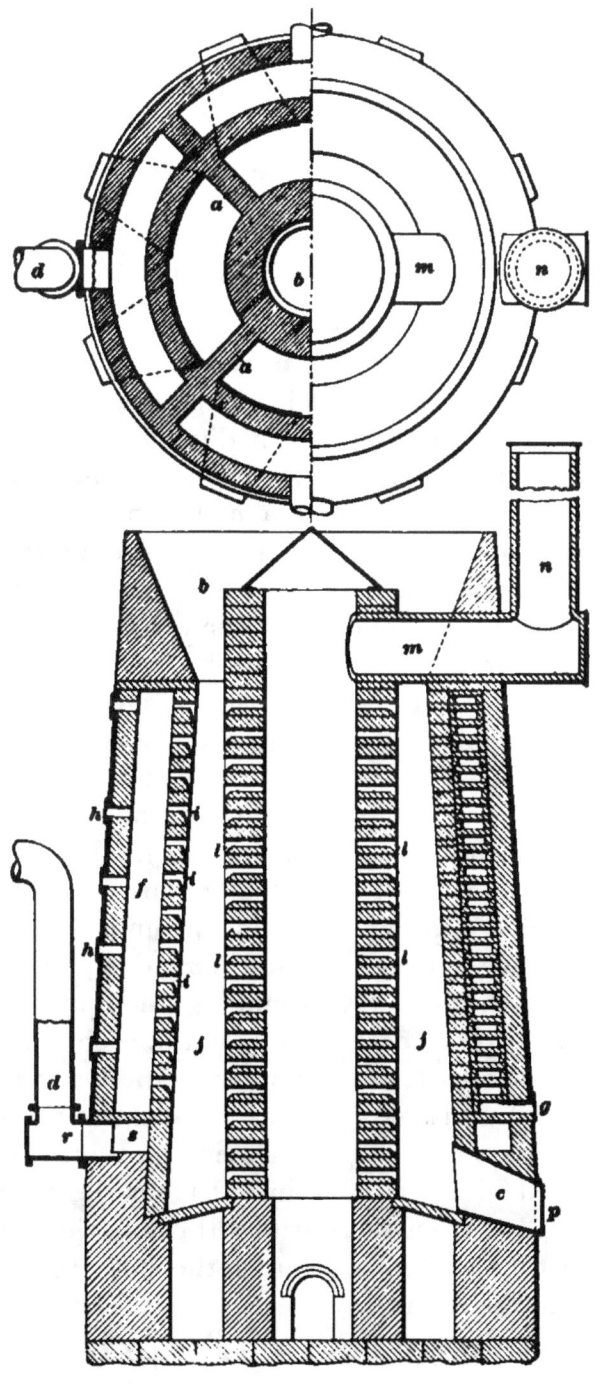

FIG. 18

this is to prevent the coal ash mixing with the ore. This precaution is necessary only with zinc ores, and even these are sometimes roasted in direct contact with the fuel. The ore is fed at c and discharged at d. The air for burning the fuel is somewhat preheated by passing through the space e (just below the hot ore that is being drawn out through d) into the ash-pit f; the draft is controlled by the damper g.

DAVIS-COLBY KILN

52. The **Davis-Colby kiln,** which is also used for iron ores, is heated by gaseous fuel. The special feature of the kiln shown in Fig. 18 consists in dividing the furnace into four independent sections by means of the partitions a. The purpose of this is to get a more uniform action than can be obtained with the ordinary construction in which the partitions a are not used. This furnace is about 20 feet high.

Ore is fed into the hopper b at the top of the furnace and is drawn off through chutes c at the bottom. Gas is supplied by the flue d surrounding the furnace and enters the combustion chambers f through branch pipes g, which are fitted with valves. Air enters the combustion chamber through the ports h, and the burning gas and air pass through the openings i into the ore chambers j. Passing through the ore, the hot gases gradually raise the latter to a red heat and are drawn through openings l into the central vertical flue. From this flue the gases pass through the pipes m into the flue n surrounding the upper part of the furnace, and from there into a suitable chimney. When the ore is hot enough, it is drawn into the lower part of the ore chamber by discharging the ore that is already there. The covers p are then tightly closed to exclude all air, and gas (unmixed with air) is admitted from the pipes r to the chambers s, and from there passes through the ore and into the exhaust flues. This reduces the hot Fe_2O_3 to Fe_3O_4, according to the equation

$$3Fe_2O_3 + CO = 2Fe_3O_4 + CO_2$$

The Fe_3O_4 is then withdrawn through the chutes c.

53. It will be noticed that the size of the ore chambers j enlarges somewhat from the top to the bottom. This is to provide for the expansion that takes place in the ore while roasting.

For calcining carbonate ores to remove carbon dioxide CO_2 or roasting such iron ores as contain small percentages of sulphur that must be expelled, the radial partitions a are not used. Also, for such ores the combustion chamber f is extended downwards to include the chamber s, because no reducing effect is needed.

CHEMISTRY OF ROASTING

54. In roasting sulphide ores, the sulphur and the metals are both oxidized, most of the former passing off in the gaseous state as sulphur dioxide SO_2, while the metallic oxides remain in a more or less porous condition.

$$PbS + 3O = PbO + SO_2$$
$$ZnS + 3O = ZnO + SO_2$$
$$Cu_2S + 3O = Cu_2O + SO_2$$
$$Cu_2S + 4O = 2CuO + SO_2$$

55. A certain amount of the sulphur is burned to sulphur trioxide SO_3. Some of this passes off in the draft, but part of it unites with the metallic oxides to form sulphates, which are not volatile, for example :

$$PbO + SO_3 = PbSO_4$$
and
$$CuO + SO_3 = CuSO_4$$

The sulphur trioxide SO_3 also acts as a very efficient oxidizing agent, both oxidizing sulphides and changing the lower to the higher metallic oxides; for instance,

$$Cu_2S + 3SO_3 = Cu_2O + 4SO_2$$
and
$$Cu_2O + SO_3 = 2CuO + SO_2$$

56. In roasting pyrite FeS_2, there are several distinct stages. One of the two atoms of sulphur in this mineral is quite readily volatilized ($FeS_2 + $ heat $= FeS + S$). The first stage in the roasting consists in the oxidation of this one atom of sulphur, which burns with the characteristic blue flame of sulphur. In proof that this atom of sulphur is actually volatilized, it is only necessary to point to the fact that by scooping out small holes in the top of an open heap of roasting pyrite and lining these holes with fine ore, a certain amount of sulphur can be collected in them. Indeed, this sulphur sometimes condenses in the upper part of the heap during the early stages of roasting to such an extent that it closes up the air passages and so hinders the roasting. Any such crust must be broken up. This same crust forms on the surface of the ore when roasting fines in hearth furnaces. Consequently the rabbling should be frequent during this period; but after the first atom of sulphur has been burned off, the rabbling need not be done so often.

57. During the second stage of roasting, the iron and the remaining atom of sulphur are oxidized, according to the following equation:

$$FeS + 3O = FeO + SO_2$$

Much of the ferrous oxide FeO formed by this reaction unavoidably becomes further oxidized to ferric oxide Fe_2O_3. The ferric oxide is undesirable in smelting, because it has to be reduced again in the smelting furnace.

58. As stated in Art. **2,** the quantity of sulphates produced is greatest when the temperature is comparatively low, when there is not too much draft, and when the bed of ore is rather deep. It is because these conditions are so well fulfilled in heap and stall roasting that unusually large amounts of sulphate are formed by those methods. The temperature at which sulphate forms most readily differs for different metals. Iron is sulphatized at a lower temperature than the other metals, yielding ferrous sulphate $FeSO_4$. As the temperature increases, this is decomposed into ferrous oxide FeO and sulphur trioxide SO_3, or sulphur dioxide and oxygen $SO_2 + O$. The sulphatizing and decomposition of the other common metals proceed in about the following order: copper, silver, zinc, and lead. It is to be understood, however, that one metal is not completely sulphatized before others begin to be; and the same may be said in regard to the decomposition of sulphates. Zinc and lead sulphates are decomposed only at very high temperatures. In the slagging of lead ores, described in Art. **14,** some sulphur is eliminated by the following reaction :

$$2PbSO_4 + SiO_2 = Pb_2SiO_4 + 2SO_2 + 2O$$

59. Arsenic and antimony behave in roasting much the same as sulphur, but a larger proportion of these are oxidized to the non-volatile condition (arsenates and antimonates) than in the case of sulphur. On the other hand, some arsenic and antimony may be volatilized as sulphide before oxidation takes place.

60. In the reducing roasts, oxygen is taken away from the higher oxides and from sulphates, arsenates, and antimonates by carbon, carbon monoxide, and sometimes by sulphur, where iron disulphide is used as a reducing agent. The reactions differ with different conditions of temperature, etc. The following examples are given of this reduction :

$$CuSO_4 + 2C \quad = CuS + 2CO_2$$
$$CuSO_4 + 4C \quad = CuS + 4CO$$
$$CuSO_4 + C \quad = CuO + CO + SO_2$$
$$Fe_2As_2O_8 + 2C = Fe_2O_3 + As_2O_3 + 2CO$$

61. In the chloridizing roast of silver ores, the principal reaction is

$$Ag_2SO_4 + 2NaCl = 2AgCl + Na_2SO_4$$

It often (and with the Stetefeldt furnace probably always) happens that the silver is not completely chloridized in the furnace; but if the hot ore is left in a heap for a number of hours, certain chemical reactions take place, which in many cases greatly improve the chloridization.

62. The roasting of carbonate ores consists simply in calcining the ore; that is, driving off carbon dioxide by heat.

$$ZnCO_3 + \text{heat} = ZnO + CO_2$$
$$FeCO_3 + \text{heat} = FeO + CO_2$$

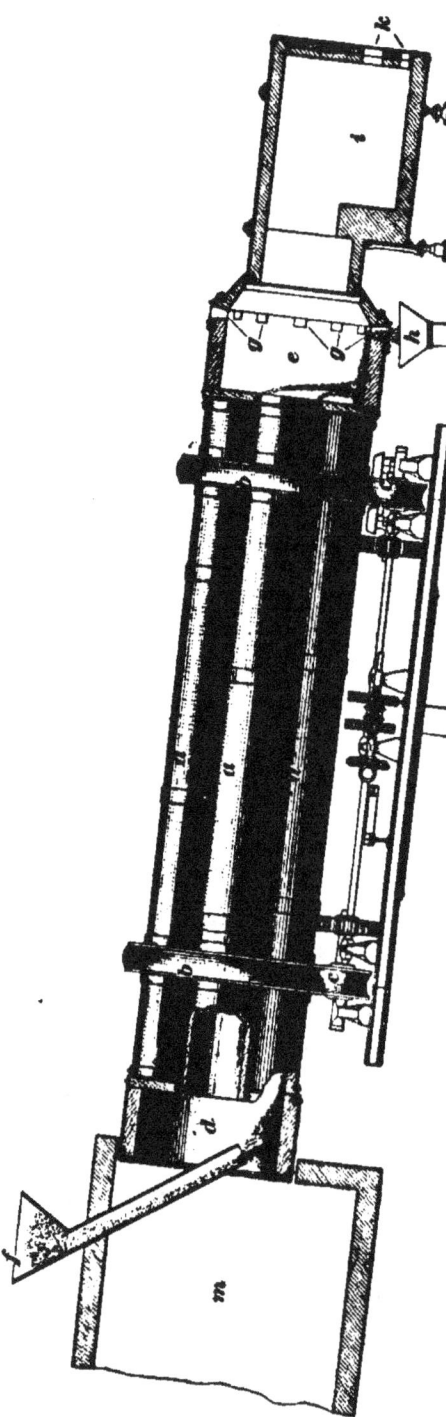

FIG. 19

In the case of iron carbonate, it is practically impossible, commercially, to prevent the FeO being oxidized to Fe_2O_3, though this is not done purposely.

In the various kinds of roasting there are, in addition to the reactions given above, various others of a more or less complicated character, which it is not necessary to discuss here.

63. Fig. 19 shows the **Argall multi-tubular roaster.** It consists of four tubes, or cylinders, a arranged side by side and held together by two heavy tires b. Each of these tires rests upon a pair of grooved friction pulleys c set so that the roaster has a little slope from the feed end d to the discharge end c. Each of the tubes a is 29 feet long and 25 inches in diameter inside the brick lining. At the feed end, the tubes are set into the hood d, into which the ore is fed from the hopper f. As the furnace is revolved by the friction pulleys c, the ore enters the tubes a. The revolution, combined with the slope of the furnace, causes the ore to gradually travel to the lower hood c, from which it discharges through the holes g into the receiving hopper h. The furnace is heated from the firebox i, which is arranged to burn oil, but can, of course, be modified to burn coal. The oil, mixed with air, is introduced through the holes k as a spray, and the hot gases pass into the hood e and thence through the tubes a of the roaster and pass off to the dust chamber m and a chimney beyond.

The object of using four cylinders instead of one is to provide a larger amount of brickwork to absorb the heat from the firebox, this heat being again given out to the ore.

The furnace has been used to roast some of the gold ores of Cripple Creek, Colorado. When treating 48 tons of ore in 24 hours in one furnace, the 2 per cent. of sulphur originally in the ore was reduced to 0.1 per cent., according to statements made. If such is the case, the roaster is entitled to take rank among the best.

64. The **Zellweger roaster** has been recently introduced at Gas City, Kansas. This furnace, while a straight-line reverberatory 135 feet long by 15 feet wide, has a rolling stirrer that moves slowly from the feed to the discharge. The stirrer consists of a heavy shaft carried by wheels, 6 feet in diameter, rolling on tracks in the depressed wheel pits on either side of the hearth. On this shaft are a number of collars carrying blades for rabbling the ore. These collars lock on the shaft when traveling from the feed to the discharge end of the furnace, and as they revolve with the shaft, the blades scoop up ore during one half a revolution and discharge it during the other half, thus gradually moving the ore forwards. During the return trip, the stirrers revolve around the shaft with the collar, but do not displace the ore more than to rake it. This furnace has external fireplaces, and is said to roast 15 tons of blende to 1 per cent. sulphur in 24 hours.

65. The roaster shown in Fig. 20 is known as the Herreshoff, although it is merely a modification of the

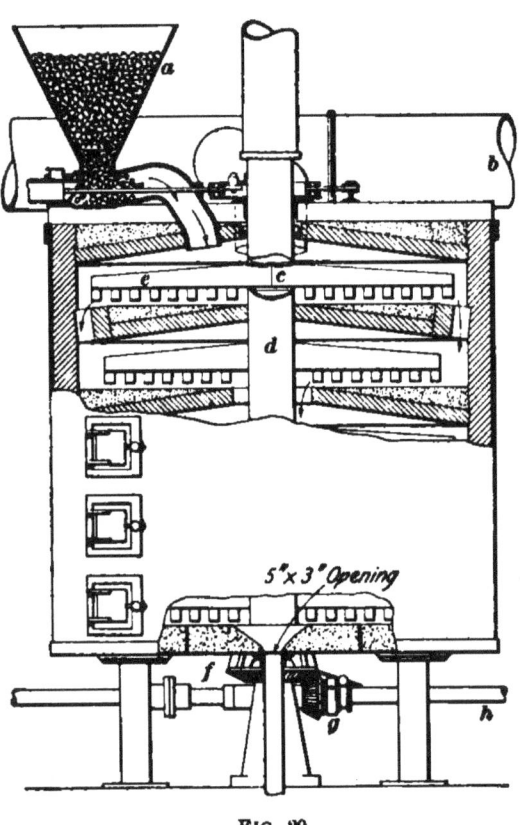

FIG. 20

McDougall. It has a central shaft d, with a number of shelves which answer for hearths placed at right angles to it. Attached to the shaft are two arms e above each hearth, which carry teeth so arranged that one set moves the ore from the center to the circumference of the furnace, where it falls on to the next lower hearth. The set of arms on the latter hearth have the teeth arranged so that the ore is moved from the circumference of the furnace to the center and discharged on to the next lower hearth, and so on to the ore discharge at the bottom of the furnace, as shown by the arrows.

Power is transmitted to the shaft d and arms e by the shaft h and gear-wheels f and g. The ore is fed on to the top hearth from the hopper a and the gaseous products of combustion pass out of the furnace through pipe b. The hollow shaft d is made large (14 inches in diameter) so that a large quantity of air is drawn up through it, this amount being increased by the sheet-iron stack c extended above the top of the furnace. Between the shelves there are cross channels passing directly through the shaft at right angles, as shown in the vertical section drawing. These cross channels are about 4 inches wide and 5 inches high and allow ample space around them for the passage of the ascending air. Into these channels or sockets the arms e are inserted, and by a groove and rib are arranged to lock when horizontal and unlock when raised at their outer ends.

By raising the outer end of the arm e about 3 inches, the rib can be pulled out. Practice has shown that these arms, weighing about 100 pounds, can be unlocked, removed from the furnace, and new ones put in and locked in place in about one minute. Each furnace requires about $\frac{1}{4}$ horsepower. The arms are the most costly part of the running expenses in regard to wear and tear; they are said to cost about $30 per year for each furnace.

COST OF ROASTING

66. The cost of roasting varies according to the degree of roasting demanded, together with the price of fuel and labor. The cost of roasting to eliminate sulphur to 2 per cent. should not, with present mechanical roasters, exceed 50 cents per ton; however, to reduce the last traces of sulphur, the cost will probably be four times this amount.

According to Doctor Phillips, sweet-roasting cost $2.18 at the Haile mine in North Carolina. The cost of roasting at the Globe smelters in Denver was $3.975 in 1887 and $2.75 in 1898 for hand-rabbled reverberatory furnaces. The cost of roasting in Brown-O'Hara mechanical roasters was $2.21 per ton of ore. The cost of roasting at the Guggenheim furnaces, Monterey, Mexico, is given as $2.43 per ton of ore in hand-rabbled furnaces.

While mechanical roasters give at first glance a cheaper product, the cost of roasting is higher when repairs and the quality of the product are taken into account.

THE CYANIDE PROCESS

(PART 1)

HISTORICAL

EVOLUTION OF THE CYANIDE PROCESS

1. Cassell Process.—Records of gold being dissolved in solutions of potassium cyanide and water extend back to 1806. The resulting solutions were first applied to gilding metallic surfaces. That gold could be dissolved from some ores by weak solutions of potassium cyanide was known in 1867, and applied in that year, but without much success from a commercial standpoint. *Doctor Cassell* perfected a process which he sold to *Messrs. MacArthur* and *Forrest*, who were granted patents in 1890 for what is known as the **MacArthur-Forrest cyanide process.** Messrs. Mac-Arthur and Forrest demonstrated the practicability of the cyanide process, although since its introduction many useful improvements have been added to make this branch of hydrometallurgy a success. The only patentable feature in the MacArthur-Forrest process was the method of precipitation by zinc shavings, and even this is denied by some.

2. Electrolytic Process.—The application of mechanical agitation to assist in dissolving gold from ores was early recognized by J. H. Rae, of Syracuse, New York, who made a test of his process in 1867. He applied to his

§ 31

For notice of copyright, see page immediately following the title page.

cyanide solution the electric current and precipitated the gold electrolytically in a bath of mercury.

Many years afterwards, *Messrs. Pelatan* and *Clerici* patented a similar process, which was more successful, however, for where Rae used an alternating current, they used a direct current. Pelatan and Clerici introduced sodium chloride into their cyanide solutions, but this was also done by J. W. Simpson, of Newark, New Jersey, in 1885. There is probably nothing in the Pelatan-Clerici process that has not been patented or used prior to its introduction in this country.

The pneumatic-cyanide process was originated about the same time in New Zealand and the United States. In this process, the agitation of the ore and aqueous potassium-cyanide mixture is accomplished by a blast of compressed air.

SCOPE OF THE PROCESS

3. Cyanide Solutions Defined. — Whenever cyanide solutions are mentioned in this work, it is to be understood that they are those made by dissolving cyanide of potassium KCN in water. The cyanide process refers to dissolving gold and silver from ore by means of cyanide solutions and afterwards precipitating the gold by some one of the methods hereinafter described. While the chemical reactions that take place between gold and cyanide solutions are known, the reactions between cyanogen and some other elements are as yet but slightly known, for which reason the cyanide process has not been fully developed and its scope is at the present time limited.

4. Treatment of Free-Milling Ores.—*Free-milling ores are most successfully treated by the cyanide process when the gold is in a fine state or subdivision.* The cyanide process has a field of its own in working tailings and concentrates resulting from wet crushing and plate amalgamation and can generally be applied to those ores that have their gold particles encased in some substance or have the gold so finely divided that it floats away over the amalgamating plates.

5. Treatment of Silver Ores.—Ores containing silver are more or less soluble in dilute cyanide solutions, but not to the same extent as gold. The chloride of silver $AgCl$ and the subsulphide of silver Ag_2S are readily soluble, but silver ores, as a rule, are slowly acted on by cyanide solutions. Sodium chloride $NaCl$ is sometimes added to cyanide solutions to hasten the reaction where much silver is in the ore, the object being to form silver chloride, which, as stated, is readily attacked by cyanide solutions. The addition of sodium chloride to the solution takes place in the Simpson and the Pelatan-Clerici processes.

6. Treatment of Base-Metal Ores. — The base-metal ores or those containing iron, zinc, lead, copper, and antimony, combined with sulphur, arsenic, or tellurium, sometimes cause a loss of cyanogen by uniting with the cyanide solution and forming soluble cyanides. This loss can be reduced materially by the use of weak cyanide solutions, as in that condition the affinity of cyanogen for gold and silver is greater than for baser metals. The action of weak cyanide solutions on lead and iron is practically nothing. The solvent action of weak cyanide solutions on copper and zinc in a metallic state is little with gold present, but when their hydrated oxides and carbonates are present in the ore, the loss of cyanogen will be so great as to render the cyanide process useless. Partially oxidized pyritous ores cause a loss of cyanogen, but they are sometimes readily adapted to the cyanide process by a preliminary treatment. Those ores that contain a quantity of antimony or tellurium generally require preliminary treatment. Arsenic seems to have no injurious effect on the cyanide solution. When gold and silver are associated with ores of copper and antimony, weak solutions of cyanide exert a very decided action on gold and silver, but do not act on copper and antimony. This fact is taken advantage of in the treatment of cupriferous ores on a commercial scale.

CHEMISTRY OF THE PROCESS

7. Cyanide of potassium is composed of the three elements potassium, carbon, and nitrogen, one atom of each combining chemically to form a molecule of cyanide of potassium. Its chemical symbol is written either KCN or KCy. It is an organic compound very active in combining with many base metals.

A large number of cyanogen compounds are formed with complex ores under different conditions of treatment. The exact nature of many of these is as yet unknown, and it will require much careful research in the laboratory by skilful chemists to determine their properties and reactions. The cyanides of the heavy metals, with the exception of gold and silver, are insoluble in water, but they are soluble in excess of cyanide. The cyanides of the alkaline metals are soluble in water.

8. Elsner's Equation.—When gold is treated with a cyanide solution, oxygen must be present or some other element that acts as an oxidizing agent. The reaction, presuming oxygen to be present, is expressed by the following equation, known as *Elsner's Equation*, as he advanced the theory in 1844:

$$2Au \ + \ 4KCN \ + \ O \ + \ H_2O \ = \ 2AuK(CN)_2 \ + \ 2KOH$$

| Gold | Cyanide of potassium | Oxygen | Water | Auric-potassic cyanide | Potassium hydrate |

The probable reaction is that the gold unites with the cyanogen, liberating potassium. Part of the potassium immediately unites with the gold cyanide, forming the double salt auric-potassic cyanide; the remainder uniting with the water liberates hydrogen and forms caustic potash or potassium hydrate. When such a solution is evaporated, it yields octahedral crystals, which show on analysis to be auric-potassic cyanide $2AuK(CN)_2$.

9. Influence of Oxygen Upon the Reaction. — In order to carry out the reaction indicated, Elsner's equation shows that oxygen should be present. That oxygen was necessary for the reaction was at first denied by some, but later experiments have shown that Elsner's theory was correct. Several devices for the artificial introduction of oxygen into the pulp have been invented.

Agitation by means of mechanical devices permits the oxygen of the atmosphere to come in contact with the pulp and thereby hasten the operation. For this purpose, mechanical stirrers, centrifugal pumps, or compressed air are employed. Sodium dioxide Na_2O_2 is also used for furnishing oxygen and increasing the time of reaction, for the sooner the atoms of gold, cyanogen, and oxygen come in contact with one another, the quicker the operation of gold extraction will be completed. Where time is not an object, the oxygen present in the water and ore, together with such oxygen as is obtained from the atmosphere during the operation, will be sufficient to carry out the chemical reaction.

10. Combining Weights of Gold and Potassium Cyanide. — Taking the final reaction given by Elsner's equation and applying the laws of chemical combination, the molecular weights will express the weights of the elements that form the auric-potassic cyanide compound. The molecular weight of the two atoms of gold is 393.4 and the molecular weight of the four molecules of potassium cyanide is 260. As the molecular weights express the combining weights, 15.12 parts of gold will require 10 parts of potassium cyanide; that is, 15.12 ounces of gold will require 10 ounces of potassium cyanide for its solution. It has been ascertained by practice that in treating free-milling ores, 20 to 40 pounds of potassium cyanide are required to dissolve 1 pound of gold. This may be partially explained by secondary reactions that are known to take place but are not fully understood.

11. Rate of the Solubility of Gold. — Researches by various scientists have demonstrated that the rate of solubility of gold in cyanide solutions depends on the strength of the solution and the supply of oxygen. A piece of gold leaf may be immersed in a strong solution of cyanide of potassium and scarcely be attacked. If, however, the proper amount of oxygen be supplied by sodium dioxide, the gold leaf will be dissolved in a comparatively short time.

Under ordinary circumstances, such as prevail when the MacArthur-Forrest process is practiced, gold will dissolve quicker in dilute cyanide solutions than in strong solutions. The maximum rate of solubility for gold in a free-milling ore, under favorable conditions, is reached with a .25-per-cent. solution. The character of the ore exerts an influence on the rate with which cyanide will dissolve the gold, and this time can be determined by laboratory experiments.

CAUSES FOR CYANIDE LOSS

12. Cyanicides.—Some of the losses of cyanide which occur when ores containing gold and silver are treated by the cyanide process may be traced to the presence of mineral acids and salts, to minerals soluble in cyanide solutions, and to a certain percentage of waste caused by washing the tailings. To these losses may be added those that are caused by evaporation or by carbon dioxide absorbed from the atmosphere.

13. Loss From the Presence of Mineral Acids and Salts.—The ordinary gangue minerals associated with gold and silver ores are largely composed of silica and silicates of the alkalis and alkaline earths. These substances are seldom of such a composition as to decompose a cyanide solution.

The metallic minerals often associated with gold and silver in quartz veins are iron pyrites, copper pyrites, zinc blende, galena, and stibnite. Iron pyrites, the most common and abundant of these, when undecomposed do not act on a solution of cyanide, but when exposed to the atmosphere in the presence of moisture, they become oxidized into ferrous sulphate and free sulphuric acid. The following equation shows the reaction:

$$FeS_2 + H_2O + 7O = FeSO_4 + H_2SO_4$$

Iron pyrites, when exposed to the elements above water level, are often changed to ferric oxide. Ferric oxide does not decompose solutions of cyanide, but its presence often causes very fine slimes, which retard the operation of draining the ore.

The ferrous sulphate and free sulphuric acid produced by the atmospheric oxidation of iron pyrites react upon solutions of potassium cyanide and may cause a loss of cyanide by the liberation of hydrocyanic acid or by the formation of ferrocyanides and ferricyanides.

14. Alkali Treatment.—*To prevent loss of cyanide in ores or tailings containing acids, iron salts, and earthy sulphates, they are treated with an alkali.* If the ore contains a large percentage of free acid, it is washed with water before treatment with the alkali. If the ore is not too acid, a quantity of caustic lime is added and thoroughly mixed with the ore before it reaches the percolating tank. If caustic soda or caustic lime is added to acid ores, it will neutralize the acids and will combine chemically with the objectionable salts.

15. Loss From Minerals Soluble in Cyanides.—Potassium cyanide acts on the oxide, sulphide, and carbonate ores of copper and the sulphides of antimony and bismuth. The loss of cyanogen, which will occur when ores of this description are treated by the cyanide process, depends on the quantity of these minerals present in the ore.

The selective action of weak cyanide solutions is taken advantage of in treating cupriferous ores. There may be sufficient copper present to decompose a 1-per-cent. solution of cyanide and give a low extraction of gold, whereas a solution containing .25 per cent. of cyanide would dissolve more gold and less copper than the stronger solution.

16. Loss From Washing Ores.—To save as much auric-potassic cyanide as possible after the ore has been treated, it is customary to wash the tailings with water. But as it is impossible to remove all the cyanide solution in this manner, some of it is lost in the tailings. Washing the tailings causes a large accumulation of dilute cyanide solutions, only a small portion of which can be used in making up new solutions, the balance being run to waste.

17. Loss by Absorption of Carbon Dioxide From the Atmosphere.—Potassium cyanide is acted on by the carbon dioxide of the atmosphere causing the formation of potassium carbonate and the liberation of hydrocyanic acid; thus,

$$2KCN + CO_2 + H_2O = K_2CO_3 + 2HCN$$

If the cyanide solution contained an alkali, the prussic or hydrocyanic acid thus liberated would be neutralized.

The loss of cyanogen that occurs by evaporation is considerable; free cyanogen CN also escapes from the solution through causes other than those described, but which have not yet been satisfactorily determined. A small loss of cyanogen takes place when wooden tanks and vats are used to retain the solution, but only when the tanks are new or leak. With iron tanks there is no loss, except in the case of leaks.

APPLICATION OF THE PROCESS

GOVERNING CONDITIONS

18. Size of Gold in Ore.—In deciding on the treatment required for any given ore, it is necessary to consider whether the gold in the ore is coarse or fine, whether the ore is neutral or acid, and what objectionable metals or minerals, if any, the ore contains.

19. Effect of Size and Condition of Gold Particles. Ores containing gold in a fine state of division are usually good cyaniding ores. If the gold is coarse, a longer time is required to dissolve it, and therefore such ore is usually subjected to some method of plate or pan amalgamation before cyaniding. When the gold is in a fine state of division, it may sometimes be dissolved by a 1-per-cent. cyanide solution in about 12 hours.

20. Time of Treatment Governed by Conditions.
In the treatment of concentrates containing pyrites, the material must remain in contact with the cyanide solution a much longer time than when tailings are being treated. This may be accounted for if there is a large quantity of gold to be dissolved, if the gold is coarse in the concentrates, or if there is some amalgam present. It takes time for a cyanide solution to penetrate the pyrite crystals to get at the gold, for the solution that can pass into the small interstices between the crystals will be held there by capillarity, resulting in a slow diffusion.

21. Acid Ores.—The **acidity of ore** is due to the decomposition of pyrites. The products of such decomposition, which consists chiefly of free sulphuric acid and soluble sulphates of iron, are destructive to cyanide solutions. The special treatment that is necessary to overcome the bad effects of acidity is discussed under the treatment of acid ores.

22. Chemical Limitations of the Process.—In some ores the gold and silver are so imprisoned in the matrix that the cyanide solution cannot reach them, even if the ore is crushed very fine. Such ores can only be partially treated unless they can be made porous by roasting, and as the latter operation is expensive, it may not be worth while. Again, there may be substances that will unite with the cyanide and prevent extraction of gold.

23. Silver Extraction.—Silver occurs usually in combination with bases that decompose potassium cyanide. When ores containing silver are oxidized or where the silver is in the condition of chloride, cyanide solutions will dissolve the silver. On the other hand, when lead, oxide of copper, or certain oxides of iron are present in the ore, the extraction is so poor as to condemn the process.

24. Slimes.— By this term is meant fine, and sometimes impalpable, pulp that floats about in the solutions and is very slow to settle. When slimes dry, they form hard lumps. If these lumps find their way into tailings and so into the vats, they will only be partially leached and consequently yield but a part of their gold and silver. Slimes are a source of annoyance from the fact that they are a hindrance to percolation and often entirely prevent it. They are not to be thrown away, however, because they frequently contain a high percentage of the precious metals.

25. Concentrates, which are derived by extracting the valuable minerals from a mass of barren rock, can be treated successfully by long contact with the cyanide solution. Agitation has been used to successfully hasten the treatment. In recent practice the concentrates are roasted. This affords an opportunity for the gold to be better attacked, as it changes the character of the ore. Sulphide concentrates sometimes arrange themselves in regular order, so that their cubes form a wall that prevents proper percolation.

LABORATORY TESTS

PERCOLATING TESTS

26. Object of Tests.—Tests are made of ores to ascertain their fitness for the cyanide process. **Percolating tests** of ore made in the laboratory are virtually similar to those carried on in practice, and consist in submerging ore in a cyanide solution, draining off the solution and finally wasting the ore. A convenient apparatus for small percolating tests is shown in Fig. 1. It consists of a funnel-shaped vessel *c*, an iron stand *s*, and a beaker *d*, into which the liquid is drained. The ore *a* rests upon a filter paper, which, in turn, rests upon a false bottom *b*

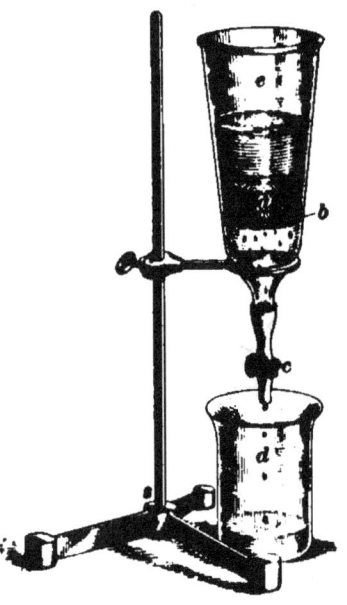

Fig. 1

in the vessel. A stop-cock c at the funnel end of the vessel e permits or prevents the liquid in the vessel escaping into the beaker d. The solution coming from the vessel may be tested for any loss of cyanide that has occurred during the percolating process and also for gold.

Whenever it is desirable to treat larger quantities of ore than the glass vessel will contain, such an arrangement of

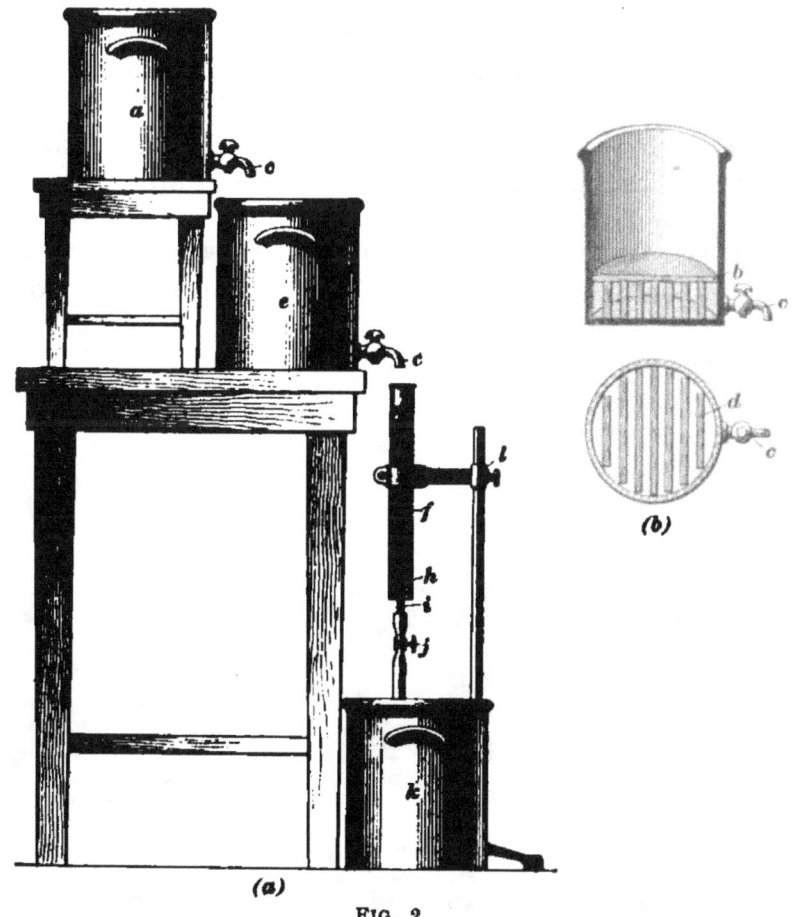

(a)

(b)

Fig. 2

glazed earthenware jars as is shown in Fig. 2 (a) will answer. A perforated false bottom b fits into the jar a above the stop-cock c, as shown in section, Fig. 2 (b). This false bottom is composed of slats d, placed ½ inch apart and surmounted by board b, upon which a piece of canvas is stretched to act as a filter. The solution is drawn from the jar *a* into the jar *e*, which answers as a gold-solution vessel. The loss of cyanide due to percolation may be ascertained from this solution. The gold in the solution is precipitated on zinc shavings, which are packed into a glass tube *f* 1.5 inches in diameter and about 18 inches long. The lower end of this tube is fitted with a perforated cork *h*, into which is inserted a glass nipple *i* provided with a rubber tube and clamp *j* for the purpose of regulating the flow into the jar *k*. The glass tube *f* is supported in a vertical position by means of an iron stand to which is attached the clamp *l*. The testing jar described has a diameter of about 12 inches and is charged with 4 or 5 inches of ore.

27. Testing Plant.—In Fig. 3 is shown an arrangement for another small testing plant composed of tubs *a*

FIG. 3

and *b*, a precipitating box *c*, and a sump *d* for receiving the filtered liquor. The tub *a* will hold 100 pounds of ore and sufficient cyanide solution to leach out the gold. From tub *a* the gold-cyanide solution is drawn into the tub *b*, from which it passes into the precipitating box *c* that contains zinc shavings. After the gold is precipitated the filtered liquor passes into the sump *d*, where it is tested for cyanide loss, then strengthened or standardized by the addition of cyanide, and, if needed, pumped back to the tub *a*.

28. Filters for Testing Plant. — A good filter bottom can be made for small testing plants by placing slats 1 inch apart across a hoop, as shown in Fig. 2 (*b*). These slats do not reach down the entire width of the hoop, but stop about 1 inch above the bottom to allow a free circulation of the liquor. A strip of canvas should be tacked between the circumference of the hoop and the inside of the tub to prevent sand from washing under the false bottom.

LABORATORY APPARATUS REQUIRED

29. The laboratory apparatus that will be required to make these tests are two burettes graduated to $\frac{1}{10}$ cubic centimeter, each having a capacity of 50 cubic centimeters and provided with floats; one burette stand; two pipettes having a capacity of 10 cubic centimeters; a graduated cylinder having a capacity of 1,000 cubic centimeters and a glass stopper; two titrating dishes having a capacity of 120 cubic centimeters; six porcelain evaporating dishes 5 inches in diameter; six iron evaporating dishes 5 inches in diameter; one wedge-wood mortar 5 inches in diameter; one iron stand with three rings; four flasks having a capacity of 8 ounces; and four flasks having a capacity of 16 ounces. A description of this apparatus is given in *Assaying*.

DETERMINATION OF FREE POTASSIUM CYANIDE IN A SOLUTION

30. Silver Nitrate Test. — Several methods have been suggested for determining free potassium cyanide in a solution. A rapid and accurate determination may be made by titrating a measured quantity of a solution with a standard solution of silver nitrate. Silver cyanide is thus formed, which will immediately redissolve if there is an excess of potassium cyanide. The reaction is as follows:

$$(1) \qquad AgNO_3 + KCN = AgCN + KNO_3$$
$$(2) \qquad AgCN + KCN = KAg(CN)_2$$
$$(3) \qquad KAg(CN)_2 + AgNO_3 = 2AgCN + KNO_3$$

31. End Point.—Titrating is finished when a permanent white precipitate of silver cyanide is produced. Silver nitrate is to be added from the burette until all the potassium cyanide has united with the silver cyanide to form a double salt of potassium silver cyanide. If more silver nitrate is added than is required to form the double salt, a permanent precipitate of silver cyanide is formed, which shows that sufficient silver nitrate has been added for the reaction.

The end reaction is more distinct when two or three drops of a 5-per-cent. solution of potassium iodide is added to the cyanide solution before titration. After all the cyanide is converted into the double salt, any excess of silver nitrate will unite with the iodide.

32. Preparation of the Standard Silver-Nitrate Solution.—A convenient standard for a silver-nitrate solution is a solution in which the quantity of silver nitrate in each cubic centimeter represents 0.1 per cent. of potassium cyanide. When such a solution is added to 10 c. c. of a cyanide solution, it will eventually form a permanent white precipitate. To prepare the standard silver nitrate, the equation of Art. **30** is taken ; it shows that it takes two molecules of KCN to unite with one of $AgNO_3$. The molecular weights are as follows :

$$AgNO_3 = 108 + 14 + 48 = 170$$
$$2KCN = 78 + 24 + 28 = 130$$

From this it may be seen that the two are in the proportion of 17 to 13, so that if 17 grams of nitrate of silver be dissolved in 1,000 c. c. of pure water, 1 c. c. of the solution will be equal to .013 per cent. of KCN. This makes an unhandy number for quick calculations, therefore the nitrate of silver solution is reduced in proportion to 10 c. c. of KCN solution as follows:

$$13 : 17 :: 10 : 13.07$$
and
$$17 : 13.07 :: 13 : 10$$

That is, the standard silver-nitrate solution may be made by dissolving 13.07 grams of silver nitrate in 1,000 c. c. of distilled water.

33. To Calculate the Percentage of KCN in a Solution.—First fill two burettes, one with the standard silver-nitrate solution and the other with the cyanide solution to be tested. Next, run into a beaker 10 c. c. of the cyanide solution; add cautiously the silver-nitrate solution from the burette until a permanent opalescent precipitate remains after thoroughly agitating the solution in the beaker. If 2 or 3 drops of a potassium-iodide solution be added to the cyanide solution in the beaker, it will assist in determining the end reaction by forming a yellow precipitate. Read from the burette the number of c. c. of standard silver nitrate used, divide by 10, and the result will be the available potassium cyanide in the solution in per cent.

To illustrate the above, suppose that 10 c. c. of the cyanide solution were used and that it took 6 c. c. of standard silver-nitrate solution to form a permanent precipitate, then $\frac{6}{10} = .6$ per cent. *KCN*.

34. To Test Strong Cyanide Solutions. — *To test a strong cyanide solution, take 3 or 4 c. c. and titrate it with silver-nitrate solution.* If 4 c. c. of cyanide solution required 6 c. c. of silver nitrate, then 10 c. c. of cyanide would require 15 c. c. of standard silver nitrate; and 15 divided by 10 = 1.5 per cent. of potassium cyanide.

In testing the strength of a strong solution in the dissolving tank, take 10 c. c. of the solution and dilute it with water to 100 c. c. Take 10 c. c. of this solution and titrate with silver nitrate, as described above. The number of cubic centimeters required of the silver-nitrate solution will be the percentage of potassium cyanide in the strong solution, for the 10 c. c. of the dilute solution only contained a tenth of the original cyanide solution, hence there is no need of dividing by 10.

35. To Test Very Weak Cyanide Solutions.—Take 100 c. c. of the cyanide solution, titrate it with the silver-nitrate solution, and divide the number of c. c. of silver nitrate required by 100; the result will be the percentage of potassium cyanide in the solution. For example, suppose that 100 c. c. of a potassium cyanide solution be used and it required 5 c. c. of silver nitrate to form a permanent precipitate, then

$$5 \div 100 = .05 \text{ per cent. } KCN$$

36. To Test Potassium Cyanide of Commerce. — Cyanide of potassium when purchased from the dealers is not pure. It may contain black carbide of iron, alkaline carbonates, and small quantities of alkaline chlorides and sulphides, for which reason it is customary to test the mixture to determine the quantity of KCN it contains.

In order to test the strength of solid potassium cyanide for the available KCN that it contains, proceed as follows: (1) Sample the cyanide to be tested by taking a specimen across the thickness of the cake. (2) Reduce the sample to a coarse powder, mix thoroughly, take a small sample of this and reduce to a very fine powder. (3) Weigh out 1 gram of the powdered sample and dissolve it in distilled water, after which add sufficient water to make 100 c. c of solution. (4) Take 10 c. c. of this solution and titrate it with standard silver nitrate. The number of c. c. of silver nirate required to produce the permanent precipitate divided by 10 will give the amount of available KCN in 1 gram of the salt. For illustration, suppose that 10 c. c. of the KCN solution required 8.5 c. c. of standard silver nitrate, then $8.5 \div 10 = .85$ KCN in 1 gram, which is equal to 85 per cent. of KCN in the crude salt.

ASSAY OF CYANIDE SOLUTION FOR GOLD
AND SILVER

37. Take 236.6 c. c., or ¼ pint, of the solution, place it in a round iron dish over a fire, and evaporate to a small bulk. As the evaporation proceeds rub the sides of the dish with a stirring rod so as to collect the salts in the bottom of the dish. To this concentrated solution add 40 grams of litharge. After mixing well, evaporate carefully to dryness. Transfer the contents of the dish to a clay crucible and add 15 grams of borax glass, 5 grams of bicarbonate of soda, and 2 grams of argol; mix the contents carefully with a spatula. Cover the contents of the crucible with a little borax glass and fuse. After quiet fusion pour the contents and allow to cool. Cupel the lead button and weigh the bead of gold and silver. If the ore contains silver, part the bead and weigh the gold.

38. Lead Evaporation Trays.—In some cyanide works the solutions are evaporated in trays made of thin sheet lead, which, with their contents, are afterwards scorified and the resulting lead button cupeled.

The lead tray, which should weigh about 20 grams, is made of pure lead foil and is shaped by laying a wooden block 2″ × 4″ on the foil and folding the lead upon the sides of the block so as to form a tray 4 inches long, 2 inches wide, and 1 inch deep. Care must be used in forming the corners of the tray so that they will not leak. For evaporation, the tray containing the solution is placed upon a piece of asbestos cardboard and heated gently by a burner underneath.

39. Gram and Grain Tables.—Consult the following tables to find the quantity of gold and silver per ton of solution. When the gold and silver are weighed with gram weights, refer to Table I; and when with grain weights, refer to Table II.

TABLE I

FOR THE ASSAY OF CYANIDE SOLUTIONS (PARK)

236.6 c. c., or ¼ Pint, of Solution Gives Fine Metal	1 Ton of Solution Gives Fine Metal			236.6 c. c., or ¼ Pint, of Solution Gives Fine Metal	1 Ton of Solution Gives Fine Metal		
Gram	Ounce	Penny-weights	Grains	Grams	Ounces	Penny-weights	Grains
.0001	0	0	5.5	.0200	2	5	20
.0002	0	0	11.0	.0300	3	8	18
.0003	0	0	16.5	.0400	4	11	16
.0004	0	0	22.0	.0500	5	14	14
.0005	0	1	3.5	.0600	6	17	12
.0006	0	1	9.0	.0700	8	0	10
.0007	0	1	14.5	.0800	9	3	8
.0008	0	1	20.0	.0900	10	6	6
.0009	0	2	1.5	.1000	11	9	4
.0010	0	2	7.0	.2000	22	18	8
.0020	0	4	14.0	.3000	34	7	12
.0030	0	6	21.0	.4000	45	16	16
.0040	0	9	4.0	.5000	57	5	20
.0050	0	11	11.0	.6000	68	15	0
.0060	0	13	18.0	.7000	80	4	4
.0070	0	16	1.0	.8000	91	13	8
.0080	0	18	8.0	.9000	103	2	12
.0090	1	0	15.0	1.0000	114	11	16
.0100	1	2	22.0	2.0000	229	3	8

CRASSE'S METHOD OF TESTING CYANIDE SOLUTIONS FOR GOLD

40. Take 236.6 c. c. of the gold-cyanide solution to be tested and add a solution of silver nitrate until a precipitate ceases to form. Add the silver nitrate a little at a time, shaking the solution after each addition. The gold will be precipitated as argentic auricyanide. Let the precipitate settle; decant off the clear solution and filter the remaining precipitate; next dry the precipitate and mix it with 30 grams

TABLE II

FOR THE ASSAY OF CYANIDE SOLUTIONS (PARK)

236.6 c. c., or ½ Pint, of Solution Gives Fine Metal	1 Ton of Solution Gives Fine Metal			236.6 c. c., or ½ Pint, of Solution Gives Fine Metal	1 Ton of Solution Gives Fine Metal		
Grain	Ounce	Penny-weights	Grains	Grains	Ounces	Penny-weights	Grains
.001	0	0	3.5	.060	0	8	23
.002	0	0	7.0	.070	0	10	11
.003	0	0	11.0	.080	0	11	23
.004	0	0	14.5	.090	0	13	10
.005	0	0	18.0	.100	0	14	22
.006	0	0	21.5	.200	1	9	20
.007	0	1	1.0	.300	2	4	18
.008	0	1	4.5	.400	2	19	16
.009	0	1	8.0	.500	3	14	14
.010	0	1	12.0	.600	4	9	12
.020	0	3	0.0	.700	5	4	10
.030	0	4	12.0	.800	5	19	8
.040	0	6	0.0	.900	6	14	6
.050	0	7	11.0	1.000	7	9	4

of litharge, 15 grams of glass, 15 grams of bicarbonate of soda, and 3 grams of argol; place the mixture in a clay crucible, fuse thoroughly, pour the contents into an iron mold, cupel the resulting lead button, part the bead, and weigh the gold. From the weight of gold calculate, from the table, the amount of gold in a ton of the cyanide solution.

Crasse's method is performed quicker than that described in Art. **37**, and gives reliable results where the value in *Au* only is required, but if it is desired to know the amount of silver in the cyanide solution, the first method must be used.

TEST FOR ORES OR TAILINGS

41. All ores and tailings should be tested in the laboratory to determine their suitableness for cyanide treatment. The principal items usually determined are the rate of percolation, the minimum extraction, the quantity of cyanide consumed, the acidity of the ore, and the neutralizing agents required. Other matters for consideration are economic, such as the strength of cyanide solutions for extraction, the time of contact for economic extraction, and the degree of precipitation from strong and weak solutions.

TEST FOR RATE OF PERCOLATION

42. Ore or tailings through which the solution will percolate at the rate of from $\frac{3}{4}$ to 2 inches or more an hour may be cyanided by percolation. When the percolation is $\frac{1}{2}$ inch per hour, or less, economic percolation is practically impossible, unless the material is very rich and will pay for the long period of contact required. The same fact is true of ores and concentrates.

TEST FOR PERCENTAGE OF EXTRACTION

43. Take 4 kilograms of thoroughly dried and mixed tailings for the test. Remove an assay sample and place the remainder into a percolating vessel. Level this charge and pour over it 4 liters of a 0.2-per-cent. cyanide solution. Close the discharge cock and soak the tailings 12 hours in the solution, after which time allow the solution to drain below the surface of the tailings. After the solution has drained $\frac{1}{4}$ hour into a vessel, close the discharge cock and pour the liquor back on the tailings, open the discharge cock and allow it to percolate into the receiving vessel again. This process is to be repeated at intervals of ½ hour for 12 hours, after which the liquor is all drained off. The tailings are next washed with 4 liters of water, then dried, and an assay sample taken from them. The cyanide solution is drained into one vessel and the wash water, which now forms a weak cyanide solution, into another vessel. Each of these solutions is assayed to ascertain the percentage of extraction. Each solution is then passed through zinc shavings, the weak solution being passed much slower than the strong solution. Assay the solutions after they have passed through the zinc shavings and the difference between the two assays will determine the economic extraction of gold. All the gold in a cyanide solution is not precipitated by the zinc, and for that reason tests should always be applied in order to remedy any decided loss.

TO TEST CYANIDE CONSUMED

44. Place in a stoppered bottle 200 grams of the ore with 200 c.c. of .5-per-cent. potassium-cyanide solution and shake for 20 minutes. Let the solution settle and then filter off a portion. Take 10 c. c. of this filtered solution and test it for the amount of cyanide it contains. The difference in the strength of the cyanide solution before and after extraction shows the percentage of cyanide consumed by leaching 200 grams of ore. If the solution contained 0.5 per cent. of cyanide before treatment and 0.45 per cent. after treatment, a consumption of 0.05 per cent. of potassium cyanide would be indicated. The consumption of potassium cyanide at this rate would be 2,000 pounds × .05 per cent. = 1 pound for a ton of ore.

The consumption of cyanide takes place almost immediately after the solution comes in contact with the ore. After 20 minutes it is safe to consider that there will be no further reduction.

DETERMINATION OF ACIDITY IN ORES

45. If the consumption of cyanide is high or above 4 pounds of KCN to the ton of ore, a portion of the material is tested for acidity or "cyanicides," by which is meant free acid, soluble metallic salts, or any other substance that will destroy cyanogen. To determine the acidity in ores and ~tailings, the following test may be made:

Place 14.5 grams of ore and 250 c. c. of water in a tall glass jar and shake well. Next fill a burette with a standard solution of caustic soda $NaOH$. Titrate the solution in the jar with the caustic soda until it becomes neutral to litmus paper or to a solution of litmus.

Each c. c. of the soda solutions used will indicate that 0.1 pound of caustic soda is to be added to each ton of ore or tailings as a preliminary wash to neutralize cyanicides before the cyanide solution is added.

46. Preparation of a Standard Solution of Caustic Soda.—Weigh 10 grams of commercially pure (c. p.) caustic soda and dissolve in 500 c. c. of pure water, after which add 500 c. c. more of pure water.

———

TO NEUTRALIZE ACID ORES

47. The test for acidity will show the quantity of caustic soda to add to each ton of ore to neutralize the acids, but caustic lime may be used sometimes to great advantage where soda is expensive and lime cheap. The quantity of lime needed in practice may be calculated from the caustic soda used in the test; for example, the quantity of caustic soda expressed in pounds multiplied by .7 will give the pounds of pure caustic lime that will be required to neutralize the acids in the ore. The caustic lime of commerce is not pure, so that a much larger quantity will be required than is indicated by the above calculation. In some mills the same number of pounds of lime are used as would be required of caustic soda.

Caustic lime is used in most cyanide works in preference to caustic soda. Caustic soda is objectionable because of its tendency to foul the cyanide solutions in the precipitating boxes and to form ferrocyanide of zinc, whenever pyritic ores are treated.

Caustic lime is usually added to the ore when it is placed in the leaching vats. There seems to be no bad effect from caustic lime on the cyanide solution nor on the zinc in the zinc boxes, although it is claimed by some that an excess will increase the consumption of zinc in the zinc boxes.

48. *If the consumption of caustic soda is more than 3 pounds per ton of ore*, it is usually advisable to first wash the ore with water before adding the alkali. This is done by covering the ore in the leaching vat with water and allowing it to percolate through the ore until the soluble acids are removed. The washed ore is then treated with a solution of caustic soda or lime of the required strength, which is determined by the amount of acid remaining in the ore.

DETERMINATION OF STRENGTH OF CYANIDE SOLUTION FOR HIGHEST EXTRACTION

49. Place 4 kilograms of ore or tailings into each of six leaching tubs or crocks, after mixing the quantity of lime required. Into the first tub place 4 liters of a 0.1-per-cent. *KCN* solution; into the next 4 liters of a 0.2-per-cent. *KCN* solution; and place cyanide solutions in each of the four remaining tubs, having increased the strength of the solution 0.1 per cent. for each tub. Let each sample of ore soak 12 hours, then precolate for 12 hours, using the solution over and over again. Test the solution drained from each tub for the cyanide consumed.

Next, add 4 liters of wash water to each tub to displace the cyanide solution left in the ore; drain off and then take samples of the tailings from each tub for assay. Compare the extraction in each case with the amount of cyanide consumed. The point to determine is the relation of the increased extraction to the increased consumption of cyanide. The proper strength of cyanide solution to select is the one that will give the highest extraction in a reasonable length of time with the least loss of potassium cyanide.

TIME FOR CONTACT WITH STRONG SOLUTION

50. To make a test for time of contact between the ore and cyanide solution, take 24 kilograms of ore and add the necessary amount of lime, if the ore is acid. Divide the ore into six parts of 4 kilograms each. Place 4 kilograms of ore into each of the six tubs and add to each 4 liters of cyanide solution of a strength determined by the preceding process to give the highest economic extraction. Let the contents of each tub soak 6 hours, then start percolation on each tub. Continue the percolation on the first tub 18 hours, the second 30, the third 42, the fourth 54, the fifth 66, and the sixth 78 hours. Displace the strong solution in each tub at the end of the respective periods for percolation with 4 liters of a cyanide solution one-fourth the strength of the original strong solution. Let this pass through the material, then wash with 4 liters of water. Drain and assay the tailings.

The assay will indicate the time beyond which the extraction is not increased. *The time of extraction will depend largely on the fineness of the gold, whether free or combined, and the presence of interfering substances.*

TESTS ON PRECIPITATION

51. To test weak and strong solutions for percentage of precipitation, the solutions from the previous tests may be taken. In some cases it is difficult to precipitate gold from weak cyanide solutions. This is one of the weakest points in zinc precipitation. The cyanide solution containing the gold is passed through a column of zinc shavings. The tube should be carefully packed with fresh zinc shavings and the solution allowed to pass through it slowly. If the precipitation takes place properly, the gold is deposited as a dark, metallic covering on the shavings in the top of the tube and the deposit on the zinc is less towards the bottom. If the deposit is gray on top and dark towards the bottom, the indications are unfavorable for good precipitation.

Take some of the solution that has passed through the tube and assay it for gold and silver. Compare this assay with what the solution contained originally, and the difference will show the degree of extraction. With good precipitation, 85 to 95 per cent. of the gold is extracted from the solution.

If the precipitation is unsatisfactory, pass the solution through two or three tubes filled with zinc shavings, and if it is still unsatisfactory, add potassium cyanide to the solution.* Pass the-strengthened solution through the tubes and determine by assaying the solution the degree of precipitation. Estimate the quantity of potassium cyanide necessary to bring the solution to the precipitating point and the cost of the additional potassium cyanide, and see how it compares with the extra quantity of gold precipitated. The addition of potassium cyanide will pay in some cases.

Imperfect precipitation can usually be corrected by increasing the length of the zinc column or by slightly increasing the strength of the solution with potassium cyanide.

PRELIMINARY ROASTING

52. Ores containing gold in combination with tellurium and antimony are benefited by a preliminary roasting operation, as the loss of cyanide is thereby reduced and the extraction of gold is increased.

In the Cripple Creek district, Colorado, most of the telluride ores are found unsuitable for cyaniding without a preliminary roast. After roasting, a high extraction of gold has been obtained, with a small consumption of cyanide. In some places, concentrates and pyritic ores that could not be

----

* Strong *KCN* solutions attack zinc and start chemico-electrical action between zinc and auric-potassic-cyanide solution.

treated economically in the raw state have been successfully treated after a dead roast. Roasting has the effect of driving off the cyanicides in the ore by volatilizing substances that destroy the cyanogen. It seems also to break up refractory ore particles by freeing the gold and to render the material more permeable to cyanide solutions.

53. After the ore has been tested by the methods described for total extraction and consumption of cyanide, if the extraction is low, try finer crushing; and if the result is still unsatisfactory, the difficulty may be corrected by a roast, as already explained.

THE CYANIDE PLANT

54. Introductory. — The cyanide process is varied to adapt it to the different metallurgical and mechanical problems that arise in different localities. There are many matters which may interfere with the successful operation of a cyanide plant, therefore it is imperative that the ore from a mine be thoroughly tested in the laboratory before the erection of a plant. Ores usually change with depth, a fact that should be remembered. There are ores that lie near the surface which yield up their gold readily, while those at certain depths on the same vein cannot be treated by the cyanide process.

LOCATING A CYANIDE PLANT

55. Character of the Plant. — In selecting the site for a cyanide plant, the items to be considered are:

1. Is the plant to treat tailings as they come from the battery or a stock of accumulated tailings?

2. Is the plant to treat raw or roasted ore?

3. The source of water supply and its sufficiency.

4. Method to be adopted for transporting ore or tailings to the leaching vats.

5.   Methods of emptying the leaching vats and transporting the tailings to the dump.

56.   Advantage of Sloping Ground.—If there is a side hill in the immediate vicinity where it has been decided to erect a plant, the leaching vats and solution tanks may be so arranged that the solutions will flow by gravitation from one set of tanks to the next.   The crushed ore can be loaded into dumping cars and run by gravitation to the leaching vats, and after it has been leached it can be sluiced from the tanks upon the lower ground or loaded into cars and run by gravity to the dump.

57.   Objects of a Cyanide Plant.—The appliances entering into the construction of cyanide plants for the treatment of ores and tailings differ both in size and design; but the general principles involved in all MacArthur-Forrest cyanide plants are the same.   The object of a cyanide plant is to treat ore or tailings containing gold or silver with a weak solution of cyanide of potassium and water, for the purpose of dissolving them and then precipitating the gold or silver from the solution in a metallic form.

58.   Differences Between Plants.—To accomplish the objects intended, a cyanide plant must have leaching vats for dissolving the gold, zinc boxes or tanks for precipitating the gold, and storage tanks for the cyanide solutions.   The important differences between plants consist in the material, size, and shape of leaching vats, the methods for discharging the tailings, and the relative positions of leaching vats, storage tanks, and zinc boxes to one another.

59.   Arrangement of a Plant.—The most convenient arrangement for a cyanide plant is to have the vats in tiers, so that each series of vats may be completely drained into the next below it.

By this means sufficient storage room is obtained to enable work to proceed from 12 to 24 hours without pumping.   Many plants have their vats on the same level, which

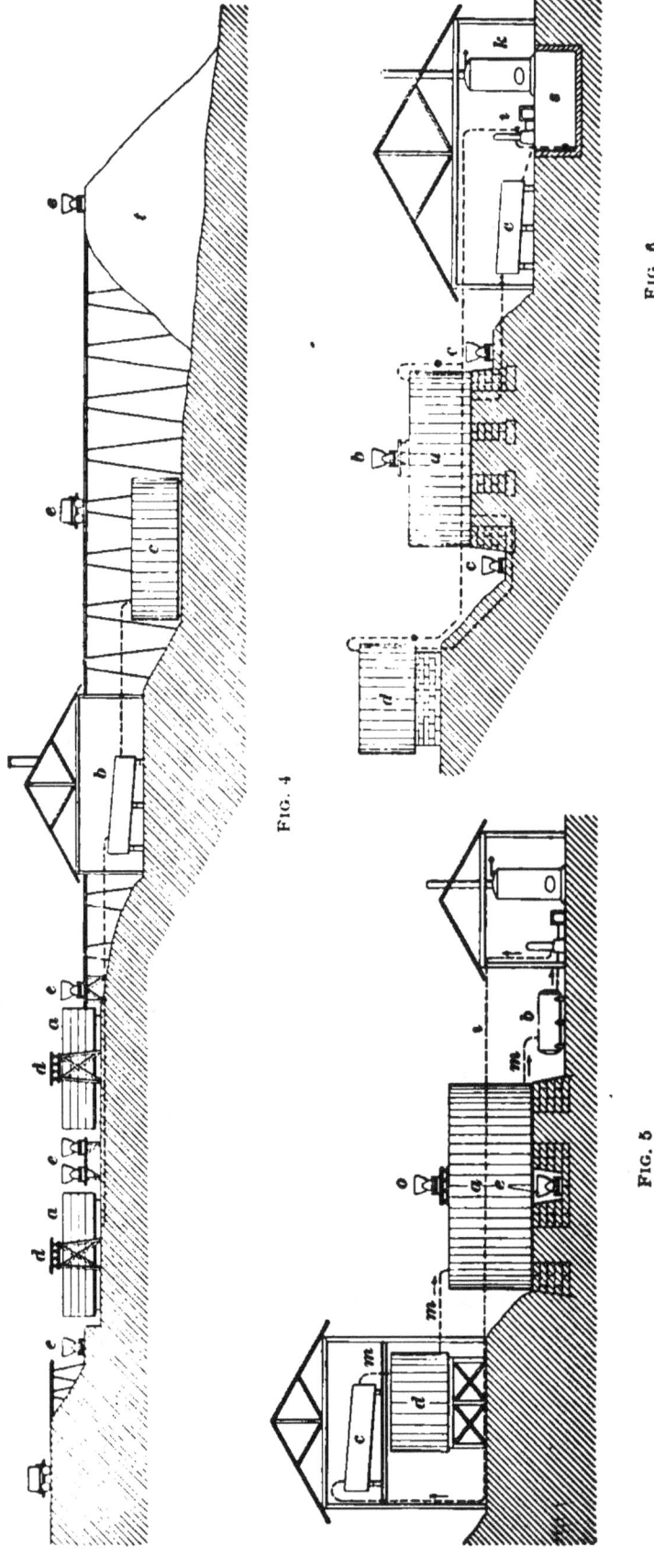

FIG. 4

FIG. 5

FIG. 6

requires that the pump be kept in operation most of the time, in order to keep the solutions in circulation.

The following figures, intended to illustrate the general designs of plants, are taken from W. R. Feldtmann's "Notes on Gold Extraction by Means of Cyanide of Potassium."

60. In Fig. 4, the leaching vats *a, a* are placed above the zinc boxes *b* and the storage tanks *c*. The solution gravitates from *a* to *c*, the latter tank, which is virtually the sump tank, acting in the dual capacity of storage and sump tank. The solutions are strengthened in tank *c* and when needed pumped to tanks *a*.

The sketch shows that the leaching tanks are loaded with ore from cars that pass over them on trestle *d* and are unloaded by shoveling the tailings over the side into tram cars *e*. After loading, the tram cars *e* are run to the tailings dump *t* and discharged. This arrangement makes an inexpensive plant, as it does away with gold-solution and storage tanks; but the method of handling the tailings increases the cost of cyaniding.

61. The plant shown in Fig. 5 is arranged so that the solution passes direct from the leaching vat *a* into an airtight sump *b*. From the sump the solution is pumped up to the zinc-precipitating boxes *c*, from which it runs by gravity into the storage vats *d*. In the storage vats it is strengthened for use when wanted in the leaching vat *a*. The tailings in this plant are discharged through a door in the bottom of the vat *a* into the cars *e*, and thence run to the tailings dump not shown in the figure. The pipe from the pump to the zinc boxes is shown by the dotted lines *i* and the pipes from the various tanks by *m*. The leaching tank is centrally charged with ore by the car *o*.

62. The plant shown in Fig. 6 is a combination of the plants shown in Figs. 1 and 2. The leaching vat *a* is centrally loaded with ore by tram car *b* and the tailings are loaded into cars *c* through doors in the sides of the vats. The solution runs from the storage tank *d* into the leaching vat *a*; from the latter tank it runs by gravity to precipitating boxes *e*, and from *e* it runs into the sump tank *s*. The dotted lines in the figure show the pipe lines from the pump *i* to the storage tank, and from the storage tank to the various vats. The boiler *k* is shown to be under the same cover as the precipitation and sump tanks, which is a good arrangement, since no one can get into the zinc boxes without the knowledge of the engineer.

63. Fig. 7 (*a*) shows the plan and Fig. 7 (*b*) the cross-section of a cyanide plant designed to treat 75 tons of ore a day. The weak-solution storage tank *a'* and the strong-solution storage tank *a* are each located on an elevation above the leaching tanks *b*. The strong-solution storage tank is connected with the leaching vats at the bottom and top by pipe *c*. Pipe *c'* connects the weak-solution storage tank with the leaching vats. Over the leaching vats is a water pipe *d* for sluicing out the tailings. The leaching vats *b* are on a level above the weak gold-solution tank *h'* and strong gold-solution tank *h*. The pipe *e* is connected with the leaching vats *b* and the vacuum pump *s*. The solution passes from the vacuum pump *s* to the weak gold-solution tank *h'* and the strong gold-solution tank *h* through the vacuum discharge pipe *u*. In ordinary leaching, where the vacuum pump is not used, the solution from the leaching vats *b* is passed through the valves *x* into weak gold-solution launder *d'* or the strong gold-solution launder *d*, depending on the strength of the cyanide solution. The weak gold-solution tank *h'* and the strong gold-solution tank *h* are connected, respectively, with the precipitating boxes *k'* and *k*. The solution passes through the precipitating boxes into the weak-solution sump *l'* or the strong-solution sump *l*. The solutions from the sump are pumped to their respective storage tanks by the centrifugal pump *w*. The solution is standardized in the strong storage tank by adding the required amount of potassium cyanide. The solution in the weak-solution storage tank is used as a wash to follow the application of the strong solution.

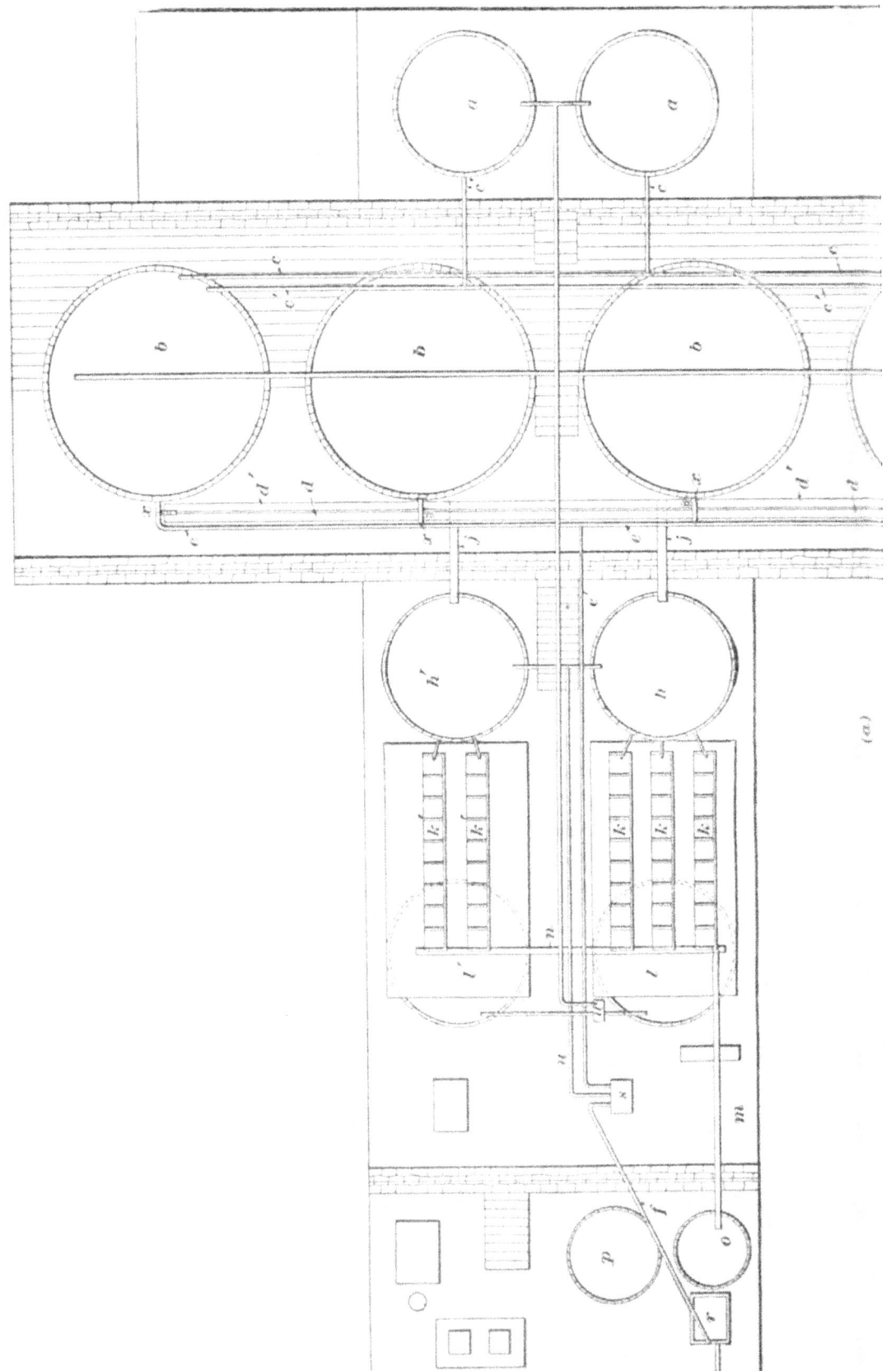

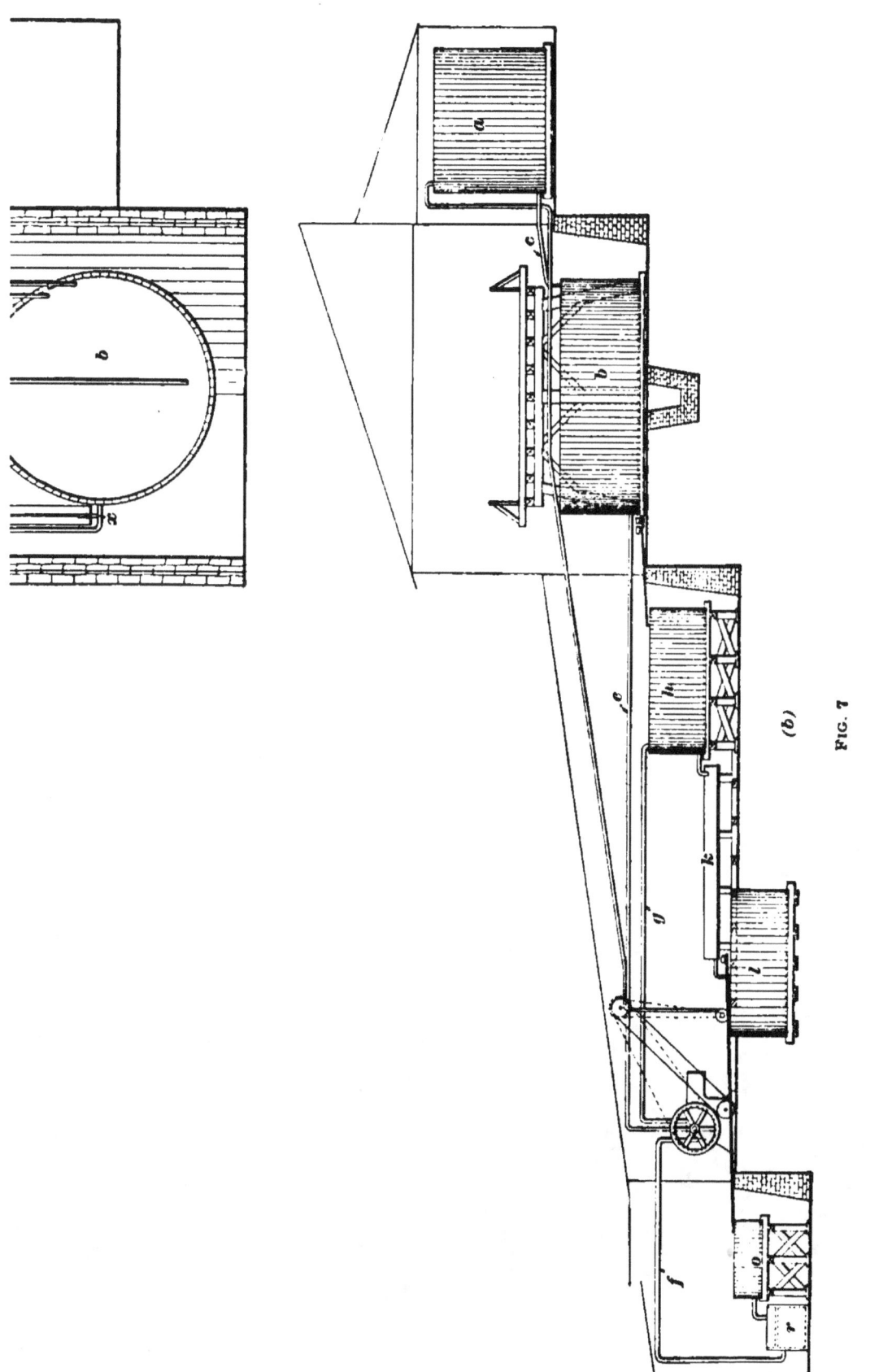

FIG. 7

The precipitated gold and silver is washed from the precipitating boxes k into a side launder communicating with the launder n, thus allowing the precipitate to pass into a movable launder m and hence into the acid tank o. The precipitate, after treatment with dilute sulphuric acid, is thoroughly washed and partially dried by suction from the vacuum pump s through the pipe f. The precipitate is washed in the acid tank o and the water decanted into the settling tank p. Any precipitate floating in the water settles to the bottom of the tank and is collected and put with the precipitate on the filter box r.

64. Capacity of a Plant. — The capacity of a plant depends on the number and size of leaching vats employed and the time required to treat the ore. The capacity of each tank should be either the same or one-half or one-third the daily capacity of the mill. The number of leaching vats will be either the same, twice, or three times the number of days required to fill, leach, and empty a vat. If a plant is designed for a 4-day treatment in four leaching vats of 75 tons each, the daily capacity would be considered 75 tons. If the leaching vats should take 6 days on some ores, the plant would not have a daily capacity of 75 tons. The capacity could be increased to this amount by adding 2 more leaching vats, each of 75 tons capacity. The remaining part of the plant would be the same as described in the specifications for a 75-ton plant.

The equipment and operation of a cyanide plant, using a 75-ton plant as a model, will be illustrated and discussed.

65. Specifications for a 75-Ton Cyaniding Plant. The following appliances will equip a plant for a daily capacity of 75 tons, with 4 days' treatment in the leaching vats. It is assumed that there are good facilities for charging the vats and discharging the tailings.

2 storage tanks (for strong and weak solutions), 12 ft. diameter × 10 ft. deep.
4 leaching tanks, 24 ft. diameter × 5 ft. deep; capacity 75 tons.
2 gold tanks (for strong and weak solutions), 12 ft. diameter × 5 ft. deep.

TABLE III

Daily Capacity of Plant in Tons	Number and Size of Storage Tanks	Number and Size of Leaching Vats	Number and Size of Gold-Solution Tanks	Number and Size of Sump Tanks	Number of Zinc Boxes of 9 Compartments Each 24" × 18" × 15"
20		4 Size, 12 ft. diam. 4½ ft. deep	2 Size, 8 ft. diam. 5 ft. deep	1 Size, 12 ft. diam. 5 ft. deep	1
50	2 Size, 12 ft. diam. 8 ft. deep	4 Size, 20 ft. diam. 4½ ft. deep	2 Size, 10 ft. diam. 5 ft. deep	2 Size, 10 ft. diam. 5 ft. deep	4 2 for strong and 2 for weak solutions
75	2 Size, 12 ft. diam. 10 ft. deep	4 Size, 25 ft. diam. 4½ ft. deep	2 Size, 12 ft. diam. 5 ft. deep	2 Size, 12 ft. diam. 5 ft. deep	5 3 for strong and 2 for weak solutions
200	2 Size, 16 ft. diam. 12 ft. deep	6 Size, 35 ft. diam. 4½ ft. deep	2 Size, 16 ft. diam. 5 ft. deep	2 Size, 16 ft. diam. 5 ft. deep	7 4 for strong and 3 for weak solutions

2 sump tanks (for strong and weak solutions), 12 ft. diameter × 5 ft. deep.

1 acid tank, 6 ft. diameter × 2 ft. deep, for the reduction of the zinc-gold slimes.

5 sets of zinc boxes, 9 in each set, 3 sets for strong solutions, 2 sets for weak solutions, making 45 boxes 12 in. high × 15 in. long × 24 in. wide, constructed of No. 12 steel or of wood.

10 screens, No. 18 wire, 8 mesh.

45 screens for zinc boxes, 15 in. × 24 in., No. 12 wire, 4 mesh.

45 screen frames for zinc boxes.

1 lathe with countershaft, for cutting zinc shavings.

1 low-service duplex piston pump (all iron), to pump solution, size 6 in. × 7½ in. × 8 in.

1 4' × 10' receiver or vacuum chamber, for exhausting moisture at ½ atmospheric pressure with 2½-inch pop safety valve and vacuum gauge and fittings.

1 6" × 8" × 12" vacuum pump. W. I. piping, valves, and fittings for solutions, water, and air, also steam pipes for pumps. Ironwork for muffle furnace, including 3 iron pans, 24 in. × 24 in. and 3 in. deep. Ironwork for 18-inch bullion furnace; 3 bullion molds, each to hold contents of No. 125 plumbago crucible.

66. Specifications for Plants of Different Capacities. By referring to Table III, a general idea can be obtained of the size of leaching vats, solution tanks, precipitating boxes, and sumps, in plants of different capacities. The general arrangement of smaller and larger plants would be the same as in the 75-ton plant described in Art. **63,** and illustrated by Fig. 7 (*a*) and (*b*).

DETAILS OF CONSTRUCTION

67. Construction of Vats.—Any material that will make a water-tight vat may be used in its construction. Up to the present time, brick and cement, wood and cement, concrete, wood, and steel have been used. Where suitable ground can be found, cisterns lined with stone and cement or brick and cement, and having their mouths level with the surface of the ground, make serviceable vats. At the Langlaagte Estate of the Transvaal, in South Africa, circular brick vats 10 feet deep and 40 feet in diameter and having a capacity of about 400 tons are constructed. For small cyaniding plants, wooden or steel vats are better suited to meet the requirements than the large vats just mentioned. The Pelatan-Clerici, the Kendall, and the Pneumatic cyanide processes use small vats having a capacity of from 1 to 5 tons, as they are able to treat ore in from 12 to 18 hours. The Pelatan-Clerici people use a wooden vat lined with cement. Masonry vats lined with cement are necessarily below ground, and in case they leak, the fact cannot be readily discovered; they are therefore not to be preferred, although they are in use at some large plants. Wooden vats lined with cement will leak in case the cement is broken, but as they are above ground the leak can be detected and the lining repaired.

68. California redwood, cypress, cedar, and white pine are trees that make suitable tank lumber. The timber trees . mentioned have soft and compressible fibres, well adapted to the tight joints necessary in cyanide tanks. Wood used in vat construction should absorb and retain a certain amount of moisture, otherwise the vats will leak. It is sometimes customary to coat wooden tanks with asphalt or paraffin paint to prevent their leaking and absorbing the gold solution. Very little leakage will occur when a well-constructed and set-up tank is soaked, but should the tank be alternately wet and dry, it will leak. Wooden tanks are purchased from dealers in the "knock-down" form for transportation. Each stave is numbered, so that any intelligent carpenter can put the parts together.

69. **Wooden leaching vats** for cyaniding usually vary in size from 20 feet to 40 feet in diameter and from 3 feet to 6.5 feet in depth. The staves a, shown in Fig. 8 (a), are about 4 inches wide by 2.5 inches thick, with a slight upward taper, so that the diameter of the vat at the top is about 4 inches less than it is at the bottom. The bottom b is made at least 3 inches thick and of the same kind of wood as the staves. The bottom pieces are fitted to each other by dowel-pins. They are then fitted into a groove cut in the staves near their lower ends, so as to leave a chime of

6 inches. The staves are held in place by steel hoops. No white lead or any packing should be used in making these vats, since if the faces of the lumber used are true, no amount of white lead can make them truer; and if they are

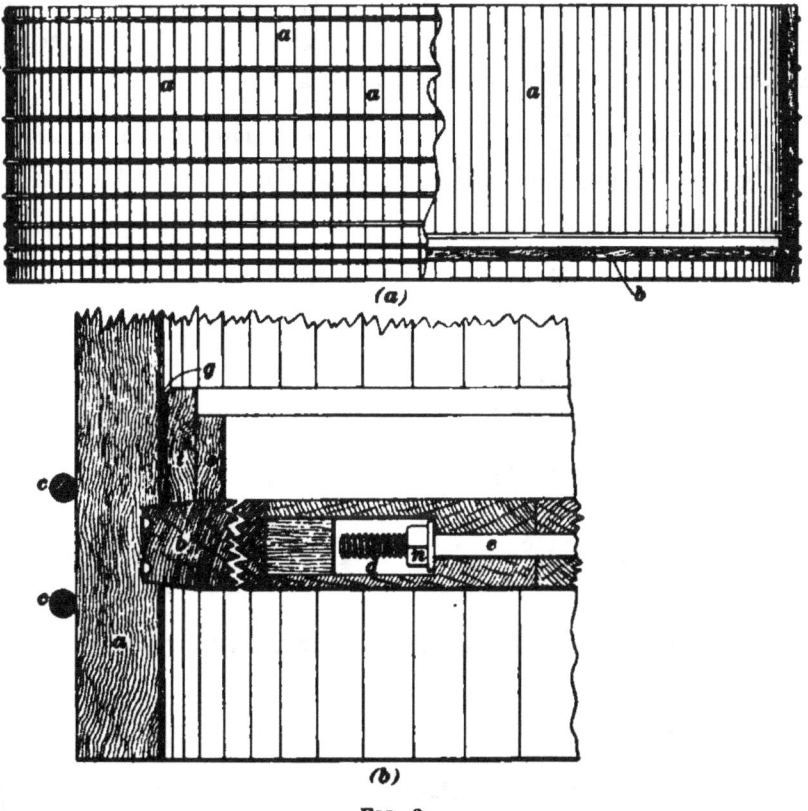

FIG. 8

not true, no amount of white lead or any other kind of packing will secure tightness. The cyanide solution, being alkaline, will unite with the oil of the white lead and remove it, which would make the vat leak worse than if no white lead had been used.

70. The **hoops** used for vats are made of iron or steel $\frac{3}{16}$ inch thick and 3 inches wide. These bands are provided with a lug at one end for the reception of a bolt made at the other end. The band is drawn tight by a nut working on the bolt and reenforced by the lug. The hoops are sometimes constructed of ⅛-inch round iron, as shown at *c*, and fastened in place the same as the flat hoops. In Fig. 8 (*b*), *a* is the side of the vat shown; *b* is the bottom of the vat, showing it to be drawn taut by countersunk bolt *e* and nut *n*. For the purpose of making a tight connection with the filter, the wooden rings *s* and *t* are made around the inside of the tub. In the space marked *g*, the filter cloth is fastened by calking it in tight with hemp packing.

71. Steel leaching vats are constructed in sections, so that they may be better transported. In Fig. 9 (*a*) is

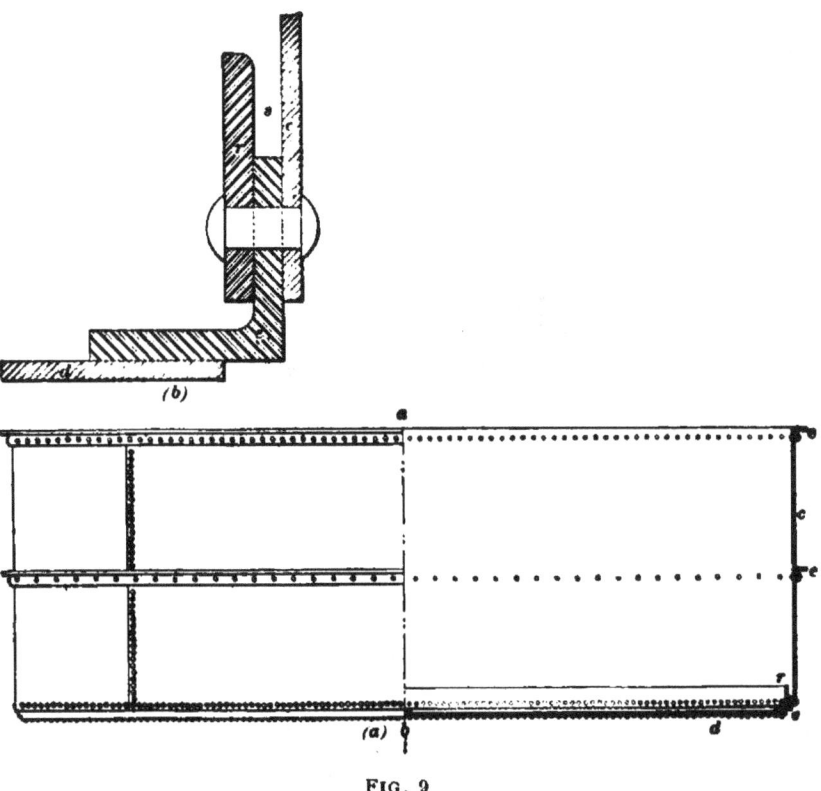

FIG. 9

shown a steel vat 20 feet in diameter and 7 feet deep. The figure is divided by the line *a b*, in order to illustrate both the exterior and interior. The side *c* and bottom *d* of the vat are of $\frac{3}{16}$-inch steel plate, riveted to $2\frac{1}{2}'' \times \frac{3}{8}''$ angle-iron rings *e*. The angle-iron rings are placed in the positions shown, in order to stiffen the steel-plate side. A ring of iron *r*, better shown in Fig. 9 (*b*), is riveted to the lower angle ring in such a manner as to form a rest for the filter and allow the filter covering to be fastened into the space *s* by rope packing.

72. Construction of Filters.—A false bottom, such as is shown in Fig. 10, is constructed for the purpose of supporting a filter upon which the ore and tailings rest. In Fig. 10, the wooden strips *a*, 1½ inches thick by 3 inches high,

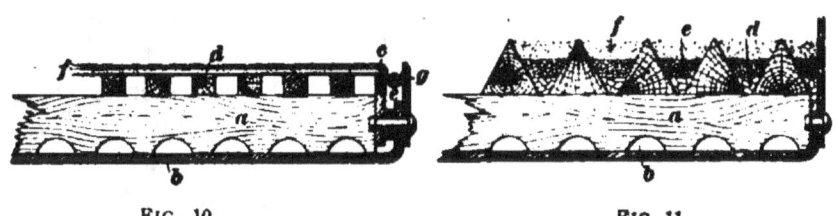

FIG. 10   FIG. 11

are arranged parallel across the floor of the vat and about 6 inches apart. That side of the strip which rests on the bottom of the vat is notched at *b* to afford a free circulation to the liquids. The ends of the slats are brought to the iron ring *c*, which is 1 inch smaller in diameter than the tank. On these strips *a* other strips *d* are nailed transversely. The strips *d*, which are 1 inch square, are nailed 1 inch apart across the tank, to act as a support for the filter cloth *f*. The filter frame in large vats is made in sections. The sections when fitted together form a circular frame. The space *c* between the filter frame and the vat permits the filter cloth to be firmly calked in its place by means of a rope *g* passing around the inside circumference of the vat and between it and the ring *c*. In vats with a bottom discharge gate, the bottom can be arranged with an incline towards its discharge gate.

73. Gravel Filter.—In Fig. 11 is shown the construction of a gravel filter. These filters are used in some places where the tailings are shoveled from the tank. In modern practice, the old gravel filter has given place to latticework and cloth filters in order to save time. The sand filter consists of a framework of wooden slats a resting on the tank bottom b. On the slats, triangular wooden pieces c are nailed. The wooden pieces c are about 3 inches high and are so nailed that there will be ½-inch spaces between them. In the **V**-shaped space thus formed, a 1-inch layer of coarse gravel d is placed; above this is placed a 1-inch layer of fine gravel e, and on top a layer of sand f.

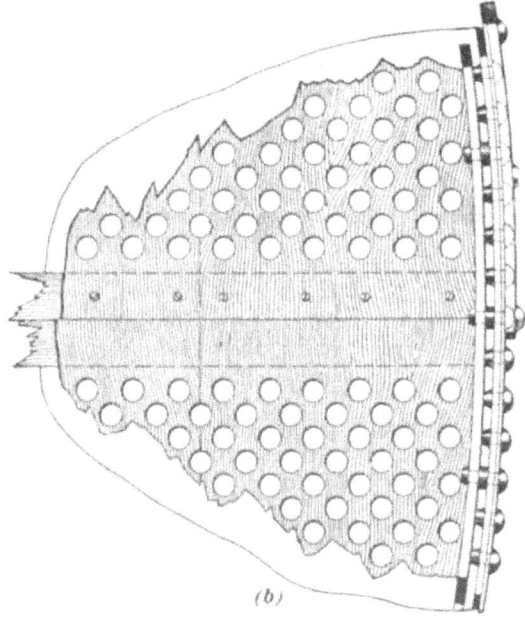

(b)

74. Board Filter.—In Fig. 12 (a) is shown the section of a perforated board filter a, which rests on grooved wooden slats b. A portion of the plan of this filter bottom is shown in Fig. 12 (b). The perforated boards a are covered with a filter cloth c, fastened in place by a calking rope g. The filter cloth is removed in Fig. 12 (b) in order to show the perforated boards.

75. Filter Cloths. — The slats or perforated boards of filter floors are covered with cocoa matting or the heaviest grade of hop cloth, the latter being the cheaper.

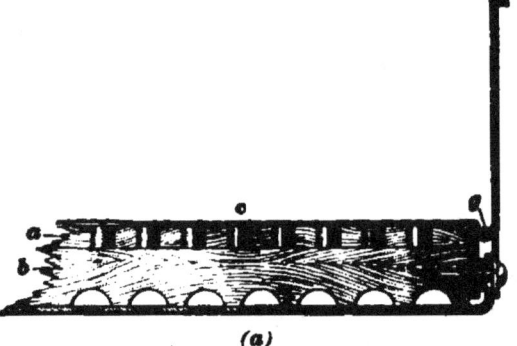

(a)

Fig. 12

This cloth is tacked on the stave ends of the slats. The covering for this cloth filter is 8-ounce canvas and is made 12 inches greater in diameter than the false bottom. The edges of the filter covering are fastened in place by pressing a 1-inch rope into the space between the side of the vat and the ring piece, as shown in Fig. 12 (a). This arrangement is very efficient in preventing the sand from washing under the filter.

76. Wrought-Iron Solution Pipes. — Each leaching vat has separate drain pipes for strong and weak solutions. The pipe a, shown in Fig. 13, is arranged to lead to the

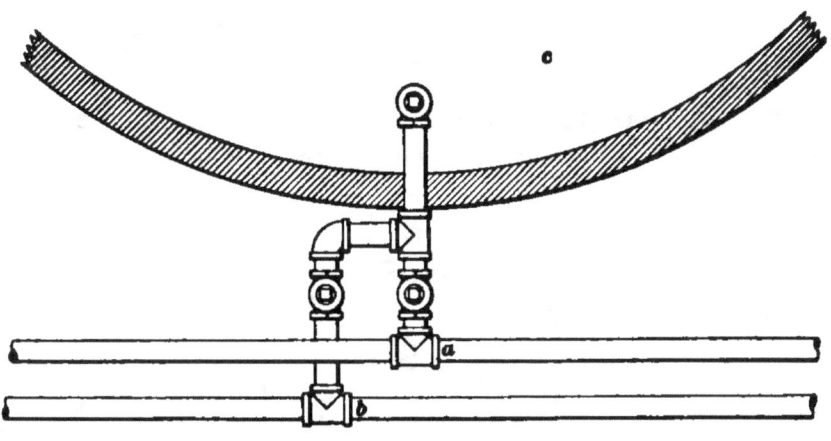

FIG. 13

strong-solution gold tank and the pipe b is arranged to lead to the weak-solution gold tank. These pipes are connected with the exhaust cylinder of a vacuum pump that is used to produce a partial vacuum under the filter bottom, in order to increase the rate of percolation.

77. Launders. — The wrought-iron solution pipes are arranged in Fig. 14, with valves a and b, so that the vacuum pipe c may drain through the pipe leading to the launder. The launder is divided into two sections d and e, each 4 inches wide and 4 inches deep. The section e is for the weak gold solution and the section d is for the strong gold solution. The valve a has a short rubber hose h attached,

so that the solution can be discharged into either section of
the launder as desired.

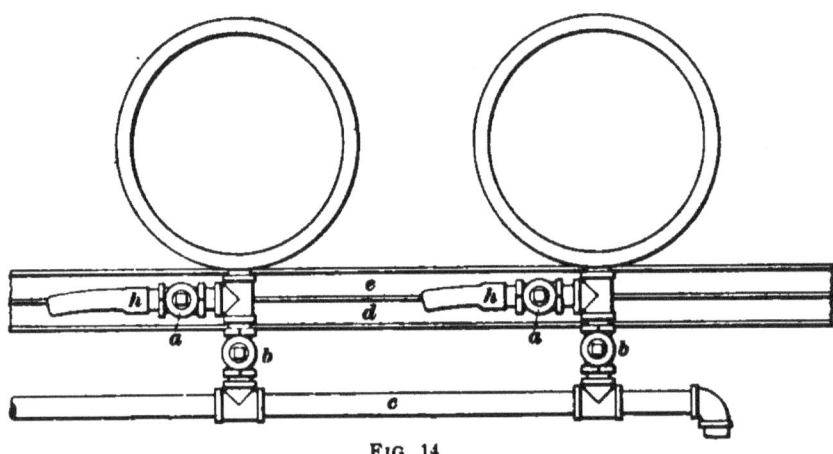

FIG. 14

78. Discharging the Vat.—There are several methods
of removing tailings from a vat, some of which depend on

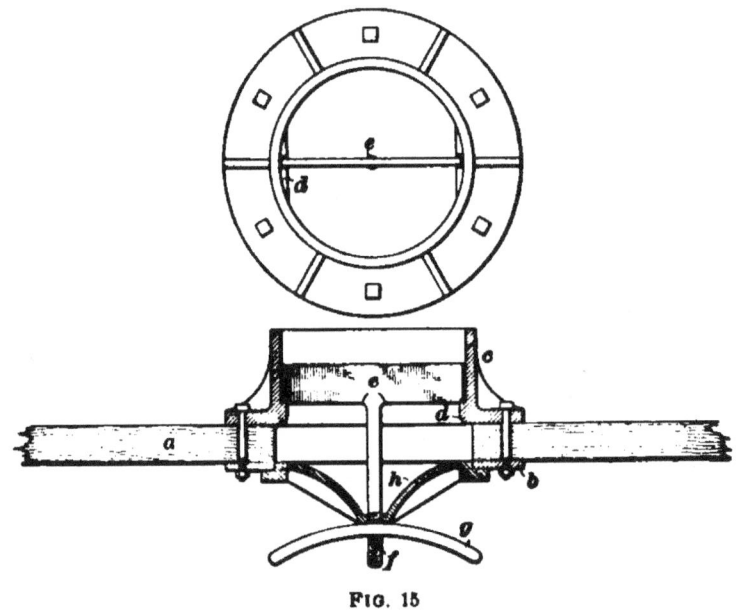

FIG. 15

the cost of labor and others on the supply of water at com-
mand. At some plants the tailings are shoveled through
gates in the bottom of the tank into dump cars. Where

there is good water pressure, sluicing out the residue is found to be cheap and quick. When the tailings contain coarse gold, the residues are sluiced slowly over amalgamated copper plates placed below the discharge hole.

When large brick leaching tanks are used, they are sometimes unloaded with the aid of traveling cranes, which lower empty car bodies into the tanks, where men shovel tailings into them. When the car bodies are filled, they are raised, placed on their trucks, and wheeled away to the dump.

79. Discharge Doors. — In the case of bottom discharging, there may be from 2 to 8 openings, depending on the size of the vat. In Fig. 15, Butter's bottom discharge is illustrated by a vertical cross-section. In the figure, a represents the tank floor, b a cast-iron ring that acts as a washer for the cast-iron cylinder c inside the tank, and a seat for the valve h. The ring b and the cylinder c are bolted together so as to draw them tight against the tank bottom a. There is a lug d inside the cylinder for the hanger e to rest upon. The hanger at its lower end is provided with a screw thread f, upon which the butterfly nut g works in order to tighten and hold in position the cast-iron valve h. The faces of the cover h and ring b should be planed perfectly smooth so as to make a tight joint.

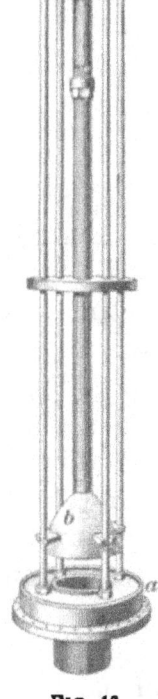

FIG. 16

80. Bottom Discharge. — A simple form of bottom-discharge valve is that shown in Fig. 16. It consists of a deep-flanged iron collar a, which is placed in a hole in the bottom of the vat and fastened to the vat floor by bolts passing through the flange and floor. The discharge valve b is operated from the top of the tank. The valve seats on a rubber ring and is lifted by a screw at c, to leave a clear opening for flushing out the tank. The opening should be about 10 inches in diameter.

81. The **side-discharge door** is shown in Fig. 17. It consists of an iron frame *a*, which is bolted to the side of the tank near the bottom. This frame supports a hinged iron door *b*. A rubber gasket is placed between the frame and the door to form a water-tight joint. The door when closed is held in place by a cross-bar *c* and the hand wheel and screw *d*. The size of the opening in this door is 10 inches × 10 inches.

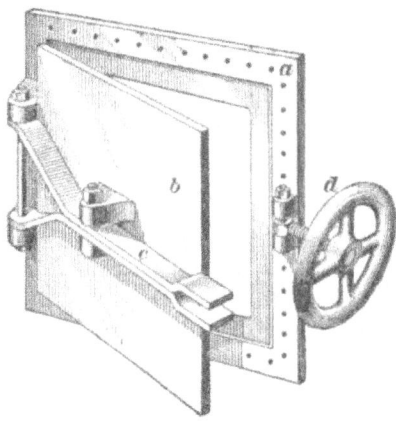

FIG. 17

82. Gold-Solution Tanks.—There are two gold-solution tanks, which should be placed low enough to receive the solution from the percolating tanks above and high enough to discharge by gravitation into the zinc boxes. The tanks act as settlers for solutions containing impurities in suspension that would interfere with the precipitation of the gold, if allowed to pass into the zinc boxes.

83. Construction of Zinc Boxes.—Zinc precipitating boxes are usually divided into compartments having double partitions. In Fig. 18 (*a*) is shown an elevation; in Fig. 18 (*b*) a plan; and in Fig. 18 (*c*) a cross-section of a zinc precipitation box. The first partition *a* does not reach to the bottom of the box *b*; the next partition *c* does not reach quite to the top of the box, and so on, to permit the solution to enter each chamber from below and pass upwards through the perforated bottom *d*, which supports the zinc shavings. The overflow from one box to the next occurs at the upper part of partition *c*.

This arrangement for upward flow permits a free flow of solution, and the gold being deposited largely at the bottom of the box falls off, leaving the passage clear for the solution. The gold and zinc slimes as they form pass through the false bottom *d* into the compartment below.

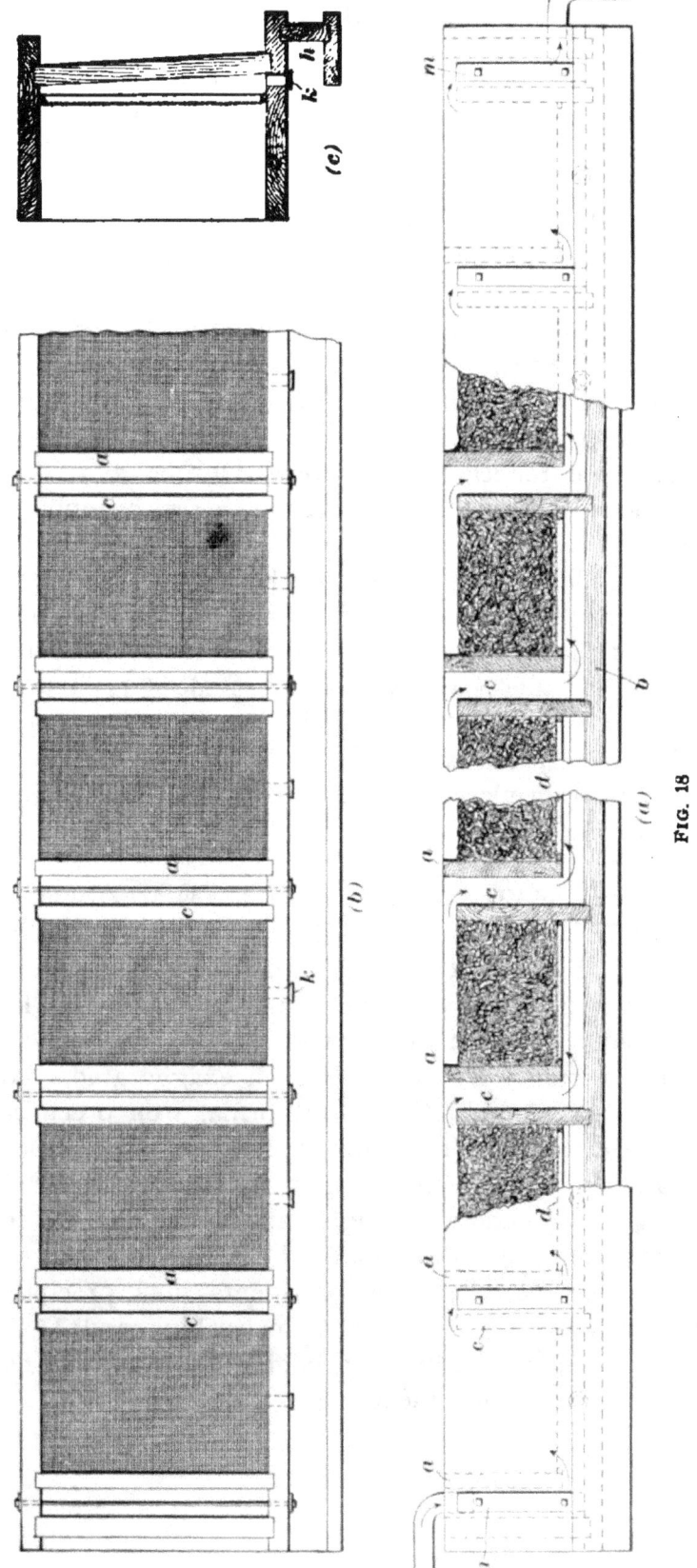

FIG. 18

In the more recent zinc boxes a permanent discharge launder h, shown in Fig. 18 (c), is fastened to the side of the box. The box in this case is provided with an inclined bottom, to facilitate the removal of the slimes. Each compartment also has a 1-inch discharge hole k near the bottom, which, when not in use, is plugged with a rubber stopper.

84. Detailed Arrangement of Zinc Boxes. — When wooden precipitating boxes are carefully constructed and put together, there is no danger of leakage. They should be made of selected planks 2 inches thick, tongued and grooved and fastened together with screws and bolts. The pipe connections for zinc boxes are made with nipples and locknuts.

The launder h, Fig. 18 (c), on the side of the precipitating box should have a tight-fitting lid that may be locked, to prevent the gold precipitates which it contains from being stolen. The baffle boards a and c are made of 1¼-inch plank. The strapping plates m are made of $2'' \times \frac{3}{8}''$ flat iron with suitable holes near each end for $\frac{3}{8}$-inch bolts. The bolts are passed from plate to plate between the baffle boards, as shown in Fig. 18 (b).

85. Size of Zinc Boxes. — Zinc boxes are usually constructed of wood, and although iron is coming into use, it is claimed by some that the galvanic action between the zinc and iron decomposes the potassium cyanide. There is undoubtedly an action of this kind between iron and zinc, but its importance has probably been exaggerated. At the Utica Works, Angel's Camp, California, iron precipitating boxes were used, and it is claimed there was no increased loss of cyanide. Iron precipitating boxes have the advantage of being easily made water-tight, and as they can be coated with paraffin paint occasionally, the galvanic action should be much lessened. Zinc compartments in a precipitating box are usually 24 inches wide, 12 inches high, and 15 inches long. The space left between baffle boards for the flow of the solution is about 3 inches.

In a 75-ton plant, three precipitating boxes of nine zinc compartments each are used for the strong solution, and two boxes of the same size are required for the weak solution. A box 18 feet long should have an inclination of 4 inches. The object in using several boxes is to have a better distribution and slower movement of the solution through the boxes to insure a better precipitation. Another advantage is that the flow of solution need not be interrupted when one of the boxes is being cleaned.

86. In Fig. 19 is an illustration of precipitating boxes with the side launders *a* connected with an ordinary launder *b* that leads to the acid tank below the floor. The

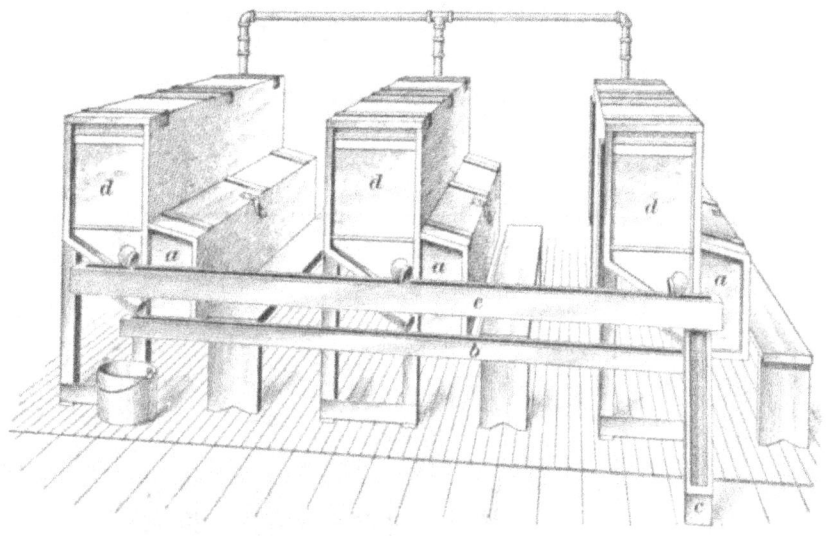

FIG. 19

cyanide solution after leaving the boxes *d* passes into a launder *e* and then through the trough *c* to the sump below the floor.

87. Protective Paints.—The outside and inside of solution vats should be protected with some kind of waterproof paint, such as asphalt, tar, or paraffin, in order to protect the timbers. This paint is also used to cover the interior of vats made of galvanized iron.

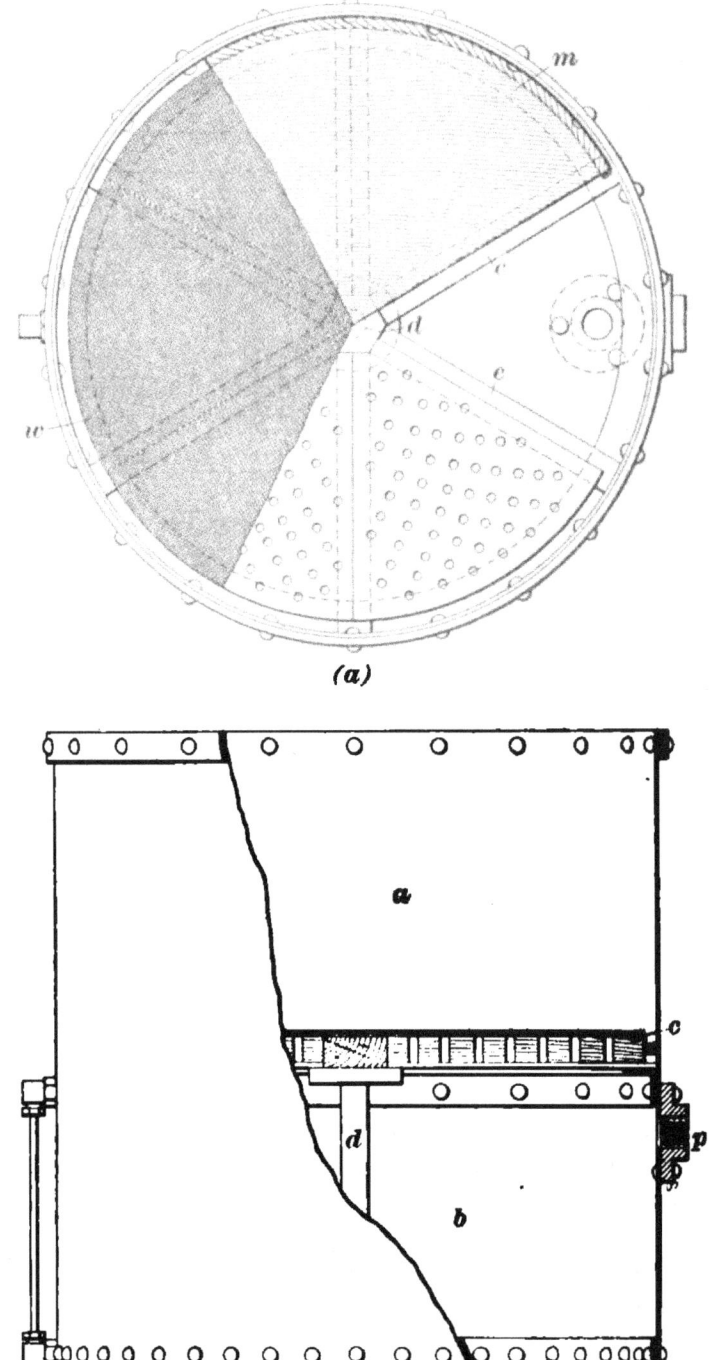

(a)

(b)

FIG. 20

VACUUM SLIME FILTERS

88. Washing Slimes.—Gold slimes are washed before refining in order to obtain as pure metal as possible. The vacuum slime filter used for washing slimes is an air-tight steel tank, as shown in Fig. 20 (*a*) and (*b*). The tank is 36 inches in diameter and 36 inches high, with the interior divided into two compartments *a* and *b* by a filter bottom *c*, which is about 12 inches from the top. The filter bottom is made of wood 1¾ inches thick, perforated with ¾-inch holes 1½ inches apart, and is supported by an iron center post *d* and bars *e*. The perforated false bottom is covered with a piece of wire cloth *w* having about 15 meshes to the inch. On the wire screen is placed a heavy mill blanket *m* of one or more thicknesses, as the case may require. When slimes are to be filtered, there is a suction created below the filter by a vacuum pump, which is connected with the filter by a pipe flange at *p* and a pipe not shown in the figure. This arrangement hastens the filtering. In the lower right-hand corner of the filter another pipe flange is shown. This connects with a waste pipe and drains the solution which passes the filter out of compartment *b*, as that solution must not be allowed to rise high enough to enter the vacuum pipe *p*. This height of solution in chamber *b* is shown by the water gauge to the left of the filter tank.

THE CYANIDE PROCESS

(PART 2)

PRACTICAL OPERATIONS

CHARGING THE VATS

1. The method used to charge percolating vats with ore will depend on whether the ore is crushed dry or wet. Dry or wet ore may be charged into the vats from iron cars that run upon trestles directly over the vats. In such cases the ore must be leveled in the vat by hand, for it is necessary to have an even surface for proper lixiviation. Large vats are sometimes filled by buckets, which are moved in various positions over the vat and dumped. The buckets in this case either run on trolleys or are moved from place to place by the arm of a traveling crane. If the ore is dry, considerable dust is raised by the operation. A third method for charging lixiviation vats is to connect them with the ore bin by means of a chute. A canvas pipe connected with the chute allows the vat to be charged evenly. This method is cheap, raises little dust, and permits an even distribution of ore about the vat.

2. Charging Vats With Tailings.—The most silicious ores when crushed wet will produce some slimes, and when clayey or earthy matter, iron or manganese oxides are contained in the ore, the quantity of slimes is increased.

Slimes, which are sometimes the most valuable part of the ore, interfere with percolation in two ways. If the slimes

§ 32

For notice of copyright, see page immediately following the title page.

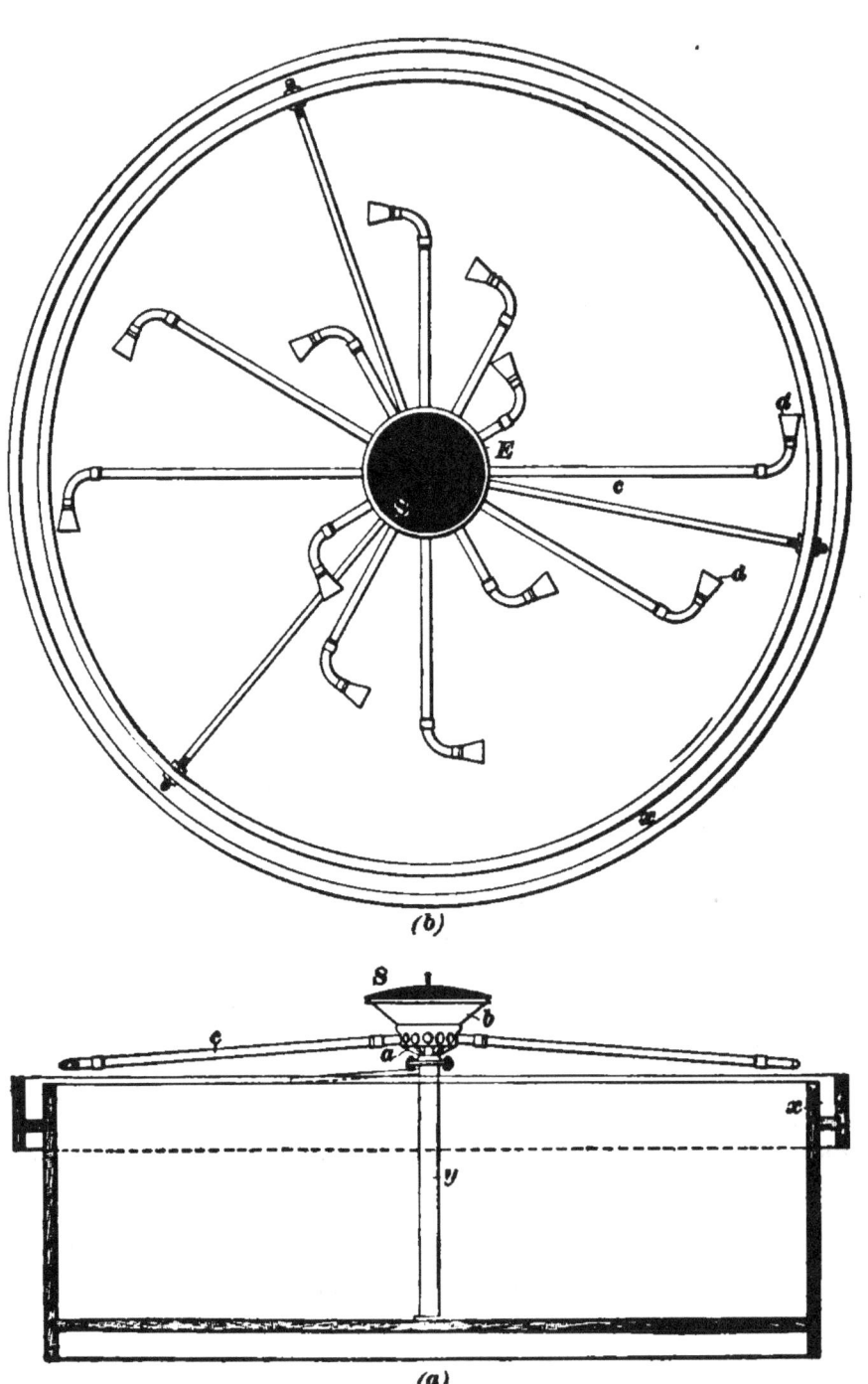

(b)

(a)

FIG. 1

are irregularly distributed in the vat, channels through which the cyanide will circulate without reaching all the ore will be formed and imperfect leaching will result. On the other hand, if they are in appreciable quantity and evenly distributed, they will retard drainage. For the above causes the slimes produced by stamp milling and amalgamation are separated as much as possible from the sands and treated separately. The tailings from a wet-crushing stamp battery are conveyed to spitzlutten, where they are classified into sands and slimes and concentrates containing some sands. There are two methods employed for disposing of the fine slimes, known as the *intermediate* and *direct* methods of charging the vats.

3. Intermediate Charging.—By the method of intermediate filling, the fine slimes are washed from the sands into a series of intermediate settling tanks. To secure an even distribution of slimes and tailings, an automatic distributor is placed in the center of the vat on an iron column y shown in Fig. 1 (*a*), which is an elevation of the distributor. There is a hopper b on top of the distributor, from which twelve or sixteen iron pipes c, with the bent ends d, shown in the plan, Fig. 1 (*b*), radiate. The iron pipes with the flat nozzles are from $1\frac{1}{2}$ to $2\frac{1}{2}$ inches in diameter, and as the stream of slimes and tailings issue from them, the escape causes the hopper and pipes to revolve slowly. The screen S is placed over the hopper b to prevent lumps from passing into the pipe arms and stopping them up.

The collecting vat is filled with water before admitting the pulp. The water flows over the side of the vat as the pulp enters, carrying with it the finest slimes. The launder x receives the water and slimes of the overflow and conveys them to the slime pit.

The settling vats contain filters, through which the water drains off, and when the ore becomes sufficiently dry it is discharged into the leaching vats through bottom discharge doors or it is loaded into cars and hauled to the leaching vats.

4. Intermediate Tanks.—The advantages of intermediate settling tanks are that the sands are collected and the slimes removed. Again, when the intermediate tanks are discharged, the sands become thoroughly mixed. Pyrites are oxidized very slightly, if at all, by this operation; consequently, the cyanide consumption is low, the extraction high, and the cost of treatment moderate.

5. Direct Charging.—This process consists in passing the pulp from the amalgamating plates into a classifier. The pulp is divided into two streams. The overflow stream carries slimes and very fine sands to the slime pit, where they are stored for future treatment. The other stream, containing mostly coarse sands with some fine sand and slimes, is carried to the leaching vat by a hose, which is moved about to give an even distribution. The excess of water passes off through discharge gates fitted inside the vat.

The advantages claimed for direct filling are that there is practically no oxidation of the pyrites in the tailings and that the tailings are handled but once; besides, the slimes are separated by the rough preliminary classification.

Some of the disadvantages of this method are that the tailings pack, and then it is very difficult to remove the water; also that the tailings are unevenly distributed, thus favoring the formation of channels during the subsequent leaching.

In some mills the water is drawn off through the filter bottom and the tailings are turned over by hand, so as to loosen and thoroughly mix them.

THE LIXIVIATING PROCESS

ORE WASHING AND TREATMENT WITH ALKALI

6. Preliminary Washing.—When ores or tailings contain sulphates, a preliminary treatment is necessary before percolation with a cyanide solution. Whenever pyritic ore is kiln-dried previous to dry crushing, soluble sulphates are formed that are destructive to cyanide. The products of the partial oxidation of iron pyrites are free sulphuric acid and soluble sulphates. Ores that contain acids and salts soluble in water are in general practice washed with water by percolation and the remaining acids neutralized by applying weak solutions of caustic soda that are allowed to remain in contact with the ore several hours. When a neutralizing solution is drawn off, a standard cyanide solution is run on. If the ore does not require more than 3 pounds of caustic soda to the ton to neutralize the acids and salts, the preliminary washing is omitted and the ore is treated directly with unslaked lime   The powdered lime is added to the ore as it is placed in the vat.

LEACHING WITH THE FIRST CYANIDE SOLUTION

7. The Weak Solution.—*The proper strength of the weak solution should be such that all cyanicides will be exhausted.* The weak solution is taken from the weak-solution tank and run into the leaching vats underneath the filter. Experiments have shown that solutions will permeate the ore sooner and make fewer channels when introduced from below than when introduced on top of the ore. The solution is allowed to pass upwards until it rises 2 or 3 inches above the ore in the tank, when it is allowed to remain in contact with the ore 2 or 3 hours. Percolation is then commenced, the liquor being drawn off into the weak-gold-solution tanks. It is not always customary to use a weak solution first, but to run on the standard or strong solution as soon as the alkali wash water has been drained off and then use a weak wash solution or merely a wash. Under ordinary circumstances, it will require about 1 hour to charge a 75-ton vat with solution, and without suction it may not drain over 3 or 4 inches per hour.

LEACHING WITH THE SECOND CYANIDE SOLUTION

8. The Strong Solution.—The time required for percolating with the strong or second cyanide solution is determined in each case by laboratory tests. The liquor is allowed to enter the vat below the filter and rise upwards through the ore. Percolation is commenced after several hours and continued until the solution coming from the vat reaches within .02 or .03 per cent. of the strength of the original cyanide solution.

In case the solution coming from the vat is less than one-half the standard solution, it is run into the weak-gold-solution tank, but if it is more than one-half the strength of the standard solution, it should be run into the strong-solution tank.

At the Standard Works in Bodie, California, the ore is soaked 20 hours in the strong solution. At the Mercur mill in Utah a series of strong solutions are used and each time the solution is drained off before the fresh is added. This practice gives higher extraction than continuous leaching on this ore. In other mills, a solution is added to cover the tailings in the vat several inches, then the solution is allowed to disappear below the surface of the ore for an hour, when more solution is added to cover the ore in the vat. A succession of changes of this description is continued until the time allotted for leaching has been exhausted. By this method of partially draining the vat at intervals, air is brought in contact with the material and the solution of gold is hastened.

9. Strong Solution Displaced by Weak.—The strong solution is completely drained off after the point of economic extraction has been reached. In a new plant, water is used to displace the strong solution until enough weak solution has accumulated. When water is used, it may continue flowing through the percolating vat until the out-flowing solutions indicate only .03 or .04 per cent. of cyanide; this solution would not contain more than 30 to 50 cents worth of gold per ton of solution. After sufficient weak solution has accumulated, it is used to displace the strong. *The strength of the weak solution is about one-third that of the standard strong solution.*

The weak solution is usually drawn off as rapidly as possible, and the faster the solution comes off, the more efficient is the washing.   The vacuum pump is often used to facilitate the drainage, and the percolation is continued until the outgoing solution is of the same strength in cyanide as the solution when it enters the percolating vat.

10. Weak Solution Displaced With Water.—Water is added to displace the cyanide solution remaining in the ore.   As little water should be used for this purpose as possible, in order to prevent the accumulation of large quantities of the weak solution.   Weak solutions accumulate on account of the moisture present in the ore from the wash water and from solutions that have passed through the zinc boxes.   The excess of weak solution is disposed of by running it to waste.

There can be no fixed rule for the exact amount of weak solution and water required to give the best results.   A careful study of all the operations is required and the effect of different washings determined by careful assays of the residues and the wash solutions as they leave the vats. These should contain no precious metal that it is possible to remove.

SAMPLING THE LEACHED TAILINGS

11. A sample of the residues in the vat is taken with a 5-foot auger introduced 20 or 30 times through the charge at different places in the leaching vat.   The auger is carefully pulled out of the material, and brings with it cores of tailings, which are collected in a bucket.   This sample is dried, quartered, and assayed.

12. Testing Solution for Cyanide.—The silver-nitrate test is used to determine when the cyanide is sufficiently washed out of the leached ore.   If a strong solution containing .2 per cent. cyanide and a gold value of about $4 per ton is diluted with the displacing liquids, so that it contains .04 per cent. cyanide, its gold value has been reduced in the same proportion.   It is clear, then, that the limit of the final displacement will depend on the cost of treatment, and this must be determined in each case by a series of tests.

N. M. III.—29

13. Discharging the Vat.—If the tailings are to be shoveled from the vat, the wash water is drained off until the tailings are dry enough for shoveling. If they are to be sluiced out, the wash water is drained off for sampling. The tailings are then covered with water, and after standing a short time the plugs are drawn and the residues sluiced out.

14. Standardizing Sump Solutions.—The cyanide solution after having passed through the various stages in the process and reached the sump has deteriorated in strength. The solution is therefore titrated with standard silver nitrate and the amount of cyanide it contains is determined, after which sufficient potassium cyanide is added to bring it to the standard strength. The solution in the sump may be raised to normal strength by placing the required amount of cyanide in the last compartment of the zinc box or by placing it in a perforated tray attached to a rope so that it can be immersed in the sump. At some works a strong solution is kept in a tank and is drawn off into the storage tanks to raise the solution to standard strength.

When pure potassium cyanide (i. e., 98 to 99 per cent. KCN) is used to strengthen the sump solution, it may be placed in a box with perforated sides and bottom and this box held under the discharge where the solution coming from the sump enters the storage tanks. When adulterated potassium cyanide is used, it should be dissolved in a small tank and the solution filtered as it enters the storage tanks, for the reason that low-grade KCN contains a number of impurities that are insoluble.

15. To Standardize a Solution with Cyanide.—Storage tanks should be provided with a float that will register the depth of the solution within. The number of pounds of cyanide to make 1 foot of standard solution in the tank should be calculated. For example, a circular tank 12 feet in diameter would contain nearly $3\frac{1}{2}$ tons of solution for each vertical foot.* If the standard solution contains .02 per

---

* See Table III.

cent. of cyanide, each ton will contain 4 pounds, and 3¼ tons, or 1 foot of solution, will contain 14 pounds.

16. Formula for Standardizing Solutions.—Having determined the strength and quantity of cyanide solution in the storage tank, the cyanide needed may be calculated from the formula

$$X = \left[A - \left(A \times \frac{B}{C} \right) \right] \times D.$$

In the above formula,

A = number of pounds of KCN in a solution whose bulk is the area of any tank multiplied by 1 foot;

B = percentage of KCN in any sump solution;

C = percentage of KCN in any standard solution;

D = number of cubic feet of sump solution to be standardized;

X = number of pounds of KCN to be added to the sump solution.

EXAMPLE.—Suppose a sump, 12 feet in diameter, contains a .21-percent. cyanide solution that measures 2¼ feet in depth. How much KCN must be added to bring the solution up to the standard strength of .25 per cent. ?

SOLUTION.—First find the number of gallons in 1 foot of the tank, thus, 12² × 1 × 5.875 = 846 gallons. This product multiplied by 8¼ will give 7,050 pounds of water in 1 foot of the tank, and $A = 7,050$ × .25 % KCN = 17.625 pounds of KCN in 1 foot of solution. Substituting these values in the above equation, the quantity of potassium cyanide to be added is found as follows:

$$X = \left[17.625 - \left(17.625 \times \frac{.21}{.25} \right) \right] \times 2.5 = 7.05 \text{ lb.} \quad \text{Ans.}$$

CHEMISTRY OF ZINC PRECIPITATION

17. Zinc Shavings.—Zinc shavings in filiform or thread-like turnings have been almost universally adopted for the precipitation of gold and silver from KCN solutions containing those metals. The shavings should be free from arsenic or antimony. A little lead in their composition is an advantage, as it promotes rapid precipitation by forming a voltaic couple with the zinc. In usual practice, 1 cubic foot of zinc shavings will precipitate the gold from 2 tons of cyanide solution. Zinc sheets, amalgam, dust, and fumes have been tried, but zinc shavings seem to have the preference, because of the ease with which cyanide solutions attack them. They also allow the free and rapid passage of the cyanide solution, and besides the screens through which the gold precipitate falls are not clogged by the zinc. The action of the zinc on the gold solution is a simple substitution of gold for zinc; according to the equation,

$$\underset{\text{auric-potassic cyanide}}{2KAu(CN)_2} + \underset{\text{zinc}}{Zn} = \underset{\text{zinc-potassic cyanide}}{K_2Zn(CN)_4} + \underset{\text{gold}}{2Au}$$

According to theory, 1 pound of zinc should precipitate about 6 pounds of gold; in practice, however, it requires from $\frac{1}{2}$ to 1 pound of zinc for every ounce of gold precipitated. The double salt of auric-potassic cyanide is one of the most stable of gold salts, but its decomposition by zinc is practically complete. The precipitated gold is not redissolved by potassium cyanide so long as there is zinc present. The potassium-zinc cyanide remains in the solution that passes to the sump tanks.

18. Preparation of Zinc Shavings.—Zinc shavings are usually cut from a cylinder composed of zinc disks. Sheet zinc of No. 9 Brown & Sharpe gauge, or .114 inch thick, is generally used for this purpose. Disks 12 inches in diameter, with an inch hole in the center, may be obtained from mill-supply houses. They weigh about $\frac{1}{2}$ pound each.

The lathe on which these filaments are cut is very simple in construction, as is shown in Fig. 2. It consists of a mandrel threaded on both ends and supplied with cast-iron disks and nuts. In practice, twenty zinc disks are placed between the cast-iron washers a and are held tight by a nut b. The disks are placed in motion and thin shavings cut from their peripheries by means of any sharp steel tool steadied on an iron rest c. The mandrel is given a speed of 350 revolutions a minute. The zinc is thus shaved off in fine threads. Guards, shown at d, prevent the shavings getting under the belt of the machine. Zinc shavings as ordinarily packed in a zinc precipitating box weigh about

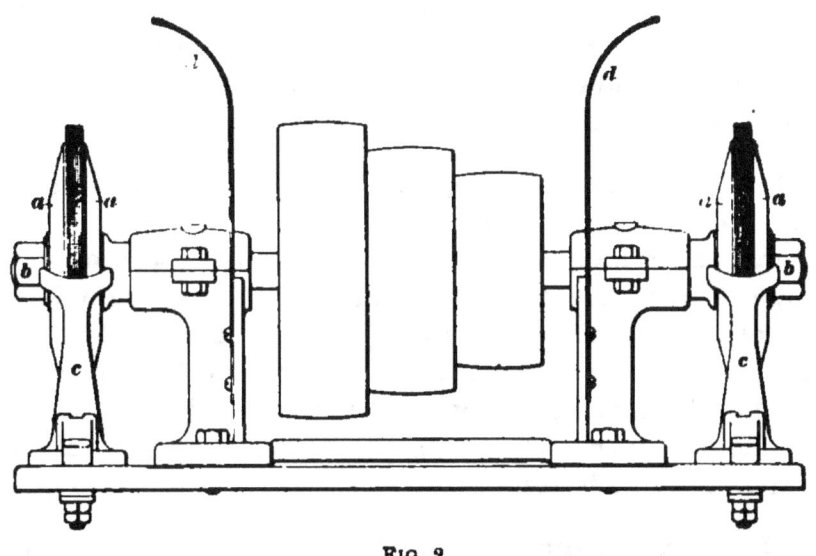

Fig. 2

6 pounds per cubic foot and should be thin enough to burn when lighted with a match.

19. Filling the Zinc Boxes.—Zinc shavings are placed in all the compartments of the zinc box except the last one, which is left empty to collect any particles of zinc and gold that may be carried from the others. The zinc shavings should be uniformly distributed and the corners of each compartment well packed to prevent the solution passing through in channels. The speed with which the solution should flow through the boxes can be determined by tests on the outflowing solution for the presence of gold.

Where gold precipitation takes place under proper conditions, the metallic deposit on the zinc is brownish black and precipitation should take place in the first compartments. In imperfect precipitation, the deposit is frequently gray or of a dull metallic color. The dry precipitate seldom contains more than 40 or 50 per cent. of gold and silver, the remainder being finely divided zinc and its impurities. The precipitation in the zinc boxes is influenced by the amount of cyanide present in the solution, as there seems to be a selective action between the zinc and metals present. It has been found that dissolved copper is precipitated faster from a weak cyanide solution than from a strong one. To overcome this, a strong cyanide solution may be allowed to drop into the first compartment of the zinc box fast enough to bring the solution to standard strength.

20. Presence of Copper.—When copper is present in the gold solution, it covers the zinc with a bright metallic copper covering. This copper deposit is observed in the lower compartments first, from which it gradually works towards the first compartment. The precipitation of the gold is very slow when the zinc is coated with copper. As previously stated, the copper may be largely kept in solution by increasing the strength of the cyanide solution before it enters the zinc boxes.

21. Preventing Copper Deposits in Precipitation. To prevent deposits of copper in the precipitation boxes, the zinc shavings are sometimes coated with lead by being placed in a 10-per-cent. solution of lead acetate. This lead-coated zinc will precipitate gold from weak cyanide solutions and leave the copper in solution. It is difficult to work ores containing much copper on account of their large consumption of cyanide and the difficulty of precipitating the gold in the presence of copper.

22. Scum on the Zinc Boxes.—In the treatment of pyritic ores and tailings, a precipitate of zinc cyanide forms occasionally on the zinc in the precipitating boxes. It is a grayish-white porous precipitate and the conditions of its formation are not thoroughly understood. As a rule, gold is being imperfectly precipitated whenever it occurs. The formation can sometimes be prevented by using lime instead of caustic soda in the preliminary wash to neutralize the acids and salts. If, however, lime is added in excess, it may form an incrustation on the zinc and also prevent satisfactory precipitation.

The presence of organic compounds will sometimes cause excessive action on the zinc, generating hydrogen so vigorously that frothing is the result. The application of an oxidizing compound to the ore, such as sodium dioxide, will often remedy this condition. If a scum forms on the zinc boxes, it should be removed and the cause of it corrected at once.

23. Care of Zinc Boxes. — Fresh zinc shavings are added daily to the last compartment of the precipitating boxes. The partly consumed zinc is brought up a step, so that the first chamber contains zinc partially consumed and rich in bullion, while the last chamber contains fresh zinc. Zinc on which bullion is already deposited is more active than new zinc; it is, therefore, advisable to replace the dissolved zinc in the upper compartments with zinc from the lower compartments and add the fresh zinc to the last compartment.

The hydrogen generated in the zinc boxes is likely to retard precipitation by polarization; to avoid this the zinc should be stirred occasionally.

24. Zinc Boxes for Weak Solutions. — A longer column of zinc shavings is necessary for weak than for strong cyanide solutions, as the gold precipitates with greater difficulty from them. The solution can be passed through two boxes of 9 compartments each, and thus come into contact with 18 compartments filled with zinc shavings.

25. Percentage of Precipitation. — When good precipitation takes place, 95 to 99 per cent. of the gold and silver should be deposited on the zinc. Some plants do not leave more than 10 to 25 cents of gold in a ton of sump solution.

REFINING THE PRECIPITATE

26. The Clean-Up. — The clean-up takes place once or twice a month, and then the cyanide solution is shut off from the zinc boxes. A current of clear water is passed through the zinc boxes to remove the cyanide solution, which is injurious to the arms. The zinc shavings are stirred with a rod or the trays holding the zinc are moved up and down, which causes the precipitate and fine zinc to pass through the perforations of the false bottom into the box below. After the gold has settled as a slimy mass, it is sluiced through the plug holes into the side launder. The slimes and zinc are allowed to run out of the launder through a 40-mesh screen, which catches fragments of zinc that are returned to the zinc boxes. Each compartment is washed out in this manner and the launder is next cleaned with a stream of water from a hose. The zinc from the lower boxes is moved up and fresh zinc added to fill the remaining compartments. The solutions containing the slimes is drawn from the launder into the acid tank, where the precipitate is allowed to settle. After the precipitate has settled the solution is siphoned into a settling tank. The precipitate is now ready for acid treatment or the calcining process.

27. Acid Treatment of the Precipitate.—The tank should be located where there is a good draft to carry away the acid fumes. In some mills, concentrated sulphuric acid is added to the mass of wet slimes in the acid tank and then an equal volume of water is put on. The mixture is thoroughly stirred and after the violent ebullition has ceased more strong acid is added, followed with an equal volume of water. Acid and water are added until no effervescence takes place on its addition; it may now be concluded that virtually all the zinc has been consumed. The solution is now allowed to stand for a few hours with occasional stirring.

28. Removing Zinc Sulphate.—After the solution has stood a few hours the tank is filled with hot water and the contents thoroughly stirred. The zinc sulphate that formed by the action of the sulphuric acid on the zinc is soluble in the water. The solution is allowed to stand until the precipitate has settled, when the clear liquid is siphoned into the settling tank and the precipitate washed from five to ten times with hot water to remove all the zinc sulphate.

29. Filtration.—After the precipitate is thoroughly washed it is removed to the filter box; the precipitate from the settling tank is collected and added to the filter box, where it is washed and dried by suction. In some works filter presses are used, into which the various dilutions of hot water may be directly pumped.

30. Drying.—The precipitate is placed in iron pans and dried in a muffle or over a furnace arranged to remove the fumes. The heat is kept low at first to drive off the moisture. It is then gradually increased to a dark-red heat and calcining carried on for about 1 hour, during which time the oxidation of the base metals, which escaped removal by acid treatment, is going on.

31. Fluxing the Dried Precipitate.—The dried precipitate is removed to the melting room, where it is broken into small pieces, weighed, and the necessary flux added. The following mixture has given satisfaction when melted in a No. 60 plumbago crucible*: Precipitate, 100 ounces; borax, 30 ounces; soda bicarbonate, 15 ounces; silica, 7 ounces. With some residues it will be necessary to modify this mixture to obtain good results.

After complete fusion the molten mass is poured into proper molds. After cooling, the slag is removed and the bullion remelted with a little borax. The slag can be crushed and melted with sufficient litharge and argol to collect all the gold and silver in a lead button, after which the button is cupeled for gold and silver.

32. Calcining Process.—The slimes are removed from the filter box and dried until just before they become dusty. They are then mixed with powdered niter, in proportions varying from 3 to 33 per cent. of their weight and gently heated in a tray of wrought iron not above a dull red. Less niter is used than is required for complete oxidation of

---

* See Table V.

all the base metals present; an excess of it would rapidly corrode the plumbago crucibles in the subsequent operations. The niter not only assists in furnishing the oxygen for oxidation, but it assists in fluxing the zinc oxide, forming zincate of potash, which is not so readily reduced as zinc oxide.

33. Melting the Oxidized Precipitate. — The dry residue is broken into small lumps, weighed, transferred to a plumbago crucible, and mixed with suitable quantities of flux. The fluxes commonly used are bicarbonate of soda, borax, and clean quartz sand. There is considerable variation in the proportion of fluxes and precipitates. If there should be much sand present (which would give a glassy but thick flowing slag), the best corrective is more soda with a little flour. When the slag is too basic (that is, a dull, lusterless one), additional borax will neutralize the base and improve the slag.

34. Fluxing Precipitates. — Some of the fluxes used are given in Table I. Any one of them may have to be increased or decreased, according to the amount of impurities present.

TABLE I

Name of Flux	Character of the Precipitate		
	Clean. Pounds	Very Zincy. Pounds	Very Sandy. Pounds
Bicarbonate of soda	15	15	20
Borax	8	12	10
Sand	5	5	0
Flour............	0	0	2

The whole of a charge as given in this table will go into two No. 35 plumbago crucibles.*

---

* See Table V.

The crucibles are placed in the fire and the contents fused until perfectly fluid. The crucibles are then withdrawn from the fire and the contents poured into molds. The metal settles to the bottom of the mold, and after cooling is turned out and the slag removed with a hammer. The bullion thus obtained is remelted with borax and run into an ingot. The second melting should be conducted at as low a temperature as possible, since gold forms a very imperfect alloy with zinc.

The slags generally contain a considerable amount of gold, and should therefore be crushed and melted with a little borax glass and poured when fluid into an ingot mold. After cooling, the slag is removed from the bullion.

MODIFICATION OF THE CYANIDE PROCESS

ACTION OF THE ELECTRIC CURRENT ON GOLD SOLUTIONS

35. Electrolysis.—When the electric current decomposes a solution of a metallic salt, the metal is carried to the negative pole, or cathode, of the electrolytic cell and deposited, while the metalloid is liberated at the positive pole, or anode. In a given time a fixed quantity of current will release and deposit a definite quantity of metal. The quantity of metal released at the anode and deposited at the cathode varies with different metals, being in direct proportion to their electrochemical equivalents. This law does not hold good for solutions containing very small quantities of the metal in solution, as in cyanide solutions, for the current does not find sufficient metal present at the electrolytes, consequently the water is decomposed. To make the precipitation as efficient as possible, the solution is kept in constant diffusion by a steady flow through the precipitating box.

36. Conditions that the Siemens-Halske Cathode Must Fulfil.—The cathode should be a material to which the gold will adhere; it should be capable of being rolled into thin sheets, which are of such a character that the gold can be recovered without loss or great expense. The cathode should be more electropositive than the anode to prevent return currents being generated when the depositing current is stopped. Thin sheet lead has been adopted as the most suitable metal for the cathode of the Siemens-Halske method of precipitation. The lead sheets are fastened to light wooden frames. Each frame contains three sheets of lead 2 ft. × 3 ft.; this gives each frame 18 square feet of surface. There are 87 frames in each precipitating box, giving an exposed surface of 1,566 square feet. The three sheets of lead in each frame weigh 3 pounds; this makes 261 pounds of lead in each box.

37. Carbon has been used for anodes, but it crumbles under the influence of the current that decomposes the cyanide. Zinc used as an anode forms a white precipitate of ferrocyanide of zinc when the ores leached contain iron. Iron anodes form Prussian blue by the reaction of oxide of iron and ferrocyanide in the solution. The iron plates are covered with canvas to prevent short-circuiting and to collect the Prussian or Turnbull blue that is formed by the ferrous and ferric salts with cyanide solutions. Iron plates are used in South Africa as anodes and lead cathode plates are suspended between them.

Cyanide can be recovered from the Prussian blue by dissolving it in caustic soda, evaporating the solution, and melting the residue with potassium carbonate.

38. Electric Current Required for Precipitation. A weak current, one with a density of about .06 ampere per square foot, is required for precipitation. It can be produced by 7 volts when the cathodes are about 1½ inches apart. The advantages claimed for such currents are that there is a firm deposit of gold and that the iron anodes are decomposed very slowly, their waste being proportional to

the current. Theoretically, $3\frac{1}{2}$ horsepower would be sufficient to run a plant having a monthly capacity of 3,000 tons;
but the amount actually required, however, was 5 horsepower.

39. Advantages of Electrical Precipitation.—Gold
may be precipitated from solutions by electrolysis independent of the strength of cyanide solutions. In the treatment of tailings, therefore, very dilute solutions can be used,
the only limit being sufficient cyanide to dissolve the gold.
Electrolysis does away with certain complications met with
in zinc precipitation, such as the formation of alumina,
lime, hydrate of iron, etc.; besides, it is cleaner and simpler,
and gives a higher grade of bullion. Its application seems
to be limited to those plants where very weak cyanide solutions can be used successfully.

40. Working Results.—At the Worcester works, South
Africa, the strong solution contains from .05 to .08 per
cent. of KCN and the weak solution about .01 per cent.
There are four precipitating boxes, 20 feet long, 8 feet wide,
and 4 feet deep. Heavy copper wires are fixed on the sides
of the boxes to convey the current from the dynamo to the
electrodes. The anodes are iron plates 7 feet long, 3 feet
wide, and $\frac{1}{8}$ inch thick. They stand on wooden strips that
are laid on the floor of the tank and they are kept in a vertical position by wooden strips on the sides of the box. To
facilitate the circulation through the box, each alternate
plate is raised 1 inch above the bottom, thus forming a
series of compartments through which the solutions must
rise and fall alternately through the successive compartments.

In a clean-up, the frames carrying the lead are removed
one at a time, the lead is removed and replaced by a fresh
sheet, and the frame returned to the box. The lead that
contains from 2 to 12 per cent. of gold is melted into bars
and cupeled.

In treating 3,000 tons of tailings, 750 pounds of lead and
1,080 pounds of iron were consumed.

PRECIPITATION WITH CHARCOAL

41. Johnson's Process.—One of the many methods proposed to take the place of zinc shavings in the precipitation of gold from auro-potassic cyanide solutions was the use of charcoal. It had been previously employed for the precipitation of gold from chlorine solutions, and there was no chemical reason why it should not be applied to a cyanide solution. The process patented by a Mr. Johnson consists in filtering a gold-cyanide solution through pulverized charcoal, from which the gold is recovered by burning the charcoal and smelting the residues with suitable fluxes. The process is considered too slow for large plants. The probable reaction that occurs may be expressed as follows:

$$2AuK(CN_2) + 2CO_2 + 2H_2O = 2Au + 2KCO_3 + 4HCN$$

42. The charcoal filter used at the South German mine, Moldon, Victoria, is shown in Fig. 3. It consists

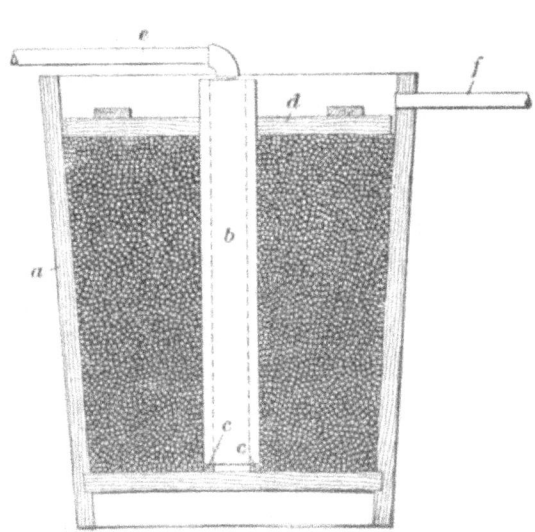

of a tub *a* 2 feet 4 inches high, 2 feet 1 inch in diameter at the top, and 1 foot 9 inches in diameter at the bottom. In the center of each tub and resting upon wooden cleats *c* is a glazed drain pipe *b* 4 inches in diameter. The pipe and tub are nearly filled with charcoal, after which the drain

FIG. 3

pipe is placed under the mouth of a pipe *e* connecting with the gold-solution tank.

The solution passes down the pipe *b* and out through the bottom, then rises through the charcoal in the tub and flows out at the pipe *f*. To prevent the charcoal rising and clogging the pipe *f*, it is confined by a board cover *d*.

There are six of these filters in a set, so arranged that the overflow from one will pass down the glazed pipe of the next, and so on until the solution is exhausted and passes to the sump. About 300 gallons of solution can pass through each filter hourly.

43. Cleaning the Filters.—The solution remaining in the tubs is poured off; then the charcoal containing the gold is removed and sent to the furnace room.

The first tub, or the one nearest the fresh gold solution, is removed after three days and its contents sent to the furnace room. The second tub is moved up to take the first tub's place, the third to take the place of the second, and finally the former first tub is made the sixth of the series by filling it with fresh charcoal. A solution that contains gold and .004 per cent. of cyanide has the gold almost completely precipitated, rarely containing more than .22 gram per ton after leaving the filter. ·

44. Recovering the Bullion.—The charcoal is next burned in a reverberatory furnace. The ash is then sifted with a 30-mesh trommel enclosed in a box, and whatever remains in the trommel is returned to the furnace for reburning. The ash in the box is fused in a graphite crucible with borax in about the following proportions: borax, 3 pounds; ash, 1½ pounds. The cost of precipitation is said to be about 25 cents per ounce of bullion.

PRECIPITATION WITH ZINC FUMES

45. Definition of Fume.—Zinc fume is a product obtained in zinc smelting. It is a blue powder containing about 90 per cent. of metallic zinc. The cyanide solution containing the gold and silver is placed in a precipitating tank having a capacity of about 30 tons. The tanks have a ¼-inch iron pipe leading to the bottom, through which air is passed at a pressure of about 15 pounds, for the purpose of

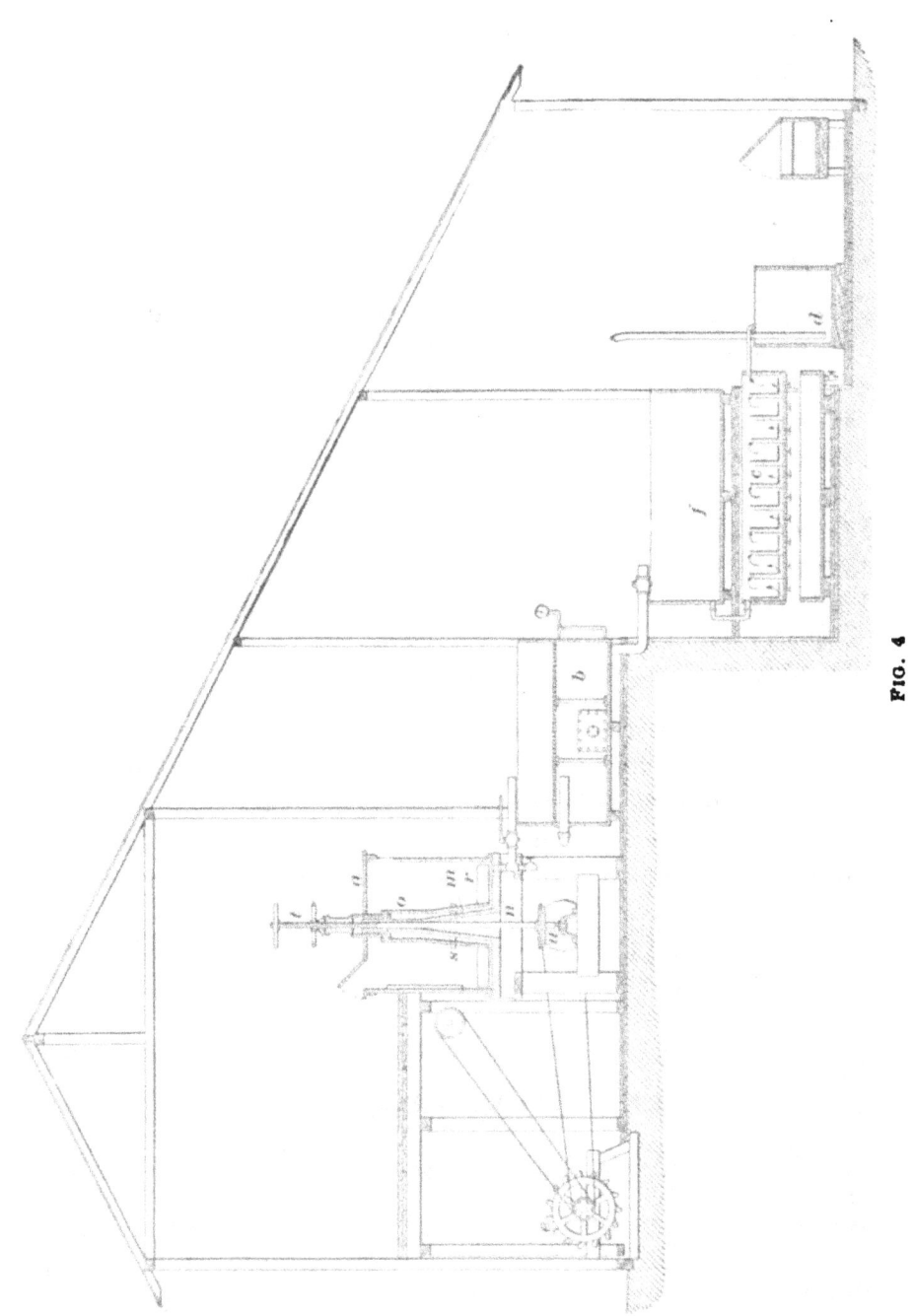

FIG. 4

keeping the solution agitated during the time the zinc dust is added.

It takes about 5 pounds of zinc dust to precipitate the gold and silver from 30 tons of cyanide solution. The zinc dust is sieved into the tank occasionally, from the time it is half full of the solution until it is full. The pulp is agitated for a few minutes after it is full of cyanide solution and all the zinc dust has been added.

46. Collecting Zinc-Fume Precipitates. — The suspended matter, or precipitate, is allowed to settle for $\frac{1}{4}$ hour. The solution is then decanted from the settled precipitate through a pipe that enters the precipitating tank about 8 inches above the bottom. As this solution contains some gold slimes, it is passed through a filter press to collect them. The pressure tanks are located below the precipitating tanks, thus allowing the solution and precipitate to be drawn quickly into the pressure tank and passed through the filter press, usually without the aid of pressure.

The precipitation of the gold and silver with zinc dust is almost instantaneous and very complete, not leaving over 20 cents in each ton of cyanide solution. The consumption of zinc is about $1\frac{1}{2}$ pounds for each ounce of gold precipitated.

COMBINATION CYANIDE PLANTS

47. Advantages of Agitation. — It has been said that by agitating the pulp, the cyanide process may be hastened, and it may be added that agitation sometimes permits slimes to be treated directly.

The plant illustrated in Fig. 4 was constructed by D. A. Schiedel for the purpose of treating slimes by cyanide and agitation. The plant, which was built of iron and steel, was composed of a tank a, a vacuum filter b, precipitating boxes c, sump tank d, and gold-solution tank f. The machinery was to be driven by a waterwheel e.

48. The Schiedel Tank. — The agitating tank is 5 feet in diameter and 5 feet high with a $\frac{1}{4}$-inch steel-plate shell and a cast-iron bottom 2 inches thick. To the bottom is cast a cone *m*, through which passes a vertical shaft *n*, which carries four arms *o* that hang down into the vat and have four ⅛-inch steel paddles *r* 6 inches wide fastened to them. These paddles are twisted like the blades of a pro-peller, and the arms to which they are attached at right angles are strengthened by a collar *s*. The shaft with the paddles can be raised by the screw spindle *t*. The driving gear *u* is placed below the tank and agitator. An opening 4 inches in diameter in the tank bottom discharges by means of a stop-cock the contents of the agitating tank through a pipe into a Schiedel patent vacuum filter *b*. A perfect separation of the gold-cyanide solution from the residues is here effected.

49. Description of Filter Box.—The filter box which is shown in Fig. 5 has ¼-inch steel plate for the sides and ⅜-inch steel plate for the bottom. It forms a rectangular box 3 feet 6 inches deep, 7 feet long by 5 feet wide. Two feet above the bottom, as shown in Fig. 5 (*a*), is a per-forated steel filter bottom *a* of ⅜-inch boiler plate, made in three movable sections, supported by angle irons *c* attached to the sides, and by the vertical supports *d*. The perfora-tions *f* shown in the horizontal section, Fig. 5 (*b*), are ½ inch in diameter and are arranged ¼ inch apart. The filter bot-tom fits closely to the sides of the apparatus; it is covered with a blanket *g*, which is kept in position by bars running along the four sides and fastened by thumbscrews *h*. A grating *i* of ⅜-inch round-iron bars, placed 3 inches apart and made in three sections, serves to protect the cloth. The space between the bars is filled with coarse sand. The filter partition divides the apparatus into two compartments, one above the other. The lower one *l* forms a closed box, which is connected with a duplex vacuum pump by the pipe *j*. By this means the air can be rarefied when the filter bot-tom is covered with pulp. The part above the filter receives the contents of the agitator. The bottom of the apparatus has a discharge pipe *k*, with a 3-inch stop-cock for running off the filtered solution into either of the two solution tanks, which are standing on the floor one step lower

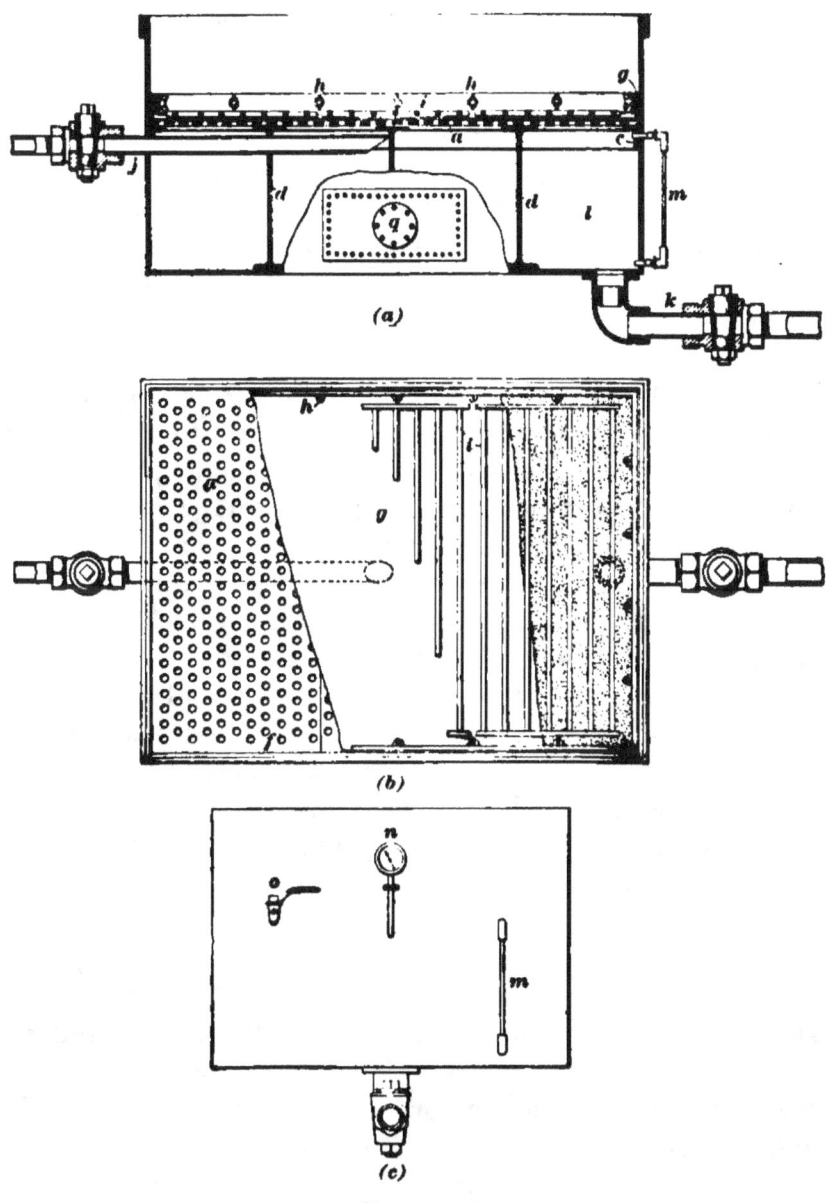

(a)

(b)

(c)

FIG. 5

than the filter. All valves and pipes are of such a size as to secure a quick charge and discharge. The filter is provided with a gauge *m*, Fig. 5 (*c*), to indicate the height of the solution within; a gauge *n*, to show the inches of vacuum; an air tap *o*, to permit an influx of air when the filtered solution is being discharged; and a manhole *q*.

50. The remainder of the process and the zinc precipitation tanks are the same as have been given in the general description. The precipitate is very slimy, and it is said that the freer it is of zinc, the more slimy it becomes. The treatment of the precipitate is the same as has been described. The steel tanks were but little, if any, affected by the cyanide solution.

51. Deductions From Combination Treatment.—The combination cyanide plant was constructed for the purpose of treating slime concentrat s from the canvas plant. Such concentrates contained a varying percentage of carbonate of lime, in some instances as much as 95 per cent., which, however, did not interfere mechanically or otherwise with their satisfactory treatment by cyanide. Such conditions would make chlorination all but impossible.

It was found that for agitation the material required an amount of solution equal to 30 per cent. of its weight and 6 hours' time for leaching. The plant described is capable of treating a much larger quantity of slimes than are usually produced per day by the canvas plant; its services are therefore only periodically required.

52. Percentage of Extraction.—The average consumption of cyanide, calculated from a large tonnage of slimes treated, amounted to 4.3 pounds per ton and cost \$2.27; the labor amounted to \$1; and the total cost of treatment by cyanide amounted to \$3.50 per ton. The average extraction amounted to 93.18 per cent. of the gold and 90 per cent. of the silver in the slimes; although as high as 96.57 per cent. of the gold has been extracted in some instances. The extraction of the gold during the agitation goes on as shown by Table II.

TABLE II

Treatment of Slimes by Agitation	Gold per Ton	Extraction per Cent.
Sample before treatment................	\$88.00	
Sample after 1 hour's agitation.........	13.00	85.23
Sample after 2 hours' agitation.........	11.00	87.50
Sample after 3 hours' agitation.........	7.00	92.05
Sample after 4 hours' agitation.........	7.00	92.05
Sample after 5 hours' agitation.........	6.00	93.18
Sample after 6 hours' agitation.........	5.00	94.31
Sample after 7 hours' agitation.........	5.00	94.31
Sample after 8 hours' agitation.........	5.00	94.31

Within the first hour 85.23 per cent. of the gold was extracted ; during the following 5 hours the increase of extraction was slow and irregular; after 6 hours no further extraction took place. For experimental purposes, Doctor Schiedel continued agitation up to 12 hours without improving on the result. The treatment of the slime concentrates by agitation was preferred on account of its quicker, cheaper, and better results, as compared with percolation.

53. Amalgam in Slimes.—Some concentrates contain a small amount of amalgam, part of which is found on the bottom of the agitating tanks; another part leaves the works with the tailings and is recovered in Hungarian riffles and on amalgamated silver plates.

54. Treatment of Concentrates. — Sulphurets, such as the concentrates of the Utica, Madison, and Eureka mines, of California, were treated on a more or less extensive scale at the same plant. The results were not very satisfactory on account of the coarseness of the concentrates. All sulphurets of the Utica mine are pure sulphide of iron. The fine canvas-plant concentrates alluded to, although less clean, gave an average extraction of 93.18 per cent., whereas vanner concentrates gave only 81.38 per cent. This later extraction, although reasonably good, could not, at the cost of $4 per ton for treatment, compete with chlorination, which yields 90 per cent. of a $50 ore at a cost of $6.50.

NOTE.—The large size of the Utica chlorination works offers special advantages and permits chlorination at this figure, which is much lower than the cost anywhere else in California.

A large number of tests proved that a high percentage of the gold is contained in the coarser particles of the sulphurets; this will account to some extent for the comparatively low percentage of cyanide extraction.

55. Cost of Plant.—The cost of this combination cyanide plant is as follows:

Grading and foundations...............	$ 200.00
Building..............................	300.00
Shafting, belting, and putting into place .	135.00
Agitator	260.00
Vacuum filter........................	165.00
Three tanks	160.00
Two zinc boxes	260.00
Two steel tanks......................	85.00
One vacuum pump....................	235.00
One liquor pump.....................	130.00
Pipes, stop-cocks, faucets, etc..........	70.00
Total	$2,000.00

THE KENDALL CYANIDE PROCESS

56. The Kendall cyanide process is based on the fact that oxygen is necessary in dissolving gold in a cyanide-of-potassium solution. The inventor of the process claims that by the addition of a certain quantity of sodium dioxide to the cyanide solution, the necessary amount of oxygen is artificially supplied and the solution of gold is hastened. The sodium dioxide is added to the ore as it is placed in the vat.

The following chemical changes are claimed by the promotors of the process to take place:

$$2KCN + Na_2O_2 + 2H_2O + 2Au = 2KOH + 2NaOH + 2AuCN$$
and $$2AuCN + 2KCN = 2AuK(CN)_2$$

The plant and the method of leaching are similar to the MacArthur-Forrest cyanide process already described. It was proposed at one time in the Kendall process to use sodium amalgam or zinc amalgam for precipitating the gold from solution. The only inducement this process has to offer is a saving of time in lixiviation.

THE PELATAN-CLERICI CYANIDE PROCESS

57. Agitator.—The process known as the **Pelatan-Clerici cyanide process** depends on agitation for quickly dissolving the gold in solution. The process also involves electrical precipitation of the gold and silver dissolved by the cyanide solution. In Fig. 6 is shown the general arrangement of the plant. The ore and weak cyanide solution are thoroughly mixed in the tank *a* by means of a cast-iron stirrer attached to the shaft *b*. From the mixing tank the thin pulp is run through a launder *c* to the lixiviating vats *d*, which hold from 1 to 5 tons of pulp and are provided with a cast-iron stirrer *e* fitted to the shaft *f*. Wooden pins *g* are inserted through the four blades of this stirrer in order to prevent the sands settling upon the bottom of the vat. The iron stirrer acts as an anode for an electrolytic bath or electrolyte, as well as an agitator, the current being passed to it through the shaft *f*. The bottom of the tank is covered with cement upon which rests a copper plate covered with live quicksilver. The plate and quicksilver form the cathode, or negative electrode, for the electrolytic bath. The anode is connected with one pole of a dynamo and the cathode with the other; the auric-potassic cyanide solution forms the electrolyte and completes the circuit.

58. Quicksilver Cathode.—The use of live quicksilver above the cathode has a twofold object: First, to catch and hold any coarse gold or silver that the cyanide solution cannot dissolve readily; second, to catch and amalgamate the fine cations of gold as they are broken up and are about to be deposited on the copper plate as free gold by electrolysis. The use of quicksilver as a means of depositing gold or silver

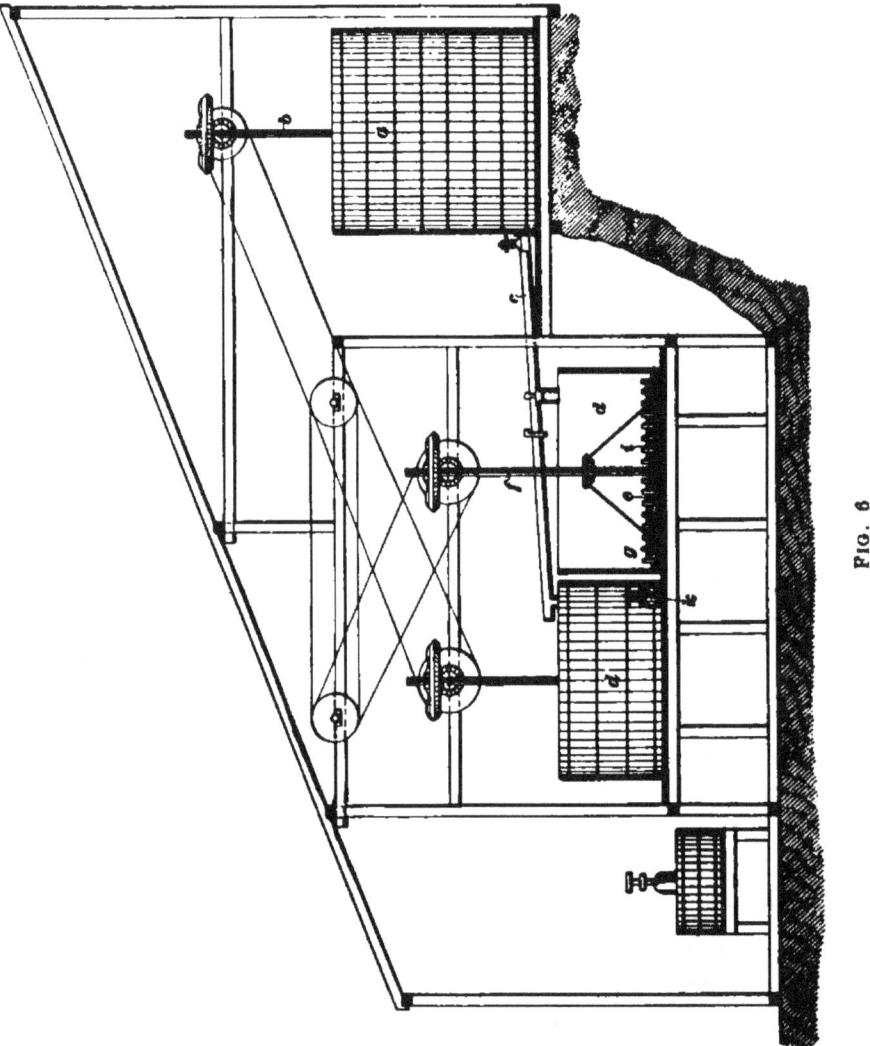

FIG. 6

from a cyanide solution was patented by J. H. Rae, of Syracuse, New York, in 1867, but owing to the employment of an alternating current, the cation became an anion before the mercury could amalgamate the free gold. The use of the direct current converts the anion into a cation, which the mercury amalgamates without trouble. The amalgam, however, is sometimes deposited very hard upon the copper plate, making the clean-up quite difficult. Another objection to the use of quicksilver in this process is due to the stirrers mixing it into the pulp, thus causing the loss of both quicksilver and amalgam in the tailings when they are run off.

59. The **theory of electroprecipitation** is based upon electrolysis. The gold is dissolved at the positive electrode by the cyanide solution in the presence of oxygen, according to Elsner's equation. It is presumed that oxygen is liberated from the water of the solution by electrolytic decomposition and that the cation of auric-potassic cyanide formed makes a tour about the bath until it nears the cathode and is attracted to it and broken up. The gold-potassium cyanide cation being broken up, cyanogen, potassium, and gold are liberated. The cyanogen and potassium probably immediately unite, forming potassium cyanide, which again goes on a tour of the vat for more gold.

60. Claims for the Process.—It is claimed for this process that it economizes in the use of cyanide, saves time and labor, and takes only from 4 to 6 hours to lixiviate and run off the tailings. The process may be adapted to slimes, as well as tailings from other processes, and to raw ore, provided they are suitable for cyanide treatment. To get rid of the tailings, a side discharge door k is opened and the stirrer worked slowly until the vat is emptied. The tank may now be washed out with a hose, the discharge door closed, and a new charge of ore immediately substituted. The liquor from the tailings is drained off and strengthened for use again in some cases; as a rule, however, it is allowed to go to waste, it being low in cyanide and practically containing no gold.

61. The Electrolytic Solution.—Salt $NaCl$ is added to the bath for the purpose of forming silver chloride, if silver is present in the ore, and then argentic-potassic cyanide, according to the following reaction:

$$2AgCl + 4KCN = 2KCl + 2KAgCN,$$

In the equation the chlorine in solution has the **same** effect on the cyanide solution as oxygen, in that it hastens reaction between the metals and cyanogen. The Pelatan-Clerici people also claim that the addition of sodium chloride forms a stable electrolyte of sufficient density for the current used.

It is claimed by some that no benefit can be derived from the use of electricity in cyaniding, while others claim that there are so few particles of gold in a cyanide solution that few become cations and anions. The fact remains, however, that under some conditions the extraction has reached 90 per cent. of the gold in the ore. The fact that this extraction was accomplished in from 4 to 6 hours, while the MacArthur-Forrest process would require as many days, shows that the use of electricity saves time and possibly values.

THE PNEUMATIC CYANIDE PROCESS

62. Basis for the Process.—The presence of **oxygen** being necessary for dissolving gold, unless some element of the haloid group, such as chlorine, bromine, or iodine, be present, agitation was adopted as a means of furnishing **a** greater supply than was contained in the ore and **water.**

FIG. 7

The **pneumatic cyanide process** agitates the pulp by means of compressed air rising through the solution, and which also furnishes oxygen for hastening the operation. The pneumatic process reduces the time to 7 hours. The tanks a shown in Fig. 7 are connected with a series of pipes b, through which compressed air is conducted to a series of pipes in the bottom of each tank. The pulp having been placed in the tanks, the air is turned on by valves c to the pipes in the bottom, which have small apertures for its escape into the solution and through which it rises, thereby causing sufficient ebullition for agitation.

DESCRIPTION OF CYANIDE MILLS .

SPECIFICATIONS FOR A 20-TON CYANIDE MILL

63. Special Construction.—Cyanide mills are sometimes designed for special purposes. A mill that was erected near Weaver, Arizona, for the treatment of tailings had the following equipment: Four leaching vats 12 feet in diameter and $4\frac{1}{2}$ feet deep; two gold-solution tanks 8 feet in diameter and $4\frac{1}{2}$ feet deep; one sump tank 12 feet in diameter and 4 feet deep; one zinc box having twelve compartments, each having a capacity of 1 cubic foot. The entire plant was constructed of wood and covered with two coats of waterproof non-metallic paint.

64. Details of Working the Ore.—The value in the tailings was largely in the dried slimes, which constituted 15 per cent. of the dump. These tailings were shoveled against a $\frac{1}{4}$-inch screen, which broke the caked slimes into fine particles that were afterwards mixed with coarse gravel. The ore was then charged into a 10-ton vat to be treated as follows:

1. The ore is leached 12 hours with a $\frac{1}{4}$ of 1 per cent. (.25 per cent.) cyanide solution, which is then drawn off.

2. The ore is leached 12 hours with a $\frac{1}{5}$ of 1 per cent. (.20 per cent.) cyanide solution run in from the top of the vat.

3. A solution containing $\frac{1}{6}$ of 1 per cent. (.166 per cent.) cyanide is run on top of the ore after No. 2 solution has been

drained off and is allowed to stand 12 hours. No. 3 solution is then drained off.

4. The wash water from a previous operation containing $\frac{1}{8}$ of 1 per cent. (.125 per cent.) cyanide is run on top of the ore and drawn off in 3 hours.

5. Water is run on the ore and after standing 3 hours is drawn off into a gold tank, to be used in the future as No. 4 solution.

All solutions pass into the gold tank, where they receive sufficient cyanide of potassium to give them a strength of $\frac{1}{4}$ of 1 per cent. (.25 per cent.) before passing through the zinc box. The solutions after passing the zinc box are strengthened and used over and over again until they become foul, when they are allowed to run to waste.

The amount of cyanide consumed is 1 pound per ton of ore. The amount of zinc consumed is $\frac{1}{8}$ pound per ton of ore. It requires six persons to operate the plant: four men to shovel, charge, and discharge the vats by wheelbarrows, one man to pump solutions by hand, and one assayer.

METHOD OF CYANIDING AT MERCUR, UTAH

65. Application of the Solutions.—The first solution, which contains .25 per cent. of cyanide, is applied at the bottom and is allowed to saturate the charge until it reaches the top of the ore, when it is shut off. This takes about 8 hours. The same strength of solution is run on from the top to cover the ore 2 inches in order to save time. It is allowed to stand 16 hours, after which time the percolation is started. Fresh solution is added when the other is drawn off and the percolation continued 24 hours.

A weak solution, which contains about .035 per cent. of cyanide, follows the strong solution at the end of the 24 hours. This solution is passed through the leaching vats from 48 to 72 hours, or until the samples taken each day show that the ore charge is ready to be washed.

Wash water is added to displace the weak solution, an operation which takes about 24 hours. It requires 6 or 7 days from the time the tank is filled with ore until it is discharged and ready to be filled again for another leaching.

66. Results from Solutions.—All solutions, whether weak or strong, flow into a common launder that leads direct to the gold-solution tanks. When a solution containing .45 per cent. of cyanide is used, the mixed solutions will contain about .35 per cent., and this solution is pumped back and used as a weak solution without the addition of cyanide. The best results are secured when the total amount of cyanide liquor is about 2 tons to 1 ton of ore.

67. Rate of Drainage.—The rate of percolation differs, but the usual rate of drainage is from ¾ to 1 inch of tank depth per hour. When the proportion of roasted ore in the charge is large, the ingredients present will make a good cementing material. This material will harden in the tanks in about 2 days, rendering sampling with an auger very difficult and greatly impeding percolation. When the percentage of limestone is high in the ore, a large quantity of lime is produced in the furnace. When the solution is added, it will hydrate, causing the material in the tank to swell perceptibly and make the leaching very slow.

68. Discharging the Tailings.—There is not sufficient water available at Mercur to flush the tailings from the vats, hence they are shoveled through the bottom gates into cars having a capacity of 3½ tons. The cars are trammed out of the building by men. Each tank has eight bottom discharge gates 15 inches in diameter, located above four tram roads, which permits the tanks to be shoveled out in from 5 to 7 hours, at a cost of from 6 to 8 cents per ton.

69. Precipitation.—The gold at Mercur is precipitated from the solution by means of zinc dust. The powder used is imported from England or Germany and contains about 90 per cent. of metallic zinc. The solution is pumped from the gold-solution tanks to three precipitating tanks, each of which holds 30 tons of solution. While the tanks are filling, air is blown into the tanks at about 15 pounds pressure through a ½-inch pipe. This is done to stir up the residues remaining in the bottom of the tanks from former precipitations, and which may contain some unconsumed zinc. Five pounds of zinc dust is added to each tank containing 30 tons of solution. The zinc dust is sifted into the solution at intervals from the time the tank is half filled until it is full. The air pipe is moved about the bottom of the tank for a few minutes to stir up all the sediment and is then removed. The suspended matter in the solution is now allowed to settle for about ½ hour, when the supernatant liquor is drawn off through an opening 8 inches above the bottom of the tank. As this solution contains some gold slimes, it is passed through the filter press. Usually the solution passes through the filter press without pressure unless the press is well filled with slimes. Precipitation with zinc dust is almost instantaneous, while its consumption is about 1⅓ pounds of zinc per ounce of gold thrown down.

70. Pressure Tanks.—Two pressure tanks, each having a capacity of 30 tons, are located below the precipitating tank. These pressure tanks are connected with the precipitating tanks by large pipes, so that the solution can be run into them in a short time. Pressure can be applied to these tanks and thus force the solution through the filter press.

The precipitating and filtering operation is continuous; while one tank is filling another is discharging and the third settling.

71. Roasting and Treating Slimes With Acid.—The clean-up takes place monthly and requires about 3 days' time. The precipitating vats are drained, as much as possible of the precipitates scooped into iron pans, and the remainder washed into the pressure tank.

The gold slimes collected are refined in a separate building. The iron pans containing the precipitate are placed in a large cast-iron muffle furnace. The cloths from the filter press are allowed to burn in the pans and the whole is brought to a dull-red heat, by which most of the zinc in the precipitates is oxidized. The roasted precipitate is passed through a ¼-inch screen, after which it is treated with dilute sulphuric and nitric acids to remove the remaining zinc, zinc oxide, arsenic, and mercury, which are present in Mercur precipitates. Nitric acid is added to assist in the oxidation of the metals present, as well as to prevent the evolution of deadly arseniureted hydrogen gas. The acid or dissolving tank is covered with a tight-fitting hood, which is connected with an exhaust fan, in order to remove the dangerous gases to the outside of the building. The acid must be added gradually to prevent the solution boiling over the top of the tank. When the addition of fresh acid does not cause further effervescence, the tank is filled with water and allowed to stand. The supernatant liquor is drawn off through a pressure tank and filter press and the slimes again washed in the same way as before. This is repeated twice to free the slimes from soluble salts, as far as possible.

72. Refining the Slimes.—The slimes are next flushed into a pressure tank, passed into the filter press, washed, dried into cakes, removed, coarsely pulverized, mixed with a flux of soda, potash, and borax glass, and smelted in two No. 300 graphite crucibles (see Table V). Before melting the product contains 60 per cent. of gold, the other 40 per cent. being largely silicious slimes. The resulting bullion is 950 fine. This is the only mill in the district refining its own bullion, all the other mills shipping to Eastern refineries. The total cost of refining is about 15 cents per ounce of gold.

CYANIDING CRIPPLE CREEK, COLORADO, ORES

73. Cripple Creek Ores.—H. Van F. Furman states that "the ores of the Cripple Creek district consist of porphyry (andesitic breccia), phonolite, decomposed granite, and quartz, and usually carry on the surface iron oxide, manganese oxide, and oxide of tellurium; below the water level the gold occurs in the minerals calaverite and sylvanite and is associated with more or less iron pyrites. The mineral fluor frequently occurs in the gold-bearing veins. While the surface ores contain free gold, they do not yield their gold contents by amalgamation, the gold usually being coated with oxide or tellurium or some substance that interferes with its extraction by this method. The extraction of gold from surface ores by potassium cyanide presents no difficulties; but the treatment of the telluride ores without subjecting them to a preliminary roast has been attended with the drawbacks of extremely fine grinding and prolonged percolation in the vats (sometimes from 12 to 14 days in order to secure a fair extraction). At present all telluride ores are roasted dead, i. e., until the tellurium is oxidized completely before being leached in the vats."

74. Cripple Creek Cyanide Mill.—The process applied at the mill of the Brodie Reduction Company, situated about 1½ miles south of the town of Cripple Creek, is typical of the method adopted for the treatment of these ores. The mill at present has a capacity of about 100 tons per day. The ore as received from the different mines is unloaded into bins, each lot being kept separate until it is sampled and paid for, the ore being purchased upon its value as determined by sample and assay, which is the invariable custom of the district. From the bins the ore is delivered by hand to a Gates crusher which reduces it to pieces not exceeding 1 inch in diameter. From the crusher it is raised by an elevator and passed through a Vezin sampler, which takes out a sample, on which the settlement as to the value of the lot is based. The crushed ore passes through a 4-tube Argall drier to a Dodge crusher, from which it is raised by an elevator to a revolving screen, the oversize going to a pair of Davis rolls, and after crushing it is returned to the screen, the undersize passing to 14" × 20" Krom rolls. The product of the Krom rolls is elevated, divided, and delivered to four 40-mesh, brass-wire,

cloth revolving screens. The undersize is carried by a
screw conveyer to the storage bins or roasting furnace, as is
desired. The oversize passes to another set of Krom rolls,
whence it is elevated to the 40-mesh screens.

75. The oxidized surface ores pass directly to the storage bins, while the unoxidized telluride ores pass to the
roasting furnace. The furnace at these works is a Pearce
turret, 40 feet in diameter, with an annular hearth 8 feet
wide, the rabble arms being water-cooled. From the storage bins the ore is drawn into tram cars, each carload
being weighed and dumped into the leaching vats. The
vats, which are circular, are constructed of No. 8 steel
(0.1285 of an inch thick), each vat being provided with
manholes for sluicing off the tailings after the charge is
leached.

76. Brodie Mill Practice.—The stock solution is stored
in steel tanks coated with paraffin paint and is kept at the
proper strength by the addition of potassium cyanide. The
solution is run in on the top of the ore to be leached and
allowed to percolate. Two solutions are used, the strong
solution containing from .5 to .75 per cent. of potassium
cyanide. The time of treatment varies from 70 to 100 or
more hours, according to the ore. After the gold is dissolved, wash water is added to displace the cyanide solution. The strong solution is allowed to percolate for about
50 hours, when the weak solution is added and afterwards
the wash water.

The solution after passing through the tank's filter bottom
is conveyed by iron pipes to the zinc boxes for the precipitation of the gold. There are two sets of zinc boxes, one set
for the strong and one set for the weak solutions and wash
water. After passing through the zinc boxes, the solutions
pass to their respective sumps, from which they are pumped
to their respective storage tanks. The zinc slimes are
washed, treated, smelted, and the resulting gold bars shipped
to the United States Mint at Denver.

DANGER IN WORKING THE CYANIDE PROCESS

PROPERTIES OF POTASSIUM CYANIDE

77. The fact that potassium cyanide is a deadly poison was at one time considered a great obstacle in the way of the successful introduction of the process. The solutions used in the process are so dilute that the hydrocyanic acid given off is of no consequence if the works are properly ventilated. All ore should be tested for acids before treatment and if necessary neutralized. This will insure safety to the men and economy in the practical operations of the process. Those working the process are not required to come into direct contact with the cyanide, either as a solid or with its solution. Even the clean-up, when properly conducted, does not require contact with the cyanide solution.

78. Symptoms of Poisoning.—Some men are very susceptible to the effects of potassium cyanide, and when the diluted solutions are brought into contact with their skin, an eruption is produced, which is not dangerous, but annoying on account of the itching. Such men should not be employed in cyanide works. If the works are not properly ventilated, the men will complain of headache, faintness, and dizziness. If it be necessary to place one's hands in a cyanide solution, they should be protected by a coating of vaseline or by rubber gloves. When the extensive use of cyanide is considered, the number of accidents is exceedingly small.

79. Sources of Poisoning.—Poisoning may occur from hydrocyanic acid being liberated from the leaching vats or tanks. In countries where the vats are not covered, poisoning from the free hydrocyanic acid liberated by mineral acids is unknown.

The poisonous gases that are liberated when the slimes are treated with acid to dissolve the zinc are very dangerous, for the slimes after washing usually contain a little

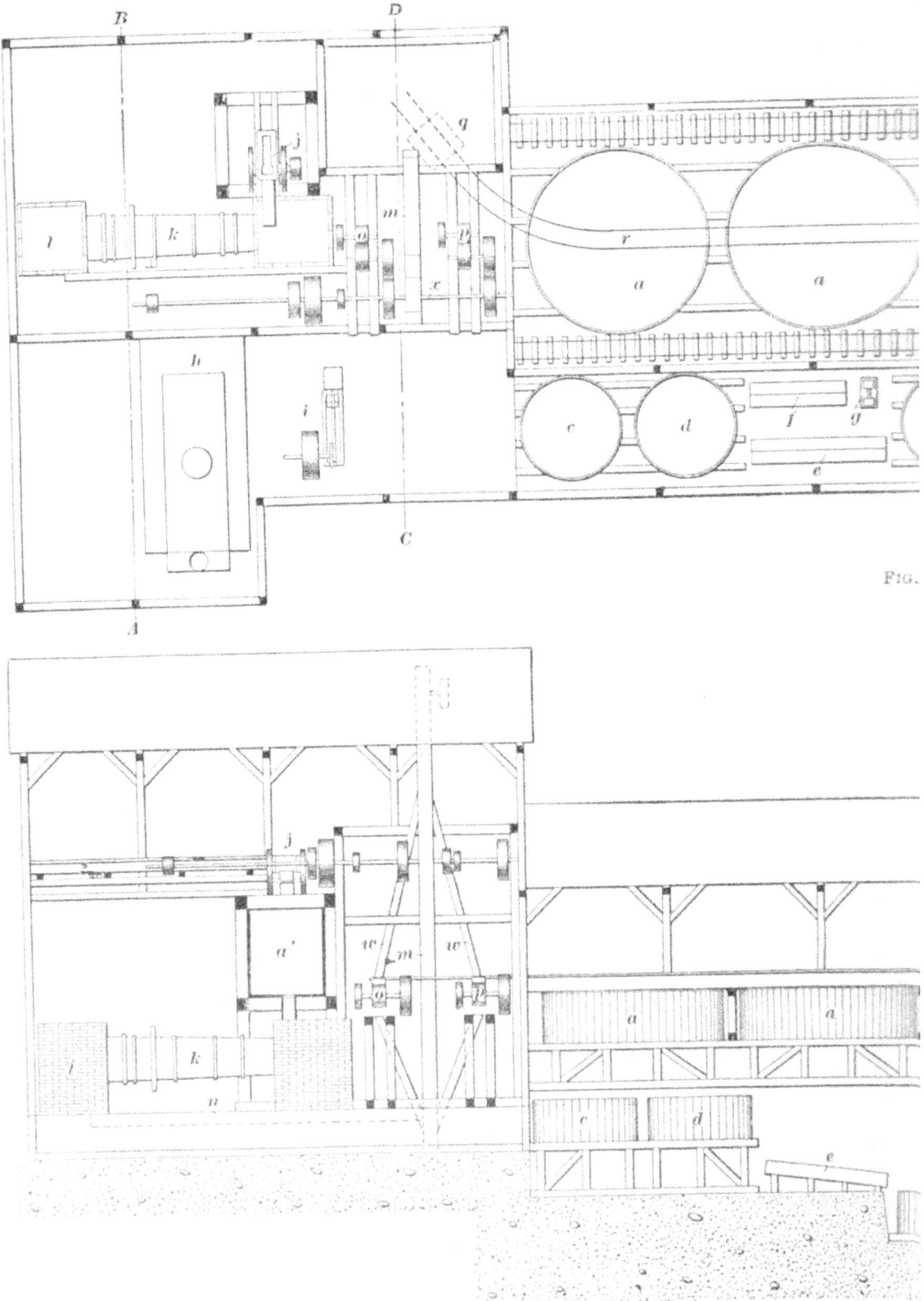

FIG.

FIG. 9

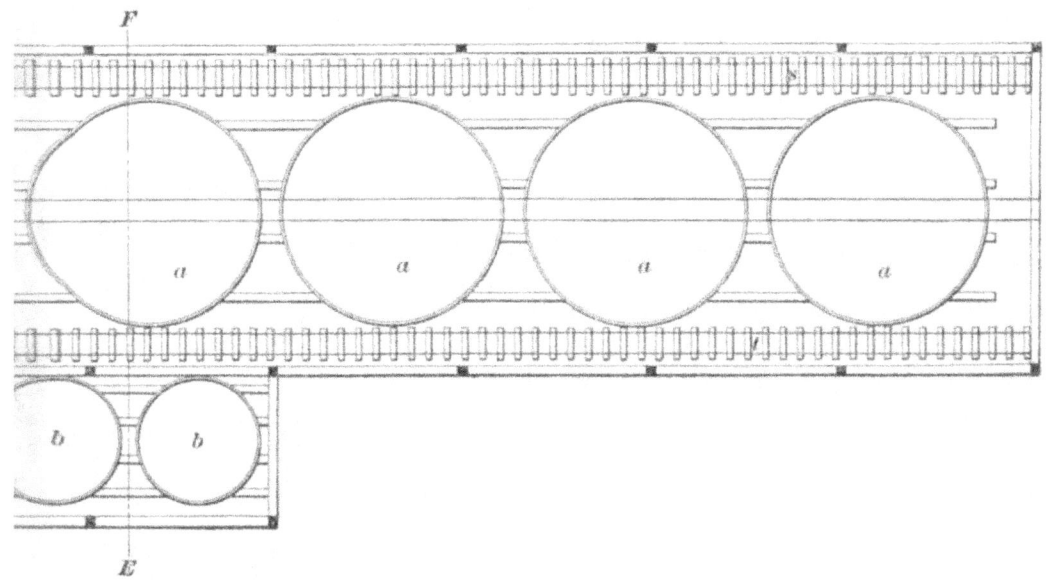

8

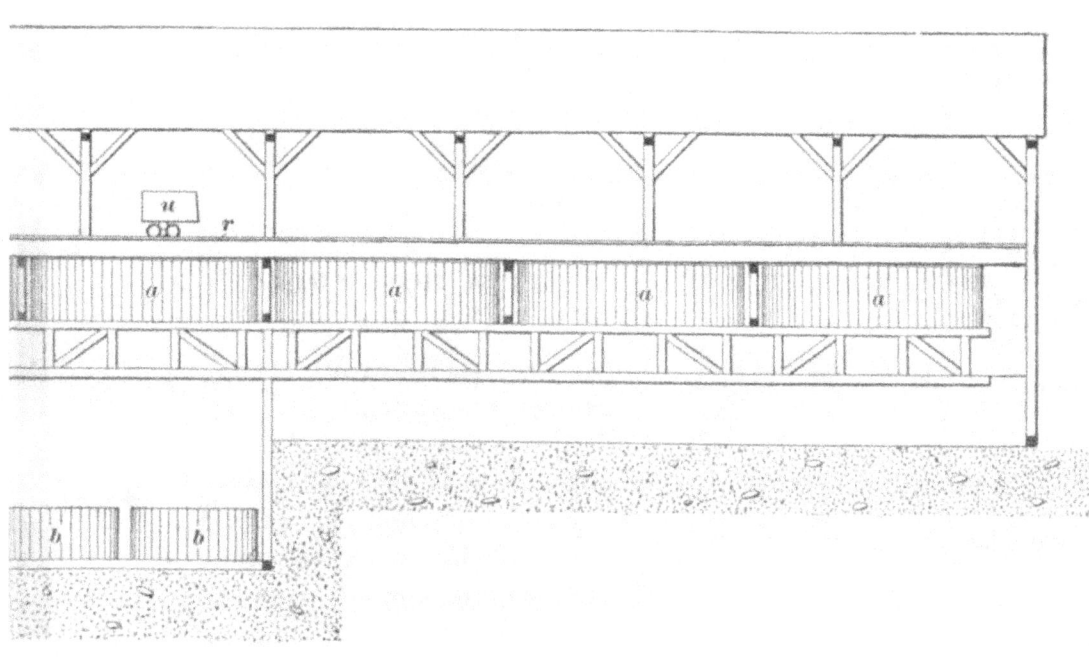

9

insoluble cyanide that will yield hydrocyanic acid upon treatment with sulphuric acid. Respirators should be worn in this case for protection.

Ores that contain arsenic are usually more or less soluble in the cyanide solution. The arsenic is precipitated on the zinc with the gold and enters the slimes. When acid is added to the slimes, arseniureted hydrogen, a deadly poisonous gas, is liberated. This gas, if inhaled, passes from the lungs into the circulation of the blood and rapidly attacks the tissue. The symptoms are first nausea, then extreme languor with pain in the legs, and finally death.

Slimes should be treated in a special chamber or cupboard connected with a chimney having a good draft when acid is used.

80. Antidotes.—In case of internal poisoning, an emetic or physical means should be used to induce vomiting.

In case of accidents, it is well to remember that *peroxide of hydrogen H_2O_2 is a powerful antidote for cyanide poisoning.* Hypodermic injections of solutions containing from 2 to 3 per cent. of peroxide have been used successfully, especially when injected at different parts of the body. At the same time the stomach was washed out with a 2-percent. solution of hydrogen peroxide. Peroxide of hydrogen H_2O_2 forms with hydrocyanic acid HCN "oxanide" $CONH_2$, which is a harmless compound, thus:

$$2HCN + H_2O_2 = 2CONH_2$$

DESIGN FOR A TWENTY-FIVE TON CYANIDE MILL

81. Plan.—The plant to be described was designed by the Allis-Chalmers Company, of Milwaukee, Wisconsin. This plant is shown by plan in Fig. 8. The six lixiviating tanks a are each 16 feet in diameter and 4 feet deep; the solution tanks c and d are 9 feet in diameter and 4 feet deep; the sump tanks b are of the same capacity as the solution tanks. Between the gold-solution tanks and the sump tank the zinc boxes *e* and *f* are shown, each tank being divided into two rows of compartments. The box *e* is for the weaker gold solution. The pump *g* is placed near the sump tanks for the purpose of forcing the solutions to the various vats, as required. The leaching vats are under a roof 106 feet 9 inches long by 23 feet wide. The remainder of the lixiviating apparatus is under a roof 53 feet 6 inches by 11 feet.

The boiler *h* and the engine *i* are separated from each other and the remainder of the machinery by partitions. The milling machinery is contained in a ground space

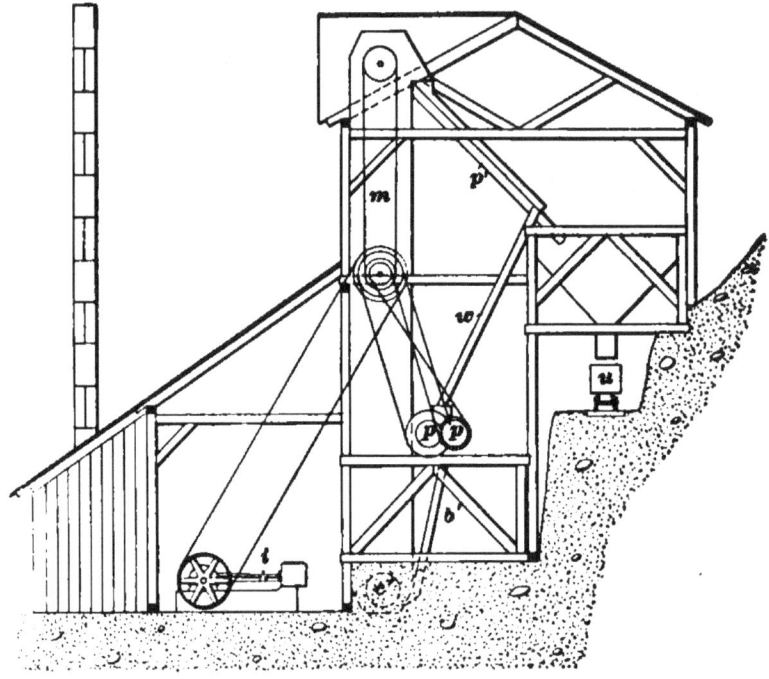

FIG. 10

44 feet 3 inches long and 26 feet wide. The crusher is represented by *j*, the ore drier by *k*, the furnace of the drier by *l*. The elevator to the screen is shown at *m*, the crushing rolls by *o* and *p*. When the ore is to be charged in a vat, it is run on to the platform scales *q* in a car and weighed; from the scales the car runs over the vats *a* on track *r* and is dumped into the proper vat. Other tracks *s* and *t* are for the cars into which the tailings from the vats are shoveled.

82. Elevation.—In Fig. 9 is shown an elevation of the 25-ton plant whose plan has just been described. The section is taken back of the line shafting *x*, in order to show the principal machines. The ore is first passed through the crusher *j* into the ore bin *a'* below. The ore runs down the ore-bin floor and is drawn off into a chute that feeds the ore drier *k*. From the drier at *l*, the ore falls into a screw line *n* and is conveyed to the boot of the elevator *m*. The elevator

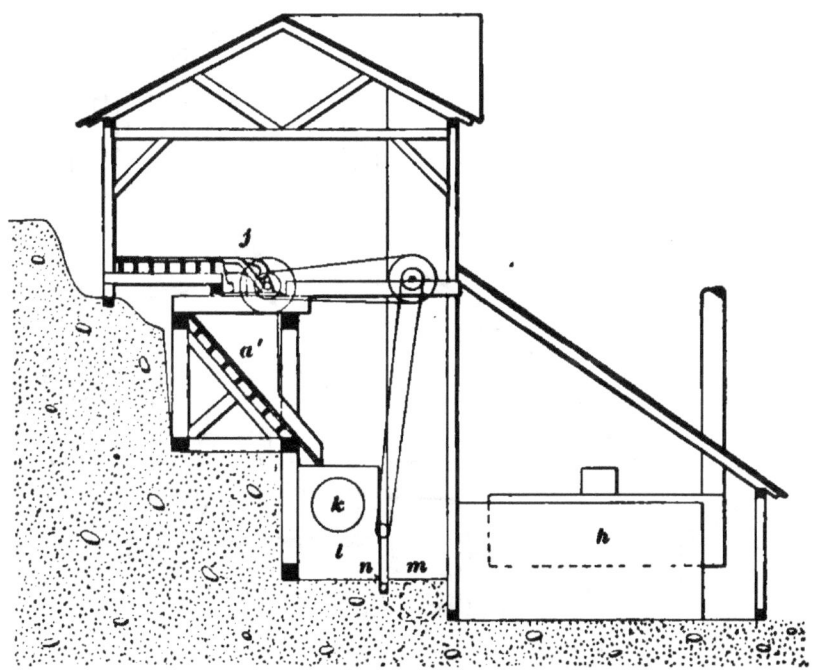

Fig. 11

raises the ore to a screen shown in Fig. 10, at *p'*, over which it passes by gravity. The coarse ore passes down the chute *w* to the rolls *p* and then down to the elevator, by which it is raised to the screen *p'* again. The fine ore from the screen passes to a bin, from which it is drawn as needed. By an examination of Fig. 9, it will be observed that advantage is taken of gravity, wherever it is thought desirable, for handling material and disposing of solutions. The lixiviating tanks are above the gold-solution tanks, the latter above the zinc boxes, and they, in turn, above the sump tanks. The charging track *r* has upon it a tram car *u* above one of the leaching vats.

83. Cross-Section.—The cross-section shown in Fig. 10 was taken through the plan on the line *C D*. It shows the rolls *p*, elevator *m*, screen *p'*, chute *w* to the rolls and chute *b'* to the elevator boot *c'*. It also shows the charging car *u* under the ore-bin chute.

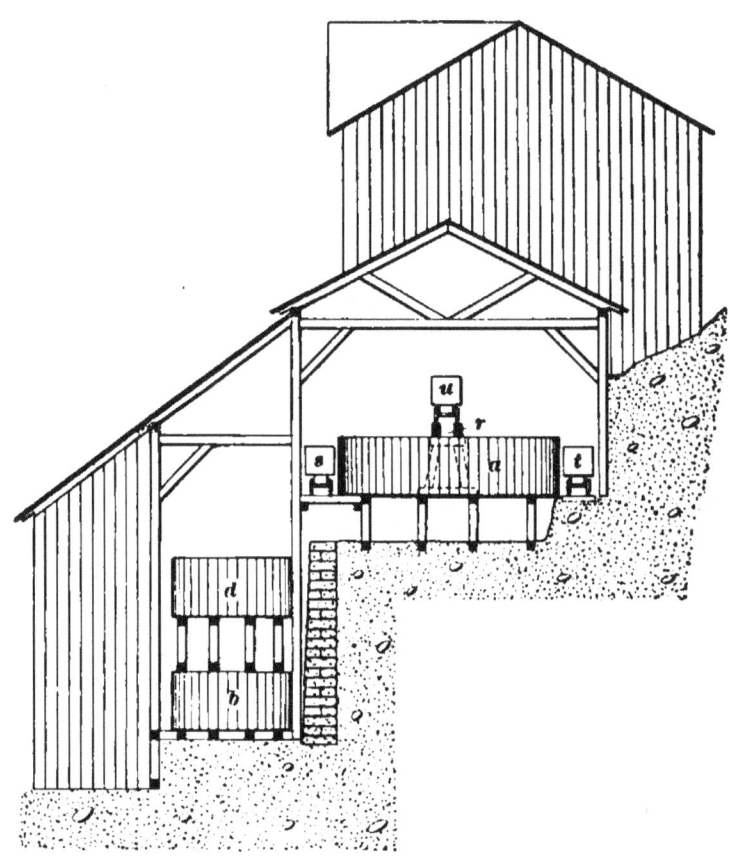

FIG. 12

84. The cross-section shown in Fig. 11 was taken through the line *A B*, Fig. 8. It shows in detail the boiler *h*, screw conveyer *n*, elevator *m*, drier *k*, furnace front *l*, crusher *j*, and ore bin *a'*, with its chute leading to the drier.

It will be noticed that the mill is built on a side hill, at an elevation that permits the ore cars to run direct to the crusher. In some cases the ore is run above the crusher, so that it is unnecessary to handle it, as in this case.

TABLE III

CONTENTS OF TANKS IN GALLONS AT ONE FOOT IN DEPTH

Diameter		Gallons. 1 Foot in Depth	Diameter		Gallons. 1 Foot in Depth	Remarks
Feet	Inches		Feet	Inches		
3		52.86	14	6	1,234.91	
3	6	73.15	15		1,321.54	
4		93.97	15	6	1,407.51	
4	6	118.98	16		1,503.62	The number of gallons in a
5		146.83	16	6	1,600.00	receptacle 1 foot in diameter
5	6	177.67	17		1,697.45	and 1 foot high is .7854 × 7.48
6		211.44	17	6	1,798.76	= 5.87 gallons. 7.48 is found by
6	6	248.15	18		1,903.02	dividing the number of cubic
7		287.80	18	6	2,010.21	inches in a cubic foot by 231,
7	6	330.68	19		2,120.34	the number of cubic inches
8		375.90	19	6	2,233.29	in 1 gallon; thus, 1,728 ÷ 231
8	6	424.36	20		2,349.41	= 7.48, or the number of gallons
9		475.75	20	6	2,468.35	in 1 cubic foot. The weight of
9	6	530.08	21		2,590.22	1 gallon of water is $8\frac{1}{3}$ pounds;
10		587.35	21	6	2,715.04	thus, $\frac{62.4}{7.48} = 8.33$. 2,000 pounds
10	6	647.55	22		2,842.79	
11		710.69	22	6	2,973.48	= 1 ton. $\frac{2\,000\ \text{lb.}}{8.33\ \text{lb.}} = 240$ gallons
11	6	776.77	23		3,107.10	
12		848.18	23	6	3,243.65	in 1 ton.
12	6	917.73	24		3,383.15	
13		992.62	24	6	3,525.59	
13	6	1,070.45	25		3,670.95	
14		1,151.21				

85. Fig. 12 is a section through the plan at EF and shows the leaching tank a, with the loading track r and unloading tracks *s* and *t*; also the gold-solution tank *d* and the sump tank *b*. It will be noted that the leaching vats and the gold-solution tanks are placed on a space leveled off on the side of the hill, in order to take advantage of gravity in drainage operations. The car *u* runs on track *r*, which is laid upon stringers over the tanks. The stringers are placed upon bents, which are between the tanks, as shown by dotted lines.

TABLE IV

CAPACITY OF LEACHING TANKS IN TONS OF DRY ORE

Diam-eter in Feet	Height of Vat						
	3 ft.	3 ft. 6 in.	4 ft.	4 ft. 6 in.	5 ft.	5 ft. 6 in.	6 ft.
10	7	8.5	10	11.5	13.0	14.5	16
12	10	12.5	15	17.0	19.5	21.5	24
14	14	17.0	20	23.0	26.0	29.0	32
16	18	22.0	26	30.0	34.0	38.0	42
18	23	28.0	33	38.0	43.0	48.0	53
20	28	34.0	40	46.0	52.0	58.0	64
25	45	55.0	65	75.0	85.0	95.0	105
30	60	75.0	90	105.0	120.0	135.0	150
40	115	140.0	165	190.0	215.0	240.0	265
50	180	220.0	260	300.0	340.0	380.0	420

86. Calculating the Capacity of Tanks in Gallons and Tons.—The capacity of a cylindrical tank may be calculated as follows: Square the diameter in feet, multiply this product by the depth in feet, and this product by 5.87, which will give the contents of the tank in gallons.

EXAMPLE.—What number of gallons will a tank 12 feet in diameter and 10 feet high contain?

SOLUTION.— $12 \times 12 \times 10 \times 5.87 = 8,453$ gallons.
Capacity in gallons $\div 240 =$ tons of water.

$8,453$ gallons $\div 240 = 35$ tons of water. Ans.

TABLE V

SIZE OF GRAPHITE CRUCIBLES

Number	Holding Capacity, Liquid Measure			Height, Outside	Diameter at the Top, Outside	Diameter at the Bilge, Outside	Diameter at the Bottom, Outside
	Gal.	Qt.	Pt.	Inches	Inches	Inches	Inches
0				2	1¼	1⅝	1¼
00				2⅝	1⅞	1⅞	1⅝
000				2½	1⅞	2⅛	1½
0000				3	2⅝	2½	1¾
1				3⅝	3⅛	3	2¼
2				4½	3¾	3⅝	2¾
3				5¼	4¼	4⅛	3
4				5⅝	4⅝	4½	3¼
5			1½	6	4⅞	4¾	3½
6		1		6½	5¼	5⅛	3¾
7		1	¼	6¾	5½	5½	4
8		1	½	7¼	5¾	5¾	4¼
9		1	¾	7⅝	6	6¼	4½
10		1	1	8	6	6½	4¾
12		2		8	6¼	6¾	5
14		2	1	8½	6¾	7¼	5½
16		2	1	8¾	7	7½	5½
18		3	1	9¾	7⅝	8	5¾
20	1			10¼	7¾	8⅝	6
25	1		1	10¾	8	8⅝	6¼
30	1	1	1	11	8⅝	9¼	6½
35	1	2	1	11⅝	9¼	9¾	7
40	2			12⅝	9¼	10¼	7¾
45	2	1		13	9¾	10½	7⅝
50	2	3		13½	10⅛	11¼	7⅞
60	3			14	10⅝	11⅝	8
70	3	1		14½	10⅞	12	8¼
80	3	2	1	15⅝	11⅛	12⅝	8⅝
90	4			15⅞	11½	12½	9
100	4	2	1	16	11⅞	13⅛	9⅝
125	4	3	1	16⅝	12½	13¾	9½
150	6	3		18¼	13¼	14⅝	10⅝
175	7	3	1	19½	14¼	15¾	10¾
200	9	3	1	20½	15	16½	11¼
225	10	1	1	20¾	15¼	16¾	12½
250	10	3		20½	15¼	17	11¾
275	11	3		22⅝	15	16⅝	12½
300	12	2		22	16¼	17½	12½
Crucibles for File Tempering							
60				17¼	9	9¼	8
70				20½	9	9¼	8
80				22¼	9	9¼	8
100				24½	9	9¼	8

NOTE.—Graphite crucibles, being carbon, oxidize more or less and should therefore be treated in a reducing and not in an oxidizing flame.

A SERIES

OF

QUESTIONS AND EXAMPLES

Relating to the Subjects
Treated of in this Volume.

———

It will be noticed that the various Examination Questions that follow have been divided into sections to which have been assigned the same section numbers as the Instruction Papers to which they refer. No attempt should be made to answer any of the questions or to solve any of the examples until the Instruction Paper having the same section number as the section in which the questions or examples occur has been carefully studied.

SURFACE ARRANGEMENTS AT REDUCTION WORKS

EXAMINATION QUESTIONS

(1) Define hydrometallurgy.

(2) What is lixiviation ?

(3) What is considered the best location for a mill near a mine ?

(4) What two classes of mill-site claims are there ?

(5) What can you say regarding the size of a mill site ?

(6) What points should be considered relative to the erection of a mill ?

(7) For what is masonry used at metallurgical works ?

(8) Referring to retaining walls, how should frost be guarded against ?

(9) When are framed bents used for trestles ?

(10) What is a scarfed joint ?

(11) What two general systems of heavy framing are there ?

(12) For what class of ores is the dry-crushing silver mill intended ?

(13) What does the term conveying and hoisting machinery include ?

§ 24

(14)   For what purpose may hoisting and conveying machinery be used at smelters?

(15)   In making use of the Hunt elevator and automatic railways, what returns the empty car to the chute?

(16)   What advantage is there in using centrifugal pumps to raise the exhausted tailings from leaching vats?

(17)   What should be the principal aim at smelting plants?

(18)   What is the quantity of water that will be necessary for concentrating machines, such as Frue vanners?

ORE DRESSING AND MILLING

(PART 1)

EXAMINATION QUESTIONS

(1) Under what conditions does it become necessary to crush ores dry ?

(2) What are the advantages of wet crushing ?

(3) What are the main points of difference between Blake and Dodge crushers ?

(4) What is the principal field for crushing rolls ?

(5) How are crushing rolls designed to resist excessive strains ?

(6) Describe the construction and crushing action of the Huntington mill; also state for what class of material it is best fitted.

(7) Give one common order for the dropping of stamps in a battery and state what is the object for dropping them in this order.

(8) What are the advantages of sectional guides for stamps ?

(9) What is a good construction for battery blocks ?

(10) What is meant by the discharge of a mortar, and how is the height of discharge regulated ?

(11) What is the objection to iron frames for batteries ?

§ 25

(12) Why are steam stamps more effectual pulverizers than gravity stamps?

(13) Tell what materials are used for battery screens and give the relative advantages of wire and punched screens.

(14) What are the advantages of automatic feeding for rolls and stamps?

ORE DRESSING AND MILLING

(PART 2)

EXAMINATION QUESTIONS

(1) What is a grizzly and at what stage of the work is it introduced?

(2) Why is close sizing necessary in the preparation of ore for concentration and should pulp for buddles or similar machines be classified, and if so, why?

(3) Describe the separating action of the spitzkasten.

(4) For what purpose is the diving board used in the spitzkasten and in the settling box?

(5) Why are settling boxes used?

(6) (*a*) What are the two general classes of concentrating machinery? (*b*) Name the principal machines of each class.

(7) What are jigs and for what are they employed?

(8) Explain the sorting action of jigs and tell how it is accomplished.

(9) If an ore containing galena and blende, with a quartz gangue, is to be jigged to concentrate both the galena and the blende separately, how many compartments should the jig have, and what would be the headings for each compartment?

(10) How is the force of water regulated in jigs?

(11) Describe the construction of a double eccentric.

(12) How do the ordinary Hartz jigs and the quick-return Hartz jigs differ in action?

§ 26

ORE DRESSING AND MILLING

(PART 3)

EXAMINATION QUESTIONS

(1) Upon what principle do pneumatic concentrators work ?

(2) Give the principle of vanners and describe their action.

(3) Give the general construction and describe the separating action of the Wilfley table.

(4) Define paramagnetic and diamagnetic and give the two general classes of magnetic concentrating machines.

(5) In gold and silver amalgamation, what are the principal causes for the loss of gold and mercury ?

(6) How is the sickening of mercury remedied when it is due to the formation of metallic oxides ?

(7) How is greasy gold made amalgamable ?

(8) How and why are amalgamating pans heated ?

(9) (a) If ore is to be amalgamated in pans, why should it be crushed as fine as practicable before being introduced into the pans ? (b) What effect has long grinding in the pan upon the mercury ?

(10) Describe the action of the Wetherill concentrator.

(11) Why are wooden pans employed when working acid ores ?

(12) How may non-amalgamable silver minerals be gotten into an amalgamable form previous to their introduction into the pans ?

(13) Describe the patio process.

ORE DRESSING AND MILLING

(PART 4)

EXAMINATION QUESTIONS

(1)   What points should be observed in selecting a mill site?

(2)   What is the essential condition of a mineral in order that it may be concentrated?

(3)   What are the principal objections to barrel amalgamation and how does it compare with the systems of pan amalgamation, such as the Boss process?

(4)   Why is it necessary to use silver-plated copper plates or to dress new copper plates with amalgam before starting up?

(5)   Describe the operation of preparing and cleaning battery and apron plates.

(6)   What is meant by sweating and skinning plates, and is the latter advisable?

(7) . What is the cause of verdigris and how may verdigris be removed?

(8)   How are apron plates arranged to secure a large amount of amalgamating surface with a limited floor space?

(9)   What are crawls and for what are they used?

(10)   For what are elevators used in mills and what are the differences of construction when they are employed for dry, wet, and hot materials?

(11)   In retorting amalgam, what dangers are to be guarded against?

(12)   What character of timbers will make good ore floors?

SAMPLING ORES

EXAMINATION QUESTIONS

(1) State the object of sampling at concentrating mills.

(2) Why are blast-furnace ores, fluxes, and fuels sampled?

(3) What is the mission of a public sampling mill?

(4) What is understood by quartering?

(5) Why should the fine ore be swept up and deposited upon the apex of the cone when quartering?

(6) What is the object of adding water to an ore when sampling?

(7) What is the object of crushing the sample or reducing the size of the lumps as it is decreased in quantity?

(8) What is fractional selection?

(9) How wide should the scoops of a Jones sampler be in comparison with the lumps of ore sampled?

(10) The fine particles of ore from a sample containing gold are sometimes richer than the lumps; will this richness apply to samples of iron ore, coke, or limestone?

(11) How are grab samples taken from ships containing iron ore?

(12) What is the object of a slotted-pipe sampler?

(13) What particular points are to be observed while sampling with a dipper or bucket?

§ 29

(14) Why should fine slimes be settled before they are poured off ?

(15) What substances may sometimes be used to hasten the settling of slimes ?

(16) How are final or assayers' samples obtained ?

(17) Give the rule for determining moisture in an ore.

(18) What objections are there to hand sampling ?

(19) How must a mechanical sampler act to give accurate results ?

(20) How does Collom's sampler resemble Topham's ?

(21) Upon what does the accuracy of sampling depend ?

(22) Why should larger samples be taken of rich than of poor ores ?

(23) Should a sample be larger when the mineral is regularly distributed than when it is irregularly distributed ?

(24) What prevents crushing fine and regularly mixing ore by machinery ?

(25) (*a*) How is smelting ore sampled ? (*b*) How may leaching ore be best sampled ?

ROASTING AND CALCINING ORES

EXAMINATION QUESTIONS

(1) Define a dead roast.

(2) Define a chloridizing roast.

(3) What is copper matte ?

(4) What objects are attained by roasting ores containing gold and silver with salt ?

(5) (*a*) What is a calcining roast ? (*b*) What is the object of drying ores ?

(6) What is the object of covering a reverberatory furnace with sand ?

(7) Why is not fresh air admitted to a furnace for a reducing roast ?

(8) When is a slagging hearth to be used in a reverberatory furnace rather than a sintering hearth ?

(9) For what purpose is silicious ore used on the slagging hearth ?

(10) Why is it advisable to use long hearths when roasting lead ores ?

(11) What roasting has the Brown horseshoe furnace accomplished successfully ?

(12) How are the rabble arms kept cool in the McDougall furnace ?

(13) (*a*) What is the Howell-White roaster intended to accomplish ? (*b*) How does it work ?

§ 30

(14) What character of lump ores are suited to shaft furnaces?

(15) Under what circumstances is it advisable to mix coal or wood with the ore?

(16) Why is it sometimes advisable to cover roasting heaps with sheds?

(17) Which is preferable, stall or heap roasting, and why?

(18) In roasting copper sulphide, what reaction occurs?

(19) In about what order does the oxidation of metals occur?

(20) (*a*) What occurs to the higher oxides during a reducing roast? (*b*) Complete the equation $CuSO_4 + 2C =$.

(21) Give the reaction for zinc carbonate and heat.

THE CYANIDE PROCESS

(PART 1)

EXAMINATION QUESTIONS

(1) What limits the scope of the cyanide process?

(2) What character of gold in ores is most successfully treated by the cyanide process?

(3) How does cyanide of potassium act with ores that contain silver?

(4) What metallic substances have more or less effect upon cyanide solutions and cause a loss of cyanogen?

(5) What means are sometimes adopted to hasten the dissolution of gold with cyanide solutions?

(6) Upon what does the rate of the dissolution of gold in cyanide solutions depend?

(7) (*a*) What action has ferric oxide upon cyanide solutions? (*b*) What action has ferrous sulphate upon cyanide solutions?

(8) How may loss of cyanide be decreased when minerals soluble in cyanide are present in the solution?

(9) Why do pyritic concentrates take more time for treatment than tailings?

(10) What are slimes?

(11) How may the treatment of concentrates by cyanide be hastened?

§ 31

(12)   What apparatus is required for testing cyanide solutions in the laboratory ?

(13)   What is a convenient standard for a silver-nitrate solution and how may it be made ?

(14)   Describe the method employed for testing the KCN of commerce.

(15)   What is the time limit for rate of percolation ?

(16)   How may the percentage of gold extraction be determined ?

(17)   Describe in what manner the proper strength of a cyanide solution for any ore may be found.

THE CYANIDE PROCESS

(PART 2)

EXAMINATION QUESTIONS

(1) (*a*) How do slimes interfere with percolation? (*b*) How are slimes gotten rid of?

(2) What should be the strength of a weak cyanide solution when it is to precede a strong solution?

(3) What are the limits that send the solutions to the strong- or weak-solution tanks?

(4) What strength should the weak solution following the strong solution have?

(5) If a gold cyanide solution contained .25 per cent. *KCN* and $6 in gold per ton, what amount of gold would it contain per ton if diluted .025 per cent. *KCN*?

(6) How is *KCN* added to sump solutions to make them standard solutions?

(7) How many tons of solution will a tank 10 feet in diameter and 5 feet deep hold?

(8) Suppose a tank 10 feet in diameter contains an .18-per-cent. cyanide solution that measures 3 feet in depth, how many pounds of *KCN* must be added to raise the solution to a standard strength of .25 per cent.?

(9) Why is not precipitated gold redissolved by a cyanide solution?

§ 32

(10) (*a*) What color has the gold on the zinc when precipitation is occurring properly? (*b*) What impurities are mixed with the gold slimes or precipitates?

(11) (*a*) What occurs when copper is in the zinc boxes? (*b*) How can the effects of copper in a gold cyanide solution be remedied?

(12) Describe the acid treatment of refining gold cyanide precipitates.

(13) Why should the second melting of bullion be at as low a temperature as possible?

(14) What properties should a metal have if it is to be used for electrical precipitation?

(15) (*a*) What is zinc fume? (*b*) How is zinc fume used to precipitate gold at Mercur, Utah?

(16) Will charcoal precipitate gold from auric-potassic cyanide solutions, and if so, why is it not generally used?

(17) On what does the pneumatic cyanide process depend?

(18) Describe the Pelatan-Clerici process.

(19) How should ores mined below water level be treated generally before cyaniding?

A KEY

TO ALL THE

QUESTIONS AND EXAMPLES

CONTAINED IN THE

EXAMINATION QUESTIONS

INCLUDED IN THIS VOLUME.

The Keys that follow have been divided into sections corresponding to the Examination Questions to which they refer, and have been given corresponding section numbers. The answers and solutions have been numbered to correspond with the questions. When the answer to a question involves a repetition of statements given in the Instruction Paper, the reader has been referred to a numbered article, the reading of which will enable him to answer the question himself.

To be of the greatest benefit, the Keys should be used sparingly. They should be used much in the same manner as a pupil would go to a teacher for instruction with regard to answering some example he was unable to solve. If used in this manner, the Keys will be of great help and assistance to the student, and will be a source of encouragement to him in studying the various papers composing the Course.

SURFACE ARRANGEMENTS AT REDUCTION WORKS

(1) Hydrometallurgy is the process of reducing metallic ores by means of liquids. See Art. **2.**

(2) Lixiviation is a process by which a soluble alkali or saline compound is extracted from an earthy mixture by washing out. See Art. **9.**

(3) A side hill. See Art. **14.**

(4) Private and public. See Art. **23.**

(5) It shall not exceed 5 acres. See Art. **24.**

(6) A suitable ore supply; the kind of power available; the location of the mill; and the character of the mill suitable for treating the ore mined. See Art. **32.**

(7) For building foundations and engine beds, though sometimes the entire plant is constructed of masonry. See Art. **37.**

(8) The back of the wall should be sloped and smoothly finished. See Art. **39** and Fig. 5.

(9) When it is not possible to drive piles and form pile bents. See Art. **43** and Fig. 13.

(10) One with which two timbers are joined without an increase of size. See Art. **48.**

(11) Heavy framing by cutting joints and heavy framing without cutting joints. See Arts. **49** and **50.**

§ 24

(12) For refractory ores that require roasting. See Art. **54.**

(13) Wire-rope tramways, conveyers, cableways, etc. See Art. **57.**

(14) For conveying ore, fuel, and flux to the furnaces. See Art. **59.**

(15) The descending loaded car picks up a cable attached to a weight and raises the weight a short distance. The weight dropping back to its former position after the car is unloaded gives it sufficient momentum to run up the plane. See Art. **64.**

(16) Large quantities of material may be handled in a comparatively short time. See Art. **66.**

(17) To simplify the handling of materials. See Art. **77.**

(18) 1.5 gallons per minute of clear water and from 1.5 to 3 gallons per minute with the pulp from the stamps. See Art. **85.**

ORE DRESSING AND MILLING

(PART 1)

(1)   When there is a lack of water; when ore is to be roasted; when ore is damp and is to be crushed by rolls and then screened.   See Art. **15.**

(2)   Less dust; no roasting required; better amalgamation where that is practiced.

(3)   One has its movable jaw pivoted at the top and the other at the bottom.   One crushes more and larger pieces of rock than the other.   See Arts. **3-5.**

(4)   For jigging processes; for lixiviation.   See Art. **15.**

(5)   By springs and by breaking cups placed to hold the loose pulley to its work.   See Art. **12.**

(6)   Huntington mill is a centrifugal roller mill, with arms from which depending rollers hang.   The rotation of a central shaft causes the rollers to fly out and press against the pan, thus grinding any ore between them and the pan. This machine is best fitted to grind middlings.   See Art. **49.**

(7)   Various orders for dropping stamps are in vogue. The chief object should be to float the ore in regular waves out of the screen and over the aprons, to prevent scouring and intermittent action.   See Art. **37.**

(8)   Sectional guides offer the advantage of using less timber, of permitting easier adjustment, and do not cause so much friction.   See Art. **26.**

§ 25

(9) To make or obtain a solid foundation and then build up the block of two-inch planks. The block should be planed and be set perfectly plumb on the mudsill, so that the mortar shall rest true and level on the battery block. The block must be tamped up and anchored. See Arts. **38** and **39.**

(10) The height of discharge is the difference in height between the tops of the dies and the bottom of the screen. The discharge is regulated by chuck blocks upon which the screen frame rests, and by replacing dies as they are worn away. See Art. **32.**

(11) Iron frames have not the same elasticity as wood and are more resonant, consequently they will arrange their particles in a crystalline order and become weakened. See Art. **25.**

(12) Steam stamps strike a heavier blow and, consequently, are better able to pulverize more ore in a given time than gravity stamps. They do not cause as much slimes as gravity stamps. They do not retain the ore in the mortar as long. See Art. **41.**

(13) Steel, iron, and brass are used for wire screens, while steel and phosphor bronze, aluminum bronze, and manganese steel are used for plate screens. See Art. **35.**

(14) By the use of automatic feeders, a uniform supply of ore is given the machines. This uniformity is not possible where hand feeding is practiced, and as a usual thing too much or too little is fed, making the product irregular and preventing good work otherwise. See Art. **56.**

ORE DRESSING AND MILLING
(PART 2)

(1) A grating made of iron or steel bars, placed usually at the top of a mill, for the purpose of separating ore which needs crushing by rock breakers from that which does not. See Art. **9.**

(2) Uniformity in the size of product is quite important, because it can be treated after sizing with better results. In the case of stamp mills, ore fed to them should not be too fine, but when the sizes are uniform crushing is facilitated so that more ore will pass through the battery screens. Leaching, roasting, jigging, and table work will be materially aided by uniform sizing. See Arts. **1** and **10.**

(3) The pulp flows in at the top and meets an ascending stream of water, against which the material settles in part. The fine material which cannot settle passes to another box, and will, if possible, settle in that. These boxes increase in size, so that if the material is not heavy enough to settle in the first box it may in some of the others. See Art. **11.**

(4) It prevents the current traveling across the box to the discharge, thus compelling the ore to meet the ascending current of water. See Arts. **11, 14,** and **15.**

(5) To settle out the fine slimes, which in some instances are more valuable than other portions of the ore. To economize in water when it is scarce, by allowing the sediment to settle so that the water may be used over again.

§ 26

The usual loss of water from this source and evaporation is about 25 per cent. of the quantity used. See Art. **17.**

(6) See Art. **21.**

(7) Jigs are concentrating machines, usually employing moving water for the separation of light from heavy particles of mineral, but sometimes air. There are two kinds of hydraulic jigs. See Art. **22.**

(8) The assorting action of jigs is accomplished in one case by having a piston which makes quick strokes force water up through a bed of crushed material resting upon a screen. In another case the screen with the bed of ore is raised and lowered in water. See Art. **22.**

(9) Three compartments. See Art. **25.**

(10) By the stroke of the piston. See Art. **25.**

(11) Two eccentrics on one shaft fastened together by the eccentric strap. The student will find it to his advantage to examine the cuts of the Blake and Dodge rock crushers, in *Ore Dressing and Milling*, Part 1, and the Hooper concentrator in *Ore Dressing and Milling*, Part 3. See Art. **26.**

(12) One has an up and a down stroke in the same length of time, the other takes a longer time on the up than on the down stroke. See Arts. **27** and **28.**

ORE DRESSING AND MILLING
(PART 3)

---

(1) The separation of the sand from the gold by a blast of air. See Arts. **24-25.**

(2) The separation of the heavier mineral from the lighter by shaking the pulp. The table is given a horizontal jerking motion by means of cams and springs or eccentrics. It is moved slowly out and returns with a jerk, which works the heavy mineral across the table in the direction of the jerk. See Art. **6.**

(3) It is a flat linoleum-covered table set on rollers and slightly inclined from front to back. The feedbox is at the back of the table and extends from end to end, being divided by a movable gate. A set of cleats $\frac{1}{4}$ inch thick at the upper end and tapering to a feather edge is nailed along the table parallel to its sides. The ore is fed in at the tailings end of the table. The lighter particles pass over the cleats and off the table with the water while the heavier minerals are carried along by the jerking motion of the table until a place is reached where the cleats are thin enough to allow them to pass over. See Art. **7.**

(4) Paramagnetic refers to substances that are attracted by the magnet. Diamagnetic, when applied to a substance, means that the substance is repelled by the magnet.

In the one class, the magnetic material is deflected from the stream of ore and deposited in a separate receptacle. In the other the magnetic material is actually picked out and carried off. See Arts. **27** and **29.**

§ 27

(5)   The loss of mercury is chiefly due to "flouring" and "sickening." Gold may be lost as float gold, non-amalgamable gold, and in amalgam.   See Arts. **37-46.**

(6)   By adding a little sodium amalgam.   See Art. **38.**

(7)   By the use of potassium cyanide.   See Art. **44.**

(8)   The heat is supposed to assist in amalgamation.   The heating is accomplished either by passing live steam into the pulp or by exhaust steam in a space between the bottom of the pan and a false bottom.   See Art. **53.**

(9)   To prevent unnecessary wear of the pans.   Moreover, long grinding flours the mercury.   See Art. **54.**

(10)   The ore having been crushed to a proper size is delivered on to a main conveyer belt, which passes through the field of the magnets.   The magnetic minerals are attracted and raised to the upper cross-belt, by which they are removed and carried to a hopper.   See Art. **32.**

(11)   The acid would corrode iron pans.   See Art. **52.**

(12)   The ore is roasted with salt to bring the silver into the form of a chloride.   See Art. **36.**

(13)   The pulp is piled on the floor of the arrastra.   When it becomes stiff enough to work, it is spaded over and salt is mixed in, after which the pile is again worked over and the magistral added.   Mercury is added with the magistral. Small quantities are also worked in from time to time.   See Art. **48.**

ORE DRESSING AND MILLING

(PART 4)

(1)   Power; nearness to ore supply; fuel supply; water supply; railroad facilities; wagon roads; working supplies; gravity assistance; side hill location, etc.   See Arts. **46** and **47.**

(2)   The mineral when crushed must separate from the gangue; furthermore, must have a distinctive specific gravity. See Art. **82.**

(3)   The general chemical principles involved are about the same in both processes, but the details and time involved are quite different.   See Arts. **1** to **18** and **12** in particular.

(4)   Muntz metal or plain copper may be used.   See Art. **22.**   There are some who claim better results are obtained from copper plates, others that they are unable to keep copper from corroding; hence it will depend in some measure upon the ore and the water used in the battery, also upon the way the battery is fed with quicksilver.   See Art. **25.**

(5)   See Arts. **25, 26,** and **27.**

(6)   The proper way to remove amalgam from the plates is to use a stiff brush if the amalgam is soft enough, if not use a rubber scraper.   Sweating is not advisable.   See Art. **28.**

(7)   Tarnished plates are usually due to acid in the ores or water which attacks the amalgam first and then the plate.   When tarnish appears, it must be immediately removed.   See Art. **29.**

§ 28

(8) Sometimes the plates are arranged so that the pulp will flow from the tail of one plate on to the head of another whose slope is directly opposite to the first plate. The plan is not generally adopted, much more depending upon the condition of the plate amalgam at the slope of plate than on its length. See Art. **32.**

(9) Overhead trolleys. See Art. **63.**

(10) Raising material from a higher to lower level. Rubber, leather, and chain belts. See Art. **58.** It would seem undesirable with link belts and sprockets as cheap as they are, to install a plant with rubber or leather belt elevators.

(11) In retorting amalgam, one must guard against the outlet of the condenser becoming clogged; also guard against giving the condenser so much heat as to warp it. See Arts. **41** and **42.**

(12) The character of timber used at mines or mills frequently depends upon the timber growing near by. Some use only the most expensive white oak, but experience has proven that beech, birch, and ash make as durable floors when the grain is placed properly. They do not wear off in splinters, but wear smoothly and gradually. See Art. **49.**

SAMPLING ORES

(1) To obtain as nearly as possible the value of the ore before treatment. See Art. **1.**

(2) In order to ascertain their composition and to proportion the different substances so that they will smelt with the least expense and give the highest extraction. See Art. **1.**

(3) They act as umpires between the buyer and the seller. See Art. **3.**

(4) Dividing a sample into four parts and discarding two of those parts. Next making a cone of the remaining ore and dividing it into four parts and discarding two of the parts, and so on until sample is reduced to proper size. See Art. **12.**

(5) Very often the richest ore is composed of fine particles, and it is for this reason that the fine ore is to be deposited upon top of the cone. See Art. **12.**

(6) To make coarse and fine ore mix better, but care should be taken not to make it too wet. See Art. **13.**

(7) To obtain correct results. See Arts. **8** and **13.**

(8) See Art. **10.**

(9) See Art. **23.**

(10) Generally not, for dirt will become mixed in with the materials; iron ore should be fairly rich; coke will have dust and possibly poor coke and slate mixed with it.

§ 29

(11)   See Art. **5.**

(12)   To sample fine material.   See Art. **16.**

(13)   The dipper must pass across the whole stream, because one side may be richer than the other.   See Arts. **17** and **18.**

(14)   It is very difficult to settle some sediments, especially float gold.   See Art. **18.**

(15)   Lime, alum, salt, soap, etc.   See Art. **19.**

(16)   The ore is passed through a sample grinder, then pulverized to pass a certain mesh screen.   The bucking board is used for the last pulverization.   See Art. **20.**

(17)   See Art. **28.**

(18)   Slow and expensive.   See Art. **29.**

(19)   Must be easily cleaned and must take the whole stream part of the time.   See Art. **30.**

(20)   They both revolve in a horizontal plane.   See Arts. **31** and **32.**

(21)   See Arts. **29** and **40.**

(22)   Because of the irregular distribution of rich particles.   See Arts. **42-50.**

(23)   See Art. **44.**

(24)   Treatment for reduction of the ore which must follow.   See Art. **45.**

(25)   (*a*) Generally by hand after crushing and drying, particularly if the ore is rich.

(*b*)   Leaching ore is best sampled by mechanical devices.

ROASTING AND CALCINING ORES

(1) The process of removing the arsenic, antimony, sulphur, and metallic oxides from an ore by a series of alternate oxidations and reductions.   See Art. **2.**

(2) The changing of metals to chlorides by roasting with common salt *NaCl*.   See Art. **3.**

(3) An artificial sulphide containing iron, formed from copper sulphide by a combination of the copper and sulphur after the excess of sulphur has been burned off.   See Art **4.**

(4) The silver is converted into the form of a chloride that is readily amalgamable.   When gold ores are to be treated by the chlorination process and oxides of copper, calcium, or magnesium remain after dead-roasting, salt is added to convert them into chlorides.   Otherwise they would absorb chlorine gas.   In this process the silver is oxidized and cannot be recovered.   See Art. **5.**

(5) (*a*) Roasting to remove carbon dioxide CO_2 and water from ores.

(*b*) The ores are more readily crushed when perfectly dry.   See Art. **6.**

(6) To prevent loss of heat by radiation.   See Art. **10.**

(7) It would supply oxygen, thus opposing reduction. See Art. **13.**

(8) When it is desired to fuse low-grade ores.   See Arts. **14** and **15.**

§ 30

(9) To prevent the hot ore from corroding the hearth proper. See Art. **16.**

(10) Because it is easier to keep the lead from sticking to the hearth and balling than in the mechanically rabbled furnaces. See Art. **20.**

(11) It has been successfully used for pyritic ores, for copper matte, and for dead-roasting zinc blende. See Art. **25.**

(12) Water is circulated through the arms and the vertical shaft. In the Herreshoff furnace there is a central air flue and the arms are not cooled. See Art. **30.**

(13) (*a*) The chloridizing roast of silver ores.

(*b*) It revolves on rollers and moves the ore towards the heat. See Art. **32.**

(14) Iron sulphides, low in sulphur and iron or zinc carbonates. See Art. **36.**

(15) When the ore contains only a small amount of pyrite or when arsenic or antimony are present in any considerable quantity. See Art. **41.**

(16) To prevent rain water from wetting the ore and forming soluble sulphates. See Art. **44.**

(17) Stall roasting; because it saves time, fuel, and labor. See Art. **49.**

(18) The copper and the sulphur are both oxidized and form a sulphate or are converted into matte. See Art. **54.**

(19) Iron, copper, silver, zinc, and lead. See Art. **58.**

(20) (*a*) Oxygen is taken from them.

(*b*) $CuSO_4 + 2C = CuS + 2CO_2$. See Art. **60.**

(21) Carbon dioxide is driven off. $ZnCO_3 + \text{heat} = ZnO + CO_2$. See Art. **62.**

THE CYANIDE PROCESS

(PART 1)

(1) The reactions which occur between cyanogen and some of the elements are not fully understood, and until they are determined it will be impracticable to treat some ores. See Art. **3.**

(2) Fine gold, free from enclosing material. See Art. **4.**

(3) With silver chloride the action is more energetic than with gold. See Art. **5.**

(4) Copper and antimony sulphides, sulphates of base metals, etc. See Art. **6.**

(5) Agitation; introduction into the solution of oxygen; or chemicals to either supply or take the place of oxygen. The latter are the halogens iodine, bromine, and chlorine. See Art. **9.**

(6) Upon oxygen and strength of the solution. Without oxygen present gold will not be dissolved, no matter how strong the solution. See Art. **11.**

(7) (*a*) Ferric oxide seems to have no effect upon cyanide solutions.

(*b*) Ferrous sulphate destroys potassium cyanide and forms potassium sulphate, which is also injurious. See Art. **13.**

(8) The loss of cyanide can be decreased by lessening the strength of the leaching solution. Some metallic oxides are

attacked by a strong solution of potassic cyanide, while with a weak solution they are but slightly affected. Also roasting decreases loss in some instances. See Art. **15.**

(9) The gold is encased by the pyrite and it cannot be reached by the solution as readily as if the ore were porous.

(10) Finely pulverized ore mixed with water. See Art. **24.**

(11) Fine crushing, agitation, and mixing with sand. See Art. **25.**

(12) See Art. **29.**

(13) In this country, 6.535 grams of silver nitrate to the liter of distilled water are used. The quantity of cyanide solution usually taken is 10 c. c., and each c. c. of the silver solution used represents 1 pound potassium cyanide to the ton of solution. See Art. **32.**

(14) See Art. **36.**

(15) $\frac{1}{2}$ inch per hour. See Art. **42.**

(16) By experiment and assay of the ore before and after percolation; or by assaying the gold solution to ascertain the gold in it and subtracting this result from the assay of the ore before leaching. See Art. **43.**

(17) This process being experimental, the reader is referred to Art. **49.**

THE CYANIDE PROCESS

(PART 2)

(1) (*a*) Slimes pack and prevent the solution from circulating freely. See Art. **2.**

(*b*) They are gotten rid of by washing them off the ore and then settling them in special tanks or ponds. See Art. **3.**

(2) When a weak solution precedes a strong solution, it has for its object the destruction of those chemical compounds that would destroy the strong solution. See Art. **7.**

(3) Solutions that are less than half the strength of the standard solution are sent to the weak gold-solution tank, but when they are more than half the strength of the standard solution, they are sent to the strong gold-solution tank. See Art. **8.**

(4) The weak cyanide solution is usually about one-third the standard solution, but it may be more or less than one-third to suit conditions. See Art. **9.**

(5) The solution would be divided by 10; hence, 60 cents is the answer. See Art. **12.**

(6) *KCN* is added to sump solutions to standardize them as follows: Stock solution of a certain strength *KCN* is added to the sump solution; or weighed lumps of cyanide of potassium are allowed to hang in baskets in the sump solution and dissolve. The baskets may sometimes be placed under the pipe discharging from the sump into the storage tank. See Art. **14.**

§ 32

(7) 12.23 tons. Ans. See Art. **15.**

(8) 10.278 lb. Ans. See Art. **16.**

(9) The alkali metal zinc precipitates gold from a cyanide-of-potassium solution. Consequently, potassium cyanide cannot dissolve gold so long as zinc is present. See Art. **17.**

(10) (*a*) The color of precipitated gold is almost black-brown when precipitation is going on properly, otherwise it takes various shades of dull gray.

(*b*) Zinc, acid, and some base-metal salts.

(11) (*a*) When copper occurs in the zinc boxes in weak-gold solutions it is deposited.

(*b*) Weak solutions containing copper and gold can have their gold deposited and the copper retained in solution by adding potassium cyanide to the solutions to strengthen them. See Arts. **19, 20,** and **21.**

(12) The object of treating gold precipitates with sulphuric acid is to dissolve out the zinc and other impurities. The sulphate of zinc formed is washed off. The clean filtrates are agitated, drained, dried, and roasted with niter. Next they are fluxed and melted into bullion. See Arts. **26** and **27.**

(13) The bullion obtained by the first smelting is impure, consequently, a further refining at a low temperature is sometimes considered advisable. See Art. **34.**

(14) Cathodes should be of a thin sheet metal, dense in character, but such that the gold will adhere to it, and not drop off after deposition. The anodes should be of a substance not easily disintegrated or readily dissolved by the solution. See Arts. **36** and **37.**

(15) (*a*) Zinc fume is the volatile substance that escapes when zinc is smelted and is afterwards condensed as "blue powder."

(*b*) Zinc fume is sifted into the solution as explained in Art. **69.**

(16)   Charcoal will precipitate gold from cyanide solutions. Professor Christy says, "The claims made by Mr. Johnson cannot be substantiated under circumstances more favorable than could ever occur in actual practice." Further, he finds precipitation quick at first, then slow, and that free *HCN* hinders precipitation, while other acids rather assist it; at least to some limited extent. To extract the gold from charcoal is expensive and tedious; consequently, the process has never found favor.   See Arts. **41-45.**

(17)   The original idea involved in the pneumatic cyanide process was to furnish agitation, to hasten the leaching, but it was found that it also furnished oxygen as a corollary. See Art. **62.**

(18)   The Pelatan-Clerici process had for its object agitation and electrical precipitation.   This would permit the treatment of slimes and coarser ore at the same time, also precipitate gold, and avoid long, tedious leaching operations. The cathode, being a bath of mercury, would catch coarse gold and retain it independent of the cyanide solution, which acts very slowly on coarse gold.   See Arts. **57-61.**

(19)   Roasted in some instances, especially sulphurets. See Cripple Creek ore treatment in Art. **73.**

INDEX

NOTE.—All items in this index refer first to the section (see the Preface) and then to the page of the section. Thus, "Agitators 27 66" means that agitators will be found on page 66 of section 27.

INDEX

INDEX

INDEX

INDEX

INDEX

INDEX

INDEX

INDEX

INDEX

INDEX

www.ingramcontent.com/pod-product-compliance
Lightning Source LLC
Chambersburg PA
CBHW081138180526
45170CB00006B/1843